Intelligent Systems Reference Library

Volume 272

The aim of this series is to publish a Reference Library, including novel advances and developments in all aspects of Intelligent Systems in an easily accessible and well structured form. The series includes reference works, handbooks, compendia, textbooks, well-structured monographs, dictionaries, and encyclopedias. It contains well integrated knowledge and current information in the field of Intelligent Systems. The series covers the theory, applications, and design methods of Intelligent Systems. Virtually all disciplines such as engineering, computer science, avionics, business, e-commerce, environment, healthcare, physics and life science are included. The list of topics spans all the areas of modern intelligent systems such as: Ambient intelligence, Computational intelligence, Social intelligence, Computational neuroscience, Artificial life, Virtual society, Cognitive systems, DNA and immunity-based systems, e-Learning and teaching, Human-centred computing and Machine ethics, Intelligent control, Intelligent data analysis, Knowledge-based paradigms, Knowledge management, Intelligent agents, Intelligent decision making, Intelligent network security, Interactive entertainment, Learning paradigms, Recommender systems, Robotics and Mechatronics including human-machine teaming, Self-organizing and adaptive systems, Soft computing including Neural systems, Fuzzy systems, Evolutionary computing and the Fusion of these paradigms, Perception and Vision, Web intelligence and Multimedia.

Indexed by SCOPUS, DBLP, zbMATH, SCImago.

All books published in the series are submitted for consideration in Web of Science.

Rohit Sharma · Gwanggil Jeon
Editors

Data Analytics for Smart Robotics and Its Applications

Springer

Editors
Rohit Sharma
Department of Electronics
and Communication Engineering
ABES Engineering College
Ghaziabad, Uttar Pradesh, India

Gwanggil Jeon
Department of Embedded Systems
Engineering
College of Information Technology
Incheon National University
Incheon, Korea (Republic of)

ISSN 1868-4394 ISSN 1868-4408 (electronic)
Intelligent Systems Reference Library
ISBN 978-3-031-87696-7 ISBN 978-3-031-87697-4 (eBook)
https://doi.org/10.1007/978-3-031-87697-4

This Springer imprint is published by the registered company Springer Nature Switzerland AG
The registered company address is: Gewerbestrasse 11, 6330 Cham, Switzerland

Preface

The edited book aims to bring together leading academic scientists, researchers and scholars to exchange and share their experiences and research results on all aspects of the Internet of Robotic Things for Smart City Infrastructure. It also provides a premier interdisciplinary platform for researchers, practitioners and educators to present and discuss the most recent innovations, trends, and concerns as well as practical challenges encountered and solutions adopted in the fields of Internet of robotic things; industry 4.0; cyber-physical systems; remote working; agriculture; manufacturing; healthcare; education; and surveillance. The edited book will focus on the state-of-the-art IoRT applications, aiming to mark their impact on several research fields, and focusing on the main open challenges of integrating robotic technologies into smart spaces. IoRT technologies and applications will also be the part to underline their influence in everyday life, inducing the need for more research into remote and automated applications.

Ghaziabad, India — Rohit Sharma
Incheon, Korea (Republic of) — Gwanggil Jeon

Organization of This Book

The book is organized into sixteen chapters. A brief description of each of the chapters follows:

Chapter "Scalable Blockchain and Robotics Technology for a Patient-Centric Next Generation Medical System" discusses current knowledge and emerging trends in blockchain and healthcare, including the integration of blockchain and robotics in healthcare. Further, this explores the role of blockchain in enhancing healthcare interoperability and its potential in personalized patient care. Finally, the paper examines the advantages and significant challenges of adopting these in healthcare and opens the way for future innovations in the sector.

Chapter "Face Mask Detection Using Multimedia Data and Deep Learning Techniques" detects whether a person is wearing a mask based on convolutional neural networks (CNNs) and VGG-16 architecture. In order to enhance the model's accuracy and computational efficiency, author used transfer learning with the VGG-16, which took advantage of pre-trained layers to classify images with various characteristics.

Chapter "Digital Twin Robots for Securing IoT Based Healthcare 4.0 Application: Key Technologies, Integration Trends and Recent Advancement" analyses the changing face of healthcare as it relates to robotics, including its origins and growth through different departments such as hospitals and rehabilitation settings, surgical procedures with related practices. It also shows the effect of teleoperation in enabling distance care. Finally, the paper concludes with a review of recent accomplishments in robotics demonstrating their significant potential to drive future healthcare delivery.

Chapter "Evolution of FinFET SRAM Cells for Smart Robotics: A Comprehensive Review" provides a comprehensive analysis of state-of-the-art FinFET SRAM cell de-signs, comparing their performance metrics across various technologies for computing and robotics applications.

Chapter "Robotics and IoT in Autonomous Vehicles: Background, Integration Trends and Recent Advances" highlights the basic classifications of IoT and discusses its various devices and how they are helping at present and will help in future. Further these devices are categorized into three parts—low-end, middle-end and high-end devices. Finally, the paper discusses robotics and autonomous vehicles, and the integration of Robotics using IoT in self-driving autonomous vehicles.

Chapter "Advancing IoT-Aided Human–Robot Interaction" begins with a brief overview of IoRT's key topic. In addition to describing the research strategy, some trends in the topics discussed are discussed. Afterwards, an overview of human-robot interaction with IoT is presented. An overall conclusion and a few recommendations for future research conclude the chapter.

Chapter "Fuzzy Frontiers: Advances in Fuzzy Systems for Real-World Applications" explores the fuzzy frontiers of computational systems, highlighting recent advances in fuzzy logic and fuzzy systems for real-world applications. Focused on bridging the gap between theoretical developments and practical implementations, the study reviews innovative methodologies and algorithms within the realm of fuzzy systems.

Chapter "Application of the Internet of Things and Multimedia Data Towards a Smart City" provides a comprehensive framework for smart city projects. The urban network helps cities gain economic power, develop unique technological solutions, and build efficient systems. As part of the smart economy and information technology, smart cities are an advanced development product, which is briefly included in this chapter.

Chapter "Machine Learning Based Solutions for Securing Robotics and IoT Systems" presents the predictive ability of ML algorithms and their roles in improving threat detection and identification, reviewing their effectiveness against different types of challenges and vulnerabilities in IoT systems. Finally, the paper comprehensively reviews the current machine learning algorithms and their effectiveness in securing IoT network, devices, and robot while also offering some valuable insights and recommendations to create more adaptive and effective security solutions.

Chapter "Internet of Robotic Things in Smart Domains: Applications and Challenges" discusses networking connectivity, robotic autonomy, cybersecurity, and interoperability of systems as key components of IoRT. IoRT architectures and frameworks are described to demonstrate their potential to revolutionize industry and society through the development of secure, efficient, and responsive systems.

Chapter "Smart Robotics and Applications Using Data Analytics—Present and Future Directions" explores the integration of smart robotics and data analytics in modern manufacturing, examining past advancements and future directions in intelligent robotic systems. With the adoption of smart manufacturing, robotic equipment has evolved from simple automation tools to sophisticated, adaptive systems powered by artificial intelligence (AI), machine learning, and data analytics.

Chapter "Machine Learning Algorithms in Smart Robotics for Sustainable Systems and Precision Agriculture" explored precision agriculture as a means of enhancing sustainability and productivity in farming through machine learning (ML) algorithms and advanced technologies. Using data-driven approaches, precision agriculture (PA) addresses climate variability, resource optimization, and environmental sustainability challenges.

Chapter "Quantum-Enhanced Lattice Cryptography: AES Encryption with Crystal Kyber and QRNG IV" delves into the fusion of NIST-approved Lattice-based Crystal Kyber Cryptography Key-Encapsulation Mechanisms (KEMs) with AES encryption, and Quantum Random Number Generation (QRNG) for initialization vectors (IVs). By incorporating NIST-approved lattice-based KEMs, the encryption process attains a higher level of security, meeting stringent standards for cryptographic protocols.

Chapter "Robot Path Planning Using Triangular Inequality" provides a comprehensive analysis of robotic task sequencing so that a robot can perform multiple tasks most effectively. The framework utilizes parent-child relationships within the node hierarchy to calculate the distance between nodes in a task sequence. For three-dimensional path planning, RRT-Connect and TRi-RRT-Connect are used in the study.

Chapter "Exploring Quantum Key Distribution: Protocol Efficiency and Implementation Challenges" provides detailed information on the blockchain concept, features, and types. The chapter also discusses a few major consensus algorithms frequently utilized in different blockchain networks. Further, the major challenges faced by existing voting methodologies have been addressed and then the paper explains that integrating the e-voting systems with blockchain technology can help overcome most of these challenges.

Chapter "Smart Factory Architecture, Infrastructure and Design of Manufacturing Systems Using Robotics" enabling technologies and their benefits are discussed to provide a comprehensive overview of the transition to smart manufacturing. In addition to discussing the challenges of implementing smart factory systems, the chapter also discusses their modularity, adaptability, and real-time capabilities, along with their technological and economic challenges.

About This Book

This book will focus on using the Internet of Robotic Things (IoRT) in business intelligence, data management, Hadoop, machine learning, cloud, smart cities, etc. IoT and Robotics emerged from the early 2000s data boom, driven forward by many early internet and technology companies. The Internet of Robotic Things is a field of research that describes the collaboration of the fields of the Internet of Things and Robotics. IoRT is not IoT aided by Robotics or Robotics driven by IoT but rather an amalgamation of the disciplines. It was slowly formed due to the realisation that the two fields had a lot of overlapping of purpose and could greatly benefit from focusing on their combination. Although IoT has existed for more than two decades and robotics even longer, IoRT is a rather novel field of research. The convergence of the two areas had been gradual, but the term IoRT was coined and the concept was formally defined in a 2014 report. IoRT is also supported by the fields of Edge Computing, Cloud Computing, Cloud Robotics, and AI in IoT. IoRT is how we can create truly autonomous networks capable of carrying out complex tasks in the physical world. The ability of robotic technology to manipulate the physical world and the capabilities of IoT to create independent and smart networks have great potential. So, the proposed book will cover all the aspects discussed in the field above.

Contents

About the Editors

Dr. Rohit Sharma is currently working as Dean (Research), ABES Engineering College, Ghaziabad. Dr. Sharma is having Post-Doctoral, from South Ural State University, Russia. Dr. Sharma is having extensive experience in academia and research. As a prolific researcher, he has authored over 70 SCIE and 122 SCOPUS publications, along with 23 books in collaboration with esteemed publishers such as Springer, Taylor and Francis, and IEEE. With more than 22 patents to his name, three of which have been granted. He currently serves as an editor for several prestigious journals including *CAEE* Elsevier, *SETA* Elsevier, *Expert Systems* Wiley, *IET Communications*, *IEEE Transactions on Consumable Electronics*, *ACM Transactions on Autonomous and Adaptive Systems*, etc. Additionally, he holds senior membership in IEEE and member of other technical societies such as ISTE, IAENG, etc. Dr. Sharma has also spearheaded 8 renowned conferences in collaboration with IEEE and Springer, and under his guidance, four scholars have awarded their Ph.D. while one other has submitted her Ph.D. synopsis.

Prof. Gwanggil Jeon received the B.S., M.S., and Ph.D. (summa cum laude) degrees from the Department of Electronics and Computer Engineering, Hanyang University, Seoul, Korea, in 2003, 2005, and 2008, respectively. From September 2009 to August 2011, he was with the School of Information Technology and Engineering, University of Ottawa, Ottawa, ON, Canada, as a Post-Doctoral Fellow. From September 2011 to February 2012, he was with the Graduate School of Science and Technology, Niigata University, Niigata, Japan, as an Assistant Professor. From December 2014 to February 2015 and June 2015 to July 2015, he was a Visiting Scholar at Centre de Mathématiques et Leurs Applications (CMLA), École Normale Supérieure Paris-Saclay (ENS-Cachan), France. From 2019 to 2020, he was a Prestigious Visiting Professor at Dipartimento di Informatica, Università degli Studi di Milano Statale, Italy. He is currently a Full Professor at Incheon National University, Incheon, Korea. He was a Visiting Professor at Sichuan University, China, Universitat Pompeu Fabra, Barcelona, Spain, Xinjiang University, China, King Mongkut's Institute of Technology Ladkrabang, Bangkok, Thailand, and University of Burgundy, Dijon, France. Dr. Jeon is an IEEE Senior Member, an Associate Editor of *Sustainable Cities and*

Society, *IEEE Access*, *Real-Time Image Processing*, *Journal of System Architecture*, and *MDPI Remote Sensing*. Dr. Jeon was a recipient of the IEEE Chester Sall Award in 2007, the ETRI Journal Paper Award in 2008, and Industry-Academic Merit Award by Ministry of SMEs and Startups of Korea Minister in 2020.

Scalable Blockchain and Robotics Technology for a Patient-Centric Next Generation Medical System

Shikhar Singh, Bharat Bhushan, Mustafa Al-Asadi, and AlaAli Hameed

Abstract In recent year, exponential increase of population creates a big challenge regarding healthcare centers to improve their services. Blockchain is a decentralized and secure ledger technology that processes transactions cryptographically in a distributed network. Its transparency and security make it ideal for addressing healthcare challenges like data management, patient confidentiality, and system compatibility. Blockchain uses smart contracts to ensure information security, patient confidentiality, and patient-controlled health data exchange. Its integration with robotics enhances real-time data collection, task automation, and secure, precise health data distribution. In this paper, we discuss current knowledge and emerging trends in blockchain and healthcare, including the integration of blockchain and robotics in healthcare. Further, this explores the role of blockchain in enhancing healthcare interoperability and its potential in personalized patient care. Finally, the paper examines the advantages and significant challenges of adopting these in healthcare and opens the way for future innovations in the sector.

Keywords Robotics · Healthcare · Electronic-health record · Decentralised technology

S. Singh (✉) · B. Bhushan
Department of Computer Science and Engineering, School of Engineering and Technology, Sharda University, Greater Noida, India
e-mail: shikhar895798@gmail.com

M. Al-Asadi
The University of Mashreq, Baghdad, Iraq
e-mail: mustafa.alasadi@uom.edu.iq

A. Hameed
Department of Computer Engineering, Istinye University, Istanbul, Turkey
e-mail: alaa.hameed@istinye.edu.tr

R. Sharma and G. Jeon (eds.), *Data Analytics for Smart Robotics and Its Applications*, Intelligent Systems Reference Library 272,
https://doi.org/10.1007/978-3-031-87697-4_1

1 Introduction

In recent year, various technologies have been integrated into blockchain, expanding its capabilities and applications. Blockchain technology was initially introduced in 2008 through the launch of the cryptocurrency Bitcoin. Bitcoin brought with it several fundamental characteristics that serves as main feature of blockchain today, which includes as decentralization, transparency, and anonymity which makes blockchain a revolutionary concept in the digital age. It serves as a robust example, proving that blockchain technology functions effectively and with nearly 400 million transactions completed as of March 19, 2019 [1]. And this has sparked discussions and proposal about the potential of blockchain technology across various other data-centric sectors like healthcare. By managing and storing data in timestamped blocks, blockchain is considered as a decentralized ledger system that works across various computing networks and linking these blocks through cryptography. The capabilities of blockchain which were first introduced in the fintech area, have a potential for wider realisations, particularly when combined with robotics which enhance functional efficiency and data management in healthcare development [2]. Robotics can facilitate real-time monitoring, improve patient interaction, complementing blockchain's secure data handling. However, applications effective solution design and venues are still underdeveloped. Blockchain assured fundamental changes across different sectors including government, fintech, health and supply chain [3]. Its potential advantages include transactions complexity and reduction in costs, improved transparency, regulation, and enhanced security.

Enduring technical improvements offers important opportunities for cost savings and biomedical innovation, but also pose challenges for the integration of emerging technology into medical care. A substantial amount of work primarily focuses on addressing useful and efficient healthcare limitations and fulfilling the growing expectations for high-quality healthcare through smart healthcare solutions [4]. Smart healthcare can be conceptualised and developed as an array of devices, tools, facilities, software and organizations integrated with conventional healthcare, smart service systems, connected applications, and biosensors [5]. IoT end nodes is the cornerstone of intelligent healthcare, encompassing a variety of medical devices and applications that connect to healthcare via the Internet.

The healthcare sector encounters security and reliability challenges which can be addressed through the modern mechanism of decentralized blockchain. Another appealing aspect of blockchain technology is the elimination of intermediaries [6]. Integration of blockchain into healthcare applications including all aspects of patient access, electronic health records, validity of safety and privacy [7]. Blockchain technology are used exponentially in this digital age in health sector. The reliability of blockchain technology renders it an efficient and effective solution across various domains, such as business, cryptocurrency, finance, advertising insurance, and copyright protection [8]. Patients remain the central focus of the health ecosystem, aiming to improve the privacy and safety of health data and their interoperability. And Blockchain looks forwards to fulfil this by enabling data sharing, management and access control for medical activities such as supply chain management, drug prescription [9]. The core principle of blockchain is a patient-focused approach, setting it apart from the traditional methods. It also aspires to provide safety, facilitate clinical trials, anti-counterfeiting drugs, and ensure interoperability with patients.

Overall, the evolution of blockchain technology in the health arena will affect secure data management, confidentiality issues of patients, and issues related to interoperability. Scalable blockchain frameworks have special capabilities to handle sensitive patient data across decentralized networks as the healthcare industry moves toward digital solutions. Blockchain immutability and transparency provide an appropriate solid basis for secure data sharing fostered with trust among healthcare providers, patients, and regulators. Results of the main contributions in this work follow:

- This paper investigates the space of blockchain in healthcare, with considerations for utility in decentralized data management, improvements to privacy and interoperability.
- The paper explores a new concept of smart contracting for patient data sharing that has been automated to increase transparency and minimize human error in health records, while also highlighting the role of robotics in facilitating these methods.
- Using blockchain applications, it presents successes in building secure and traceable drug supply chains that avoid issues related to counterfeiting and, therefore, enhance effective pharmaceutical distribution, with robotics aiding in logistics and monitoring.
- It discusses consensus protocols, such as PoS and DPoS, in the health context and scalability, data consistency, and reliability with several health networks across healthcare networks, while considering how robotics can enhance operational efficiency in these networks.
- Compare and contrast blockchain frameworks, for example, Hyperledger Fabric and Ethereum, are delineated to ensure secure storage of health data and also on interoperability issues on patient's records, supported by robotics systems that assist in data collection and management.
- It demonstrates how blockchain enable the real-time sharing of health data between health care service providers, enhanced by robotics that streamline data collection and improve patient care as well as making possible their up-to-date accurate medical history using authorized providers.
- The paper discusses predictive modeling with AI in the context of hospital patient monitoring and emergency response, emphasizing the integration of robotics and blockchain in critical situations to maintain patient outcomes.

The remainder of the paper is organized as follows: Sect. 2 covers blockchain technology and robotics, including blockchain introduction, types, consensus algorithms, robotics introduction, and benefits of blockchain and robotics in healthcare. Section 3 focuses on the healthcare domain, discussing its challenges and the need for blockchain solutions. Section 4 addresses critical challenges in healthcare and explores blockchain solutions for data collection and storage, data sharing and interoperability, financial records, and pharmaceuticals. Finally, Sect. 5 reviews recent advancements.

2 Background Knowledge

From areas of electronic data security to actual care delivery, blockchain and robotics are changing healthcare. Secure, decentralized management of sensitive patient data is just one of blockchain's critical functionalities, which ultimately found its importance with the advent of telemedicine and remote monitoring-a place where issues related to integrity and privacy of data come into focus. Robotics, powered by AI and cyber-physical systems, automates routine tasks and improves patient care in both hospitals and home environments. So each of these technologies offers health care solutions where the needs of a patient are at center stage, reducing the burden on healthcare professionals and ensuring effortless contact between healthcare professionals and patients at a distance.

2.1 Blockchain Technology

Blockchain is a cutting-edge technology that makes significant influence on modern society due to its decentralization, transparency, and security features. It can also be defined as a sequence of digital blocks that are linked and interconnected with one another and works as open distributes ledger. Initially, it was used to record digital currency transactions, but later on it began to be applied for other purposes other than payments and currencies [10]. However, it gained its recognition in recent years when it was used in crypto currency called "Bitcoin" to record transactions. The blockchain technology have been explained in the subsections below. Blockchain technology is also known as Distributed ledger technology (DLT) is intended to support verification-drive transaction services in an ecosystem that is typically untrusted [11]. Blockchain technology assures that data or records kept on the ledger is immutable by design, which means no one can alter, delete, or even add any record without the consent of other network user. Blockchain technology holds significant potential for various industries, including supply chain management, financial services, academia, real estate, health care and more [12]. Its inherent properties make it naturally fit to enhancing the transparency, maintaining records and reduce illicit activities in the art market. The attributes of blockchain technology, namely decentralization, non-tampering, openness, transparency, and traceability, greatly aid in enhancing the legitimacy of electronic data [10]. Blockchain-based deepfake data supervision has a wide range of applications. Many nations, including China, now formally acknowledge that blockchain technology is an efficient way to store electronic evidence [13]. Blockchain technology was introduced just over a decade ago, created by the anonymous individual or group behind the cryptocurrency Bitcoin, using the pseudonym Satoshi Nakamoto. However, Blockchain Technology was first described in 1991 by W Scott Stornetta and Stuart Haber as a cryptographically secured chain of blocks [14]. In 1998, computer scientist Nick Szabo worked on a decentralised digital currency called 'bit gold'. Theory of cryptographically secured chains was further expanded in 2000 by Stefan Konst for implementation. Nakamoto implemented the first blockchain in 2009 by using bitcoin as a public ledger for transactions made. And in 2014, Blockchain technology was separated from cryptocurrency and its potential in other financial and inter-organisational transactions was explored [15].

2.1.1 Types of Blockchain

Blockchain technology comes in two broad classifications: public blockchain and private blockchain. Based on other criteria and analysis, it further divided as consortium blockchain and consortium blockchain. Blockchain basically involves a network of nodes or we can say blocks working on the peer-to-peer (P2P) network system. Each block in the network contains a copy of the shared ledger and is timely get updated [16]. The nodes verify transaction, initiate and receive them. And due to these characteristics, it is classified into four types.

- **Public Blockchains**: Public Blockchain are truly transparent, secure, and decentralized distributed ledger system. And in this type of technology, user is authorized to access anyone's current or past records. Two types of models are used in Public Blockchain, one is proof-of-work and the other is proof-of-stake, and the verified person receives the transaction rewards. In Public blockchain, participants don't need to worry about authenticity unlike of private blockchain. And there is no need of knowing other nodes and so there is no risk of fraud in transaction [17]. And these features, give birth to mining and exchanging cryptocurrencies. Most common examples of this type of Blockchain are- *Bitcoin, Ethereum, Litecoin.* And important features of Public Blockchain are- Open and Adaptable Environment, Full Transparency, Anonymity, High Security and Privacy, no regulations and Rigid Policies, Distributed.
- **Private Blockchains**: Private Blockchain are restricted, closed, and offer access features. They are typically used within organizations or enterprises where selected members participate. System administrators grant transaction permissions. These blockchains include authorizations, permissions, security, and accessibility. Popular examples are Multichain, Hyperledger, and Corda. With fewer nodes, transactions per second are faster, but third-party access can reduce security and increase hacking risks.
- **Hybrid Blockchains**: Hybrid Blockchain combine public and private blockchain features, managing both centralized and decentralized systems in a closed environment. They utilize a permission-based private system and a permission-less public system as needed. Only selected data sections are public, while the rest

remains confidential, giving users control over data access. Hybrid technology's flexibility allows easy integration with multiple public blockchains, enhancing security and transparency.
- **Consortium Blockchain**: Consortium Blockchain are semi-decentralized, managed by multiple organizations that act as nodes, facilitating information exchange and mining. One key advantage is their high customization, combining public permission-less and private permission-based systems [18]. Users can view and select portions of records from the ledger while keeping the rest secure. Their adaptability allows easy integration with private blockchains, enhancing security and transparency.

2.1.2 Consensus Algorithms

Consensus algorithm is used to maintain the integrity of blockchain by ensuring that any node or group of nodes can't manipulate the network. By stopping hostile actors from seizing control of the network and guaranteeing legitimate transactions and efficient network functioning, it offers security [19]. Consensus Algorithms avoid centralization by ensuring that all nodes reach an agreement about transaction authenticity, which aids in the decentralization process. Consensus algorithms facilitate transparency by enabling simple tracking and prevention of fraudulent activity by making all transactions accessible on the blockchain. By enabling nodes to swiftly concur on transaction legality and promptly add fresh blocks to the blockchain, they increase efficiency [20]. Different blockchain uses different algorithms that are demanded by the platform. The intention and requirement of the different blockchain platforms tend to differ. For example, proof-of-work, proof-of-stack, proof of authority are used for consensus among nodes in a network. Some of them are as follows –

- **Proof of work (PoW)**: It was first introduced by Bitcoin's creator Satoshi Nakamoto, secures the network and prevents double-spending. Miners confirm transactions and append new blocks to the blockchain through solving complex hash problems. A miner, therefore, solves a problem first then turns out to be the one that generates a new block which gets appended to the blockchain if agreed upon to by other miners [19]. This process requires significant processing power and rewards the solving miner with cryptocurrency. PoW boost security by complicating to alter blockchain data, as modifying a block requires redoing the proof-of-work for that block and all subsequent ones, preventing fraud and manipulation.
- **Proof of stake (PoS)**: It is a consensus method that could be more energy-efficient than Proof of Work (PoW), which is used in Bitcoin and consumes significant energy. In PoS, the probability of mining a new block depends on the amount of cryptocurrency a miner holds, rather than computational power. Nodes stake a certain amount of cryptocurrency to qualify as candidates for validating new blocks. An algorithm selects one of these candidates to validate the block and

receive transaction fees. The selection process incorporates factors such as stake amount, coin age, and randomness to ensure fairness across the network.

- **Practical Byzantine Fault Tolerance (PBFT)**: It is a replication algorithm designed for withstanding byzantine faults. PBFT is a consensus method getting utilised by Hyperledger Fabric as it can handle up to 1/3 malicious byzantine replicas. In each round, a new block is introduced, and a primary, according to specific rules, would be chosen and is responsible for ordering the transaction. The whole process can be segmented into three phases: pre-prepared, prepared and commit. If the node had votes from more than 2/3 of all the nodes in any phase, then the node enters into the next phase [21]. So, PBFT makes a condition that every node must know the network. SCP is also the Byzantine agreement protocol like PBFT. In SCP, participants are given the right to select a set of other participants to trust while in PBFT, every node needs to query all the other nodes. The dBFT (delegated byzantine fault tolerance) implementation of Antshares is based on PBFT. In dBFT, to record the transactions, some other professional nodes vote.
- **Delegated Proof of Stake (DPoS)**: PoS and DPoS differs mainly as PoS operates as a direct democracy whereas DPoS functions as illustrative democracy. To create and varify blocks, representatives are chosen by stakeholders. Blocks can be verified quickly with significantly fewer nodes, resulting in rapid transaction confirmation. Meanwhile, Delegates can modify network settings, such as block timing, block capacity, etc. User can vote easily and so they need not to worry of dishonest delegates. The backbone of Bitshares is DPoS.
- **Ripple Protocol**: Ripple is a consensus protocol. Within the greater network, collectively-trusted sub-networks are used. Nodes, in the net of work, have been classified into two categories, and for attaining consensus process we are using a server, whereas for transferring funds we are using a client. Every single server has an Unique Node List (UNL), and it has to be a server [22]. The server will communicate with the nodes in UNL while deciding to send the transaction into the ledger. If the agreement received is above 80%, then the transaction is included in the ledger. As long as the percentage of faulty nodes in UNL is less than 20%, the ledger for a node will be correct.
- **Tendermint Protocol**: Tendermint is a Byzantine fault-tolerant consensus algorithm. On every round, a new block is selected and distributed by a proposer as an unverified block. Tendermint consists of three phases: the Prevote Phase: validators decide to send a prevote for the proposed block; the Precommit Phase: if a node receives more than two-thirds of prevotes, it sends a precommit for the block, and a node enters the commit phase once it receives more than two-thirds of precommits; the Commit Phase: after checking the block, a node broadcasts a commit for that block, accepting it if it receives at least two-thirds of commits. Unlike PBFT, nodes should lock their tokens as validators. For any dishonest validator there is penalization [13]. Table 1 shows comparison between different consensus algorithm.

Table 1 Typical consensus algorithm comparison

Attribute	PoW	Ripple	PBFT	PoS	DPOS	Tendermint
Node Identity Management	Unrestricted	Unrestricted	Restricted	Unrestricted	Unrestricted	Restricted
Tolerated Adversary Power	$< 25\%$ Processing power	$< 20\%$ Compromised units in UNL	$< 33.3\%$ Faulty nodes	$< 51\%$ Ownership	$< 51\%$ Validators	$< 33.3\%$ Byzantine voting strength
Energy Efficiency	None	High	High	Moderate	Moderate	High
Example	Bitcoin	Ripple	Hyperledger Fabric	Peercoin	Bitshares	Tendermint

2.2 *Robotics*

This is the realm where the field of robotics changes healthcare into automation of very vital work, leading to efficiency in both the hospitals and homecare settings. AI-robots like Lio can move around their respective healthcare settings without human control as they assist patients and pick or manipulate objects, relieving the workload of often besieged healthcare professionals in patients' rooms with mundane tasks. Equipped with a rich set of sensors, soft materials, and features with collision detection, these robots ensure safe interaction with their patients and staff. In Healthcare 4.0, robotics becomes fully integrated into CPS, the result of which is to apply IoT and cloud computing in the environment of virtual and real in order to ensure smooth-running operations for both physical and digital healthcare environments [23]. This has been found to be efficient in remote patient monitoring, as well as for delivering homecare support, since constant, real-time observation can be provided without the need for constant human intervention. These functions are particularly crucial to patients who are elderly or have chronic diseases. They keep such patients at home by providing them with individualized care. Robotics should also be mentioned in the case of telemedicine for distant rehabilitation and performance of tele-operated procedures without physical contact, which is the expanded access to health care without unnecessary physical contact-another important advantage during a pandemic or outbreak. Apart from the efficiency of operations, the use of robots in health care deals with important issues such as nursing shortages and the increasing demands of an aging population. Scalable patient-centric care solutions through robotics allow for more resilient and adaptive healthcare systems, much better suited to meeting the needs of patients as well as healthcare providers. Benefits and challenges in AI-Healthcare-Robotics Scenario is depicted in Fig. 1.

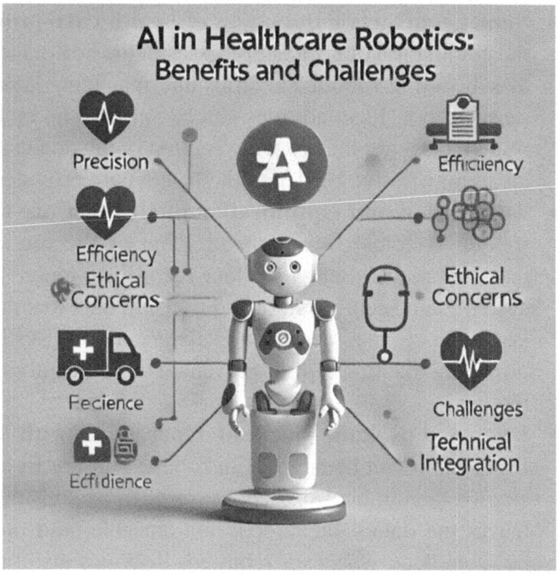

Fig. 1 AI in healthcare robotics

2.3 Benefits of Blockchain Technology and Robotics in Healthcare

- **Improved Data Security and Privacy for Robotics-Driven Systems**: The most sensitive patient data being accessed by the robotics systems have a safe, decentralized repository in blockchain. This means that all records obtained for the care of health, sensor data, or other sensitive information compiled by Lio and other robotic assistants will be encrypted, tamper-proof, and accessed only by qualified people. This reduces the rate of data leaks and is far more private when robots deal with personal and medical data.
- **Increased Interoperability of Health Care Systems and Robotics**: Augmenting the interconnectivity of the different health technologies, such as robots, blockchain offers an open and secured network with standardized data transfer between the parties. In healthcare, robots play the role of personal assistants or even assist the surgeon in surgeries, and they can be used for information flow with other systems, for instance, electronic health records or remote monitoring tools, ensuring an efficient and safe flow of information.
- **Efficient and Transparent Supply Chain Management for Medical Robotics**: The use of blockchain technology will be able to track the provenance of robotic parts such as sensors, parts, and medical instruments used in health care robotics, so only genuine and certified components are used in robotic systems, thus eliminating counterfeit and low-quality materials from any robotic system, which boosts patient safety [24].

- **Smart contracts automation of health care processes**: Some of these can be the automation of critical processes in the health industry involving robotics such as scheduling robotic surgeries and managing cases involving procedures done on the robot. Its conditions can be verified, and payments can be automatically issued to trigger the proper execution of the processes so that these will be less of an administrative burden to the health care providers and the patients.
- **Automation and Optimization of Health-Care Operations**: For example, the robotic systems can automate some of the routine activities like monitoring patients, medication, and data recording. Upon integrating blockchain, these systems can securely store and transmit data to enhance efficiency in operations. For instance, in a medication delivery robotic system, the dosage or administration of medication error is prevented through automatic updates in the blockchain records.
- **Improved Robotic Surgical Operations with Blockchain Authentication**: Blockchain can be adapted in robotic surgery to log and verify each and every step of the surgical procedure, right from consultation to final operation. This leaves the data safeguarded, non-erasable, and provides an audit trail that can be referenced whenever required. Robotics also offers the precision needed for minimal invasive interventions and minimizes recovery times while improving the patient's outcomes.
- **Real-Time Remote Patient Monitoring and Telemedicine**: Robotics and blockchain combine to facilitate sophisticated telemedicine solutions. Robots can monitor patients from remote locations, record real-time vital signs, and provide those results to the medical practitioner. Blockchain is critical for storing robotically gathered data so that it can be accessed at all locations and points; such data provides the basis for especially the aged and chronic patients requiring extended care.
- **Personalized and Adaptive Healthcare Using Robotics**: Robotic homecare systems, such as personal assistants, can be programmed to learn the individual requirements of patients for proper care, like renting mobility aids for them, reminding patients of their daily routine, health status monitoring among others. Blockchain will ensure that all interactions and data collected by the robots will be logged in a safe environment, while patients' healthcare information will be updated in real-time, hence making the healthcare more relevant and time-friendly [25].
- **Efficient and Secure Coordination in Homecare**: The machines can help patients with daily activities in homecare; the blockchain ensures that every action carried out by the systems is safe and traceable. For example, every time a robot dispenses medication or helps with the physical therapy, it will be logged by the blockchain, ensuring that the right treatment gets to the patient while being verified at real-time by their health provider.

2.3.1 Blockchain in Healthcare

The inability to securely manage and retrieve personal health information (PHI) in timely manner is a major challenge encountered by the healthcare industry. For healthcare practitioners to improve patient-physician communication, increase the use of healthcare-related data, and gain a comprehensive understanding of a patient's well-being, efficient handling and retrieval of patient data is essential. Blockchain technology is seen as a key breakthrough that can address these challenges. Blockchain has the ability to greatly enhance healthcare information handling by offering a clear and secure system for manging and PHI access. Better patient care, more effective data processing, and improved overall healthcare outcomes might result from its deployment [26].

- Healthcare organizations can benefit from using blockchain in several ways at once. Initially, it pertains to patient medical records. EHRs may be safely kept in a decentralized system that is impervious to hacking and manipulation using blockchain technology. Second, choices on patient care may be made more quickly by hospitals and other healthcare institutions. The fact that all patient data is kept in a single, round-the-clock system makes it feasible. Quick data access made possible by blockchain technology also expedites the process of issuing medical certifications to healthcare workers.
- What matters most to patients? Most likely, it has to do with protecting their private medical information. That is made possible by blockchain, which grants people control of their medical records. Moreover, no one is able to access patient data without consent. Consensus mechanisms, which are technologies provided by blockchain, ensure that only the patient may read medical information. Wearable technology is also covered by blockchain-powered data security [27]. Patients may securely send their doctors the data that these gadgets capture.

2.3.2 Robotics in Healthcare

Robotics in healthcare transforms patient care, optimizes clinical tasks, and supports overburdened medical staff. Robots are used in a vast variety of applications, particularly in surgical support, interactive roles with patients, and logistics in hospitals. High-precision interventions are made possible by advanced robotics, such as through robotic surgery, as in the example of the use of image-guided systems in minimally invasive interventions. Robots are also used in the care of patients through the provision of companionship, monitoring the vital signs of patients, and assisting them in their daily activities, which is very essential as far as geriatric care is concerned. Other mobile robots disinfect rooms with UV light, deliver supplies, and monitor body temperature, thus enhancing hygiene and safety measures considerably and even in pandemic situations [28]. With AI assistance, onsite data processing through healthcare robots largely minimizes dependence on cloud services, has protected patient data, and ensured being up to date with very strict healthcare regulations. These innovations contribute.

- Robots help in very precise surgery procedures and minimally invasive surgeries.
- Mobile robots are used in hospitals to perform logistical tasks such as disinfection, delivery, and monitoring temperatures.
- An autonomous healthcare robot would be capable of completing the task without relating it to cloud data, thereby providing enhanced privacy for the patients.
- For instance, these sophisticated systems are supposed to carry sensors that improve safety while taking care of patients while interacting with humans.
- Personal assistant robots like Lio help to assist the patient in most day-to-day activities and provide companionship in elder care.
- Robots that possess scene understanding enabled with artificial intelligence could navigate through more complex environments in healthcare settings and then adapt to changes.as much to the efficiency of the setting as a more patient-centric attitude [29].

3 Healthcare

Our health requirements are addressed by healthcare systems and as the population ages, there are even more needs for productivity. In addition, we can appreciate the requirement for the most correct and up to date data such as in the case of COVID 19. For patients recovering from a COVID 19 infection, information regarding a patient's condition or temperature is very useful. With the elderly and the chronic illnesses, patients tend to be more aware of their health and are in search of better health care services [30]. Rather the other way, these days people are embracing patient centric approach instead of focussing on the healthcare profession. Over the years, paradigm changes, where there was a focus on the patient and health care users in the hospital, have decreased in favour of information, computer, and mobile technology advanced universal health care. Recent market trends, like Internet of Medical Things (IoMT) devices allow you to acquire almost any type of market intelligence almost instantly and economically [31]. The processes have been rendered cheaper and more effectual by the presence of ubiquitous and mobile devices. However, more issues have been introduced with the advent of web based IoMT devices.

3.1 Definition

There are three primary parts of healthcare system: (i) Main suppliers of medical treatment services, including physicians, nurses, medical technicians, and hospital administrators (ii) Emergency support services, and (iii) Consumers of health and health-related services, such as individual patients [32]. This study explores health maintenance, which includes remotely managed, technology-driven services delivered by dedicated service providers in order to promote, preserve, or recover beneficiaries' health. Breaches in security and privacy continue to increase in the medical field. While in 2017, over 300 breaches were confirmed, from 2010 to 2017, up to 37 million information might have been compromised [33]. The healthcare system is made up of many interwoven parts that make it a complex ecosystem. The people who offer medical care services, such as physicians, nurses, technicians, and hospital managers, are its main constituents. These experts are essential in the identification, management, and avoidance of diseases. The healthcare system includes not just medical treatment services but also emergency services and consumers of health and health-related services, including individual patients [34]. Critical care is provided in emergencies using services such as ambulances and the emergency room. The patient, therefore, remains the focus of the health system, as s/he represents the final beneficiary of its services, with attention shaped to their experiences and needs.

3.2 Challenges of Healthcare

- Securing and effectively managing the large volumes of on health information generated by routine health operations and service delivery is one of the considerable challenges still facing the industry [34]. Wearable and other health monitoring technologies also produce large volumes of personal health information. Much of the data remains inaccessible, inconsistent across systems, and hard to interpret, utilize, and share. This is derived from multiple sources and maintained in centralized IT systems, making it cumbersome to share and coordinate [35]. Extraction, submission, reception, and consolidation of patient information require significant resources and a lot of time. Ensured access and proper management of information enable health care systems to establish comprehensive views of patients, thereby raising the quality of care, enhancing communication, and maximizing health outcome.
- The healthcare field also struggles with interoperability issues, inaccessible medical records, and a shortage of secure, overarching population health information. Recent public health crises highlight the inadequacies of the current healthcare system. In the end, data security is one of the main issues confronting the healthcare industry [36]. The main store for vital health data is a frequent component of many healthcare organizations antiquated legacy IT infrastructure, leaving it open to ransomware and other breaches.
- Concerns on data security and patient information security in recent times have been exacerbated by security problems with respect to patients' medical records. They profile Beazley, an international cyber insurance provider, which indicates that healthcare institutions accountant for 45% of all ransomware incidents in 2017 [37]. There are rising instances of medical data fraud and theft of medical records. According to the HIPPA Journal, there were over 350 breaches reported in 2017 up from less than 20 in 2009. In the year 2018, the DHHS reported that

over 13 million medical records were compromised during data breaches within the healthcare system [38]. In 2016, there was an average of at least one health data breach each day, and the number of unique patient records affected reached a total of over 27 million. In the year 2018, hackers had unauthorized access to 1.4 million patient records of the UnityPoint Health network, which at that time, represented the largest instance of medical data theft in the USA. Among those were also lab documents of exposed test results, therapy, social security number of the patient and insurance [39].

3.3 Need of Blockchain in Healthcare

High-performance health facilities with the most advanced, contemporary technology are a must in the fast-paced development of healthcare. Blockchain will be a highly crucial feature transforming the sector of healthcare [40]. The health system today also concentrates more on a patient-centric approach by addressing two major issues: (i) well available and accessible services, and (ii) ready availability of relevant medical resources at any given time. Blockchain helps healthcare organizations deliver quality patient care and good services [41]. Health information exchange is yet another time-consuming, repetitive exercise that further inflates healthcare costs; blockchain can make the task more efficient. Through blockchain, citizens can take part in health research programs, and the richer the research and data sharing, the better the outcomes for community health. Traditionally, a centralized database is used to manage the entire healthcare system and its organizations. However, interoperability, data sharing, and protection due to population health management persist [42]. Blockchain technology directly meets all these issues, supporting the capabilities of real-time updates, secure access, data exchange, interoperability, and integrity. Data privacy concerns are major issues, especially in regard to wearables and personalized care. Blockchain offers patients and healthcare providers secure and straightforward methods for capturing, transferring, and accessing data across networks without risking security breaches [43]. This technology will efficiently solve many of the problems in the health care sector, including interoperability, security of data, and accelerating health processes through a fully transparent platform, decentralized, and secure. Probably one of the biggest advantages of blockchain is a feature that lets patient data be shared securely and efficiently, which thereby endows medical professionals with a whole view of a patient's history, effectively upgrading care quality. Another area where blockchain will help in reducing healthcare costs is by reducing redundant data entry and hence administrative burdens [44].

4 Blockchain Based Solutions

More than $1.7 trillion per year is expended by the U.S. healthcare sector, the largest globally. The average yearly cost of healthcare per American is $10,739, surpassing that of citizens in any other country. Close to 18% of U.S. Growth Domestic Product (GDP) is allocated to healthcare. And this share is projected to reach about 20% of U.S. GDP by 2027 without any charge [45]. Strategies and tactics is implementing in healthcare industry for improving quality of care and to address the expansion of pharmaceutical and medical. The healthcare industry is rapidly evolving due to critical challenge, issues, and opportunities arising by technology advancement. The traditional alliance among big pharma, trained medical professionals and big government is no longer the dominant force [46]. Tech giants are strategically entering the healthcare industry to tackle rising medical and pharmaceutical costs, as well as improve access and quality of care. These tech giants are investing heavily in research and development (R&D) to improve consumer health. The current healthcare system primarily focuses on treating symptoms rather than addressing underlying causes of illness. The COVID-19 pandemic might be accelerating the shift towards continuous healthcare, which relies on data-oriented decision-making. Tech companies like Apple, Google, Amazon, and others have introduced wearables that consistently track personal health metrics, like pulse rate, body temperature, respiratory sounds, step count, cycle length variability [47]. The data collected by these wearables can be analysed to understand the impact of factors like diet, sleep, and behaviour on individual health. The following sections explain how blockchain technology can be helpful in solving problems in health care data management. Table 2 highlights the potentials of blockchain for healthcare.

4.1 Data Collection and Storage

In healthcare monitoring, the production of huge amounts of personal data relevant to well-informed, data-driven decisions by these technologies has led to inadequate access levels to some of that data as well as inefficient management thereof, which affects health care. The various periods of a patient's life mandate the collection of a number of items that form a health care information, which is spread out across a large number of systems, making the resultant records incomplete and even inconsistent. Providers are generally set to maintain dominant stewardship of patient information to the detriment of the patients' access [1]. It severely restricts patients and healthcare professionals from seeing the full picture of patient health, which limits possibilities for coordinated care and stagnates improvement in healthcare outcomes through information.

Table 2 Blockchain potentials for healthcare

Classification	Potential applications	Primary advantages
Patient	• Empowering patient • Track health records • Verify recent prescriptions • Secure information exchange with healthcare providers	• Increase trust and transparency • Better collaboration • Personalised patient experience • Higher efficiency, lower costs • Global access to medical records and prescriptions
Regulation and oversight	• Monitoring adherence to standards • Verification via smart contract	• Real-time audit tracking • Automatic privacy regulation enforcement • Data-sharing monitoring without revealing data
Cross-organization processes	• Fund transfers • Medical devices distribution network • Temperature-monitored supply network • Resources	• Automated smart contract payments • Quicker payments and transaction transparency • Universal access to medical records and prescriptions
Management and service centers	• Financial management	• Effective Monitoring • Reduces admin costs • Reliable audits and faster transactions
Pharmaceuticals	• Verify drug provenance • Industry-wide information source	• Tracks and traces pharmaceuticals • Anti-counterfeiting proof • Prevents counterfeit product distribution • Detect pharmaceutical complications

4.1.1 Current Condition of Medical History and Record-Keeping in Healthcare

Attributes of the present healthcare information collection system and records keeping are as follows.

- It heavily depends on face-to-face interactions between patients and physicians.
- The current health care system is able to take advantage of the data to improve patient care.
- It results in an elongated and time-consuming process for seeking healthcare.
- Most healthcare systems lack adequate care for patients because access to information is very limited.
- Due to insufficient data availability, many healthcare systems fail to provide appropriate treatment to patients.

- Availability of exact information is low and it makes the management system of healthcare less effective.
- It provides insufficient security and reliability for healthcare information.

4.1.2 Blockchain Solutions for Healthcare Record Keeping

One of the advantages of blockchain technology is that it favours managing all kinds of digital data where shared write access among several entities is required, and verification of integrity of data and consensus is of utmost importance [48]. All critical medical information can be kept safe through a blockchain. Blockchain technology proves especially beneficial for the purpose of logging the increasingly consistent flow of transactions due to its various uses, which can be of help to the record-keeping in health care. It would be particularly apt in an open space of consumer exchange, where data is continuously generated and older entries lose importance [39]. Currently, the applications of blockchain are seen in personal data protection, DNA data, medical records, and vital medical history data. Using blockchain technology, a healthcare provider can maintain the information of patient records, which are accessible to both patients and physicians at anytime from anywhere with the network.

4.2 Data Sharing and Interoperability

Challenges in identifying patients and the issue of information blocking—where healthcare providers place excessive restrictions on patient data or electronic health information sharing—are problems arising from poor interoperability. The absence of universally recognized patient identifiers significantly hampers efficient healthcare practices [49]. Information blocking practices further disrupt effective healthcare processes. Interoperability becomes especially crucial during crises like the COVID-19 pandemic.

- *Existing data sharing*: Centralized IT systems store all the medical records digitally. Thus, it acts as a major impediment to seamless sharing. Processes for requesting, transfer of data, receiving, and consolidation of patient information become slow and cumbersome and very costly. Fragmented health records, along with non-compatible backend systems, and also due to regulatory constraints, technological advancements relating to managing medical data have been obstructed. Isolated system protocols concerning various healthcare storage systems act as a barrier to an efficient flow of data, retrieval, and analysis. It further imposes limitations on patients' access to their health history through isolated databases.
- *Blockchain based data sharing solutions*: Blockchain technology offers an answer in overcoming interoperability difficulties that exist in healthcare, thus resulting in smooth processes in data sharing. In a permissioned healthcare blockchain,

patient-centered services are issued a unique identifier, or hash ID, with a guarantee of security and user privacy [50]. This occurs most notably through control in data sharing as decryption keys are sent to designated health care providers to ensure security, privacy, interoperability, and even patient-centeredness. It benefits the most both the patient and the health care provider with accurate, up-to-date, and comprehensive medical records. It even creates promising advancements in financial transactions and record management.

4.3 Blockchain Solutions for Financial and Record Management

Problems in billing related to errors in billing and fraud make the entire process cumbersome for the healthcare industry. It is here that blockchain offers an answer by decentralizing the record-keeping process, thereby making the payment processes smoother and rejecting fraudulent transactions. One of the most impressive features of blockchain technology is that it can be used to write self-executing agreements among parties, known as smart contracts, in which terms are written directly into lines of code on the blockchain. These smart contracts are stored on the blockchain at some particular addresses, and if cryptocurrency is received at an address, it activates the contract. The consensus protocol would validate it, meaning less third-party interference, higher traceability, and greater security [51]. This can be very helpful in healthcare for developing an open, efficient, value-based payment system for health care, reducing such old mistrust between physicians and payers. With blockchain records, financial penalties or reimbursements can be tied to specific health parameters, ensuring minimum human error while enforcing value-based payments. Also, blockchain offers clear visibility to insurance companies, which creates trust as it provides information about what action was taken and the claims received. Data from several sources trusted by the blockchain ecosystem can also be collected reliably, eliminating the possibility of data tampering, thus improving the claims process significantly. Blockchain can be the most secured and relied upon way to ensure the integrity and privacy of personal, sensitive health information, thereby giving power and control to both healthcare providers and patients in medical insurance [52].

4.4 Blockchain Solution for Pharmaceuticals

- **Drug traceability**: Counterfeit drugs are a problem in the health care systems of most countries around the world. An estimated 15% of the drugs sold in developing countries are counterfeit. It results in a huge loss of financial value; in fact, the pharmaceutical industry loses $200 billion per annum due to counterfeit drugs. This is an issue that Blockchain can solve by coming up with a drug-tracking system that is transparent and secure within a supply chain. Blockchain will ensure the validity of drugs due to proof offered, like end-to-end tracking [4]. The entire drug-from production to delivery-will be tracked using blockchain and recorded in case of a successful journey, which helps keep the quality and genuineness of drugs intact. Patients will have a chance to validate the genuineness of drugs before a purchase-which will mean no circulated adulterated medicines. The third application of blockchain is to record effects of drugs on patients for continuous analysis and can trace the outcome over time, therefore providing more insightful judgments about the safety and efficacy of drugs.

- **Healthcare supply chain**: Blockchain technology is increasingly being adopted in the pharmaceutical sector, tracking, at all levels, the journey of raw materials, ingredients, and products through the supply chain, to ensure transparency and security. It enables far greater tracing back to the source all the way to the patient [53]. In pharmaceutical research, the technology is also used to protect sensitive information about medical supplies and the integrity of the supply chain. This improves traceability of origin and possibility of authenticating medical products, which turns out to be extremely important for ensuring the safety of the supply chain and preventing its fraud. It is especially relevant in the struggle against the misuse of controlled substances and illegal pharmaceutical trade. For instance, Modum.io is an example of applying blockchain in practice: a Zurich-based start-up that connects blockchain with sensors to enhance the efficiency of pharmaceutical supply chains. The company, Modum.io, in collaboration with SAP Software Solutions and Swiss Post, develops MOD-Sense. The blockchain solution traces temperature-dependent pharmaceutical products based on temperature monitoring [54]. This ensures the quality of drugs along the entire supply chain process. Not only does this preserve the integrity of drugs, but it also assures efficiency in the supply chain, saving on costs while generally improving the efficiency of the pharmaceutical industry.

5 Recent Advancement

Arbabi et al. [42] presented a thorough review of the current developments that provide blockchain-based solutions for the health sector along with a framework that classifies these solutions taking into account the interactions, functional parts, challenges, and benefits. Ramzan et al. [2] provided detailed overview of blockchain technology in healthcare, including an introduction, history, technical details, blockchain technology types and motivations behind the leading healthcare projects using this technology. Wang et al. [33] proposed a PHS framework combining blockchain technology with ACP methods in an enhanced process of diagnosis and treatment. Blockchain creates links of patients, hospitals, and other health organizations for secure sharing and auditibility of their data. A prototype of gout treatment demonstrates how the framework works. Pradhan et al. [55] reviewed blockchain IoT, health care, and relevant recent developments since 2018 till 2023. The current study focuses on different types of blockchain IoT, applied technologies, and various health-care-related applications such as the remote monitoring of patients, safety and security of medical records, and disease prediction. Key challenges, solutions, and comparative insights on metrics have also been discussed.

In another work, Zarour et al. [54] introduced a novel model to evaluate the impact of different blockchain models on Electronic Health Records (EHRs) in the healthcare industry, though Fuzzy Analytic Network Process (F-ANP) and Fuzzy-TOPSIS techniques to provides a scientific approach to select the most suitable blockchain model for ensuring secure and trustworthy EHRs. Ren et al. [56] introduced a blockchain-based intelligent health system based on AI and IoT integration to enhance data security and model reliability. Given heterogeneous healthcare data, the proposed tensor meta-learning model could appropriately process them, and even further improvements in performance would be achieved when labeled data are limited. Palanisamy et al. [57] reviewed the development of blockchain applications in healthcare through 50 studies that appeared in the literature between 2015 and 2020, discussing emerging trends, technical issues, and challenges in adoption. Private and Ethereum-based blockchain systems dominate, and future directions include integration with AI, cloud-based solutions, and parallel architectures for blockchain. Subramanian et al. [58] provided the blueprint and execution of a diabetes blockchain consortium that would help health care providers to effectively prioritize what the diabetic patient requires from health care system during a pandemic, such as vaccination, oxygen beds, telemedicine, diabetes reimbursement, and other related records. Sumaidaee et al. [59] presented a framework that focuses on the InterPlanetary File System (IPFS) as an example of a distributed storage system and it provides use case to elaborate the application of the framework and includes a set of technical indicators. Badr N. et al. [60] proposed a Blockchain-based platform for Bangladesh's healthcare sector using Hyperledger Fabric to uphold the integrity of data as well as confidentiality. The solution will be accommodating in engaging public and private providers in secure collaboration, and it supports national data warehousing for analytics purposes on the healthcare front.

Additionally, Ajayi et al. [61] presented a blockchain-based solution that supports the scalable and secure exchange of EHRs between health-care entities while providers store and manage their respective records in completely isolated blockchain networks; the system detects and prevents unauthorized access to a patient's data and even limits insider threats through stopping compromised EHR uploads into the blockchain. Geng et al. [62] proposed an integrated healthcare service system that is based on blockchain technology to break data silos among health care providers and enhance the secure sharing of information. The system provides protection for patient privacy, guards against data tampering, and ensures improved quality of service using the nonsensical alteration and tracing properties of blockchain, thus sounding proper with intensive experiments that verify its efficiency in bettering health delivery. Myrzashova et al. [63] proposed a blockchain-based framework of federated learning to improve the privacy and security of smart health applications. The proposed approach uses block chain-managed edge nodes and FL for decentralized safe data usage across IoMT, EHR management, and digital health care systems, highlighting the main advantages and future challenges that need to be addressed. Panda et al. [64] presented a blockchain-based emergency detection system using encrypted physiological information for patients, which does not require decryption for action. The system automatically initiates rapid responses to deliver healthcare services through nearby hospitals. Chukwu et al. [15] reviewed blockchain in healthcare, with a focus on the frameworks and implementations of Electronic Health Records (EHRs). This study engages blockchain as a new avenue for trust, security, and privacy issues but discusses performance-related trade-offs in terms of storage and cost.

Similarly, Haddad et al. [65] proposed a solution that is based on the integration of Machine Learning and Blockchain. The rationale behind the combination is that machine learning enables the extraction of the essential information, while blockchain with consensus protocols guarantees transaction security and data authenticity. Dhingra et al. [40] survey blockchain applications within healthcare supply chains, with attention to related topics, such as health record management transparency and the challenge of drug counterfeiting. To this study, the applications that appeared in EHR management were identified and pointed out to be growing but theoretical so far. Vidap et al. [66] developed an electronic healthcare records storage and sharing system based on the Hyperledger Fabric framework, with a permissioned blockchain-based solution to achieve secure and efficient data management. Mišeikis et al. [67] Lio, Developed for Human–Robot Interaction and Personal Care: Fully Autonomous Navigation, Real-Time Data Processing, and Enhanced Safety for Healthcare Setting-Hospital Setting, Especially in COVID-19 Prevalent Areas. Yang et al. [68] presented a new vision for CPS-based homecare robotic systems by reviewing the latest advancements in AI, sensing, materials and communication, and by discussing future challenges in the field of Healthcare 4.0. Sahu et al. [69] introduced the open-source toolkit ENTRI, which combines AMBF with 3D Slicer to offer real-time visualization and control within the image-guided framework of robotic surgeries. Maki et al. [70] conducted a review of recent trends in digital and facility management for healthcare, suggested a causal loop diagram to clarify interactions among key variables, and conducted a survey within Qatar's healthcare sector to assess the level of adoption of digital tools. Sharma et al. [71] provided a model that enables secure data sharing and effective data management in healthcare using event-driven smart contracts, medical record data, and off chain data storage which promotes adoption of blockchain based solution for healthcare. Mallick et al. [64] introduced BCGeo: Blockchain-enabled geospatial service for smart healthcare systems. The framework deals with important challenges, including data privacy, scalability, and patient prioritization to enhance healthcare delivery for newcomers in Bhubaneswar, India. Al-Marridi et al. [72] presented a blockchain-based optimization framework that utilizes local computational resources to enhance the latency, security, and cost-effectiveness of every participating node. Table 3 summarizes these recent advancements, providing a comprehensive overview of their major contribution.

Table 3 Summary of recent advancements

References	Year	Major Contribution
Arbabi et al. [42]	2023	Blockchain solutions for healthcare with a framework on interactions, functions, challenges
Ramzan et al. [2]	2023	Overviewed history, types, and healthcare applications
Wang et al. [33]	2018	PHS framework combining blockchain and ACP for secure patient-hospital data sharing
Pradhan et al. [55]	2021	Blockchain IoT for patient monitoring, record security, and disease prediction
Zarour er al. [54]	2020	F-ANP and Fuzzy-TOPSIS to evaluate blockchain models' impact on EHRs
Ren et al. [56]	2023	AI-Integrated blockchain system for secure data processing in healthcare
Palanisamy et al. [57]	2024	Healthcare blockchain trends, focusing on AI and clod integration
Subramanian et al. [58]	2021	Diabetes blockchain consortium to prioritize healthcare needs in pandemics
Sumaidaee et al. [59]	2023	IPFS-based distributed storage framework for healthcare
Badr N. et al. [60]	2019	Hyperledger blockchain for secure data sharing in Bangladesh's healthcare
Ajayi et al. [61]	2020	Blockchain solution for secure EHR exchange across healthcare providers
Geng et al. [62]	2024	Blockchain system for breaking data silos and securing healthcare information sharing
Myrzashova et al. [63]	2023	Federated learning blockchain for secure smart healthcare applications
Panda et al. [64]	2022	Blockchain based emergency detection system using encrypted patient data
Chukwu et al. [15]	2020	Blockchain EHR frameworks, focusing on security and privacy
Haddad et al. [65]	2021	Integrated machine learning with blockchain for secure transaction processing in healthcare
Dhingra et al. [40]	2024	Blockchain potential in healthcare supply chains and EHR transparency
Vidap et al. [66]	2023	Hyperledger-based EHR system for efficient data management
J. Mišeikis et a. [67]	2020	An autonomous healthcare robot with enhanced safety and COVID-19-specific features
G. Yang et al. [68]	2020	CPS-based homecare robots, reviewing advancements in AI and sensing for Healthcare 4.0
M. Sahu et al. [69]	2024	ENTRI, a toolkit for real-time control in robotic surgeries integrating AMBF and 3D Slicer

(continued)

Table 3 (continued)

References	Year	Major Contribution
O. Maki et al. [70]	2022	Digital healthcare management trends and surveyed Qatar's adoption of digital tools
Sharma et al. [71]	2024	Smart contracts and off-chain storage for secure healthcare data sharing
Mallick et al. [73]	2023	BCGeo for geospatial healthcare services addressing privacy and scalability
Al-Marridi et al. [72]	2024	Blockchain optimization framework for latency, security, and cost

6 Future Research Direction

Scalable blockchain technology has the potential to introduce an overhaul in healthcare data management and safe handling across multiple domains including patient records, medical supply chains, and emergency response systems. Its potential is especially exciting for applications that require strong data integrity and privacy guarantees, such as patient electronic health records, and pharmaceutical logistics. Scalable blockchain networks can also cater to real-time data-sharing requirements of any connected health application-including telemedicine and the Internet of Things-for patient monitoring systems-to make healthcare delivery faster and more accurate. Scalable blockchain solutions can hence achieve further interoperability, security, and operational efficiency in the healthcare systems by enabling smooth data integration and secure, decentralized exchanges of data across a wide healthcare ecosystem. For a scalable and efficient blockchain-integrated health care system, future research directions include the following:

- Improving blockchain scalability for large data is still a matter of great concern. Research into innovative consensus algorithms as well as data partitioning techniques might bring about the efficient processing and storage of health care data on blockchain that would not degrade performance in handling high-volume medical data applications.
- There is a need for the building of powerful data integrity and privacy protocols in blockchain for health care. Research can talk about the advanced forms of cryptographic techniques and permissioned blockchain models that might stop unlawful access with respect to strengthening the protection given to patient data.
- On this, smart contracts are established to reduce the possibility of error and fraud healthcare billing, thus providing an easy reimbursement process. More effort should be brought into value-based billing frameworks in the future studies with clear references on blockchain transparency and security as a basis for making effective payment systems.

- The pharmaceutical supply chain will authenticate the drugs and prevent counterfeiting with blockchain-based traceability solutions. Drugs can be traced from production to patient for improved trust and transparency, especially those that are sensitive to temperature, which require proper handling.
- Semantic interoperability in blockchain-enabled EHRs should be addressed for effective sharing of cross-platform data in healthcare. Stenciled protocols for data formats, access, and transfer can be developed across different healthcare systems to make them more compatible and accessible.
- Smart-contract-based incentive systems may be formulated to incentivize health care providers and patients to share data with each other, where incentives would be received by the data providers for the studies and public health purposes in a secure manner, building collaborations with consideration to privacy.
- The combination of machine learning with blockchain can produce predictive insights without compromising data security and integrity through the blockchain's decentralized model that provides verifiable and tamper-proof healthcare analytics.
- Further research is required to be designed to such a way as to support blockchain architectures enabling low-latency data access, hence boosting transaction processing speeds and then decreasing wait times for accessing healthcare records across networks, especially in real-time applications.

7 Conclusion

The combination of blockchain and robotics presents transformational potential in the health arena, where blockchain may solve some of the major concerns: issues of data security, patient privacy, and interoperability between systems. Since blockchain is decentralized in nature, it provides a safe framework for storing and managing sensitive health data and reduces the chances of data breaches and unauthorized access. Blockchain lets patients own and manage their health data, while smart contracts with data authenticity ensure secure information exchange among health care providers while maintaining security, clarity, and reliability in medical record-keeping. Robotics improves healthcare through greater accuracy in treatment, reduces manual repetition, and helps in the possibility of remote operations for wider accessibility and efficiency in patient care. Blockchain and robotics can go hand in hand to develop a healthcare system that coalesces assured, real-time data alongside sophisticated automation. Robotic systems with blockchain can be used to record procedural data in a secure manner, enabling an indispensable, tamper-proof record that would be extremely useful for analysis and follow-up care. Such technologies still face major challenges, such as scalability issues, regulatory compliance, interoperability, and high implementation costs, before they can be widely applied in healthcare. Only interoperability among various health care systems would ensure that blockchain solutions could be applied universally through standardized frameworks. Although these challenges are substantial, improved data safety, better care for patients, and greater efficiency in operations are some benefits associated with the fusion of blockchain and robotics in healthcare. Continued research and collaboration between healthcare providers, technology developers, patients, and policymakers may shape blockchain and robotics to extensively revolutionize the landscape of health care by building a safe, efficient, and more responsive system. This integration is a step toward a future where the healthcare system would be technologically advanced, and its adaptability to the changing requirements of medicine can be improved.

References

1. M. Prokofieva, S.J. Miah, Blockchain in healthcare. Austral. J. Inform. Syst. **23** (2019b). https://doi.org/10.3127/ajis.v23i0.2203
2. S. Ramzan, A. Aqdus, V. Ravi, D. Koundal, R. Amin, M.A. Al Ghamdi, Healthcare applications using blockchain technology: motivations and challenges. IEEE Trans. Eng. Manage. **70**(8), 2874–2890 (2023). https://doi.org/10.1109/TEM.2022.3189734
3. A. Anjum, M. Sporny, A. Sill, Blockchain standards for compliance and trust. IEEE Cloud Comput. **4**(4), 84–90 (2017). https://doi.org/10.1109/MCC.2017.3791019
4. P.G. Shynu, V.G. Menon, R.L. Kumar, S. Kadry, Y. Nam, Blockchain-based secure healthcare application for diabetic-cardio disease prediction in fog computing. IEEE Access **9**, 45706–45720 (2021). https://doi.org/10.1109/ACCESS.2021.3065440
5. P. Sundaravadivel, E. Kougianos, S.P. Mohanty, M.K. Ganapathiraju, Everything you wanted to know about smart health care: evaluating the different technologies and components of the internet of things for better health. IEEE Consum. Electron. Mag. **7**(1), 18–28 (2018). https://doi.org/10.1109/MCE.2017.2755378
6. S. Khatri, F.A. Alzahrani, M.T.J. Ansari, A. Agrawal, R. Kumar, R.A. Khan, A systematic analysis on blockchain integration with healthcare domain: scope and challenges. IEEE Access **9**, 84666–84687 (2021). https://doi.org/10.1109/ACCESS.2021.3087608
7. B. Bellaj, A. Ouaddah, E. Bertin, N. Crespi, A. Mezrioui, Drawing the boundaries between blockchain and blockchain-like systems: a comprehensive survey on distributed ledger technologies. Proc. IEEE **112**(3), 247–299 (2024). https://doi.org/10.1109/JPROC.2024.3386257
8. M. Maslin, M. Watt, C. Yong, Research methodologies to support the development of blockchain standards. J. ICT Standard. **7**(3), 249–268 (2019). https://doi.org/10.13052/jicts2245-800X.734
9. Z. Zheng, S. Xie, H.-N. Dai, X. Chen, H. Wang, 'Blockchain challenges and opportunities: a survey.' Int. J. Web Grid Serv. **14**(4), 352–375 (2018)
10. J. Yang, C. Ma, D. Li, J. Liu, Mapping the knowledge on blockchain technology in the field of business and management: a bibliometric analysis. IEEE Access **10**, 60585–60596 (2022). https://doi.org/10.1109/ACCESS.2022.3179714
11. K. Yaeger, M. Martini, J. Rasouli, A. Costa, 'Emerging blockchain technology solutions for modern healthcare infrastructure.' J. Sci. Innov. Med. **2**(1), 1–7 (2019). https://doi.org/10.29024/jsim.7
12. M.N.M. Bhutta et al., A survey on blockchain technology: evolution, architecture and security. IEEE Access **9**, 61048–61073 (2021). https://doi.org/10.1109/ACCESS.2021.3072849
13. S. Wang, L. Ouyang, Y. Yuan, X. Ni, X. Han, F.-Y. Wang, Blockchain-enabled smart contracts: architecture, applications, and future trends. IEEE Trans. Syst. Man Cybern. Syst. **49**(11), 2266–2277 (2019). https://doi.org/10.1109/TSMC.2019.2895123
14. G.R. Gray, Understanding blockchain, in *Blockchain Technology for Managers* (2021), pp. 11–41. https://doi.org/10.1007/978-3-030-85716-5_2

15. E. Chukwu, L. Garg, A systematic review of blockchain in healthcare: frameworks, prototypes, and implementations. IEEE Access **8**, 21196–21214 (2020). https://doi.org/10.1109/ACCESS.2020.2969881
16. S. Ozcan, S. Unalan, Blockchain as a general-purpose technology: patentometric evidence of science, technologies, and actors. IEEE Trans. Eng. Manage. **69**(3), 792–809 (2022). https://doi.org/10.1109/TEM.2020.3008859
17. Z. Xing, Z. Shi, A fast and deterministic algorithm for consensus set maximization. IEEE Access **6**, 26175–26180 (2018). https://doi.org/10.1109/ACCESS.2018.2835302
18. M. Belotti, N. Božić, G. Pujolle, S. Secci, A vademecum on blockchain technologies: when, which, and how. IEEE Commun. Surv. Tutor. **21**(4), 3796–3838 (2019). https://doi.org/10.1109/COMST.2019.2928178
19. Y. Shang, Finite-Time weighted average consensus and generalized consensus over a subset. IEEE Access **4**, 2615–2620 (2016). https://doi.org/10.1109/ACCESS.2016.2570518
20. J. Yli-Huumo, D. Ko, S. Choi, S. Park, K. Smolander, Where is current research on blockchain technology?—A systematic review. PLOS ONE **11**(10) (2016). https://doi.org/10.1371/journal.pone.0163477
21. Z. Zheng, S. Xie, H. Dai, X. Chen, H. Wang, An overview of blockchain technology: architecture, consensus, and future trends, in *2017 IEEE International Congress on Big Data (BigData Congress), Honolulu, HI, USA* (2017), pp. 557–564. https://doi.org/10.1109/BigDataCongress.2017.85
22. D.T. Dang, N.T. Nguyen, D. Hwang, A quick algorithm to determine 2-optimality consensus for collectives. IEEE Access **8**, 221794–221807 (2020). https://doi.org/10.1109/ACCESS.2020.3043371
23. R. Mohan, J. Arce, S. Mokhtar, D. Cattaneo, A. Valada, Syn-mediverse: a multimodal synthetic dataset for intelligent scene understanding of healthcare facilities. IEEE Robot. Autom. Lett. **9**(8), 7094–7101 (2024). https://doi.org/10.1109/LRA.2024.3418272
24. M. Dyck et al., Toward safe and collaborative robotic ultrasound tissue scanning in neurosurgery. IEEE Trans. Med. Robot. Bionics **6**(1), 64–67 (2024). https://doi.org/10.1109/TMRB.2024.3349626
25. R.H. Taylor, A perspective on medical robotics. Proc. IEEE **94**(9), 1652–1664 (2006). https://doi.org/10.1109/JPROC.2006.880669
26. S. Ma, S. Wang, W.-T. Tsai, Delay analysis of consensus communication for blockchain-based applications using network calculus. IEEE Wirel. Commun. Lett. **11**(9), 1825–1829 (2022). https://doi.org/10.1109/LWC.2022.3183197
27. R. Ben Fekih, M. Lahami, Application of blockchain technology in healthcare: a comprehensive study, in *Lecture Notes in Computer Science* (2020), pp. 268–276. https://doi.org/10.1007/978-3-030-51517-1_23
28. B. Thamo, V. Voulgaridou, H. Wood, J. Stone, K. Dhaliwal, M. Khadem, Toward robotics-assisted endomicroscopy in percutaneous needle-based interventions. IEEE Trans. Med. Robot. Bionics **6**(1), 110–119 (2024). https://doi.org/10.1109/TMRB.2023.3337869
29. R. Chengoden et al., Metaverse for healthcare: a survey on potential applications, challenges and future directions. IEEE Access **11**, 12765–12795 (2023). https://doi.org/10.1109/ACCESS.2023.3241628
30. M. Akay et al., Healthcare innovations to address the challenges of the COVID-19 pandemic. IEEE J. Biomed. Health Inform. **26**(7), 3294–3302 (2022). https://doi.org/10.1109/JBHI.2022.3144941
31. S.K. Nagarajan, G. Narasimhan, A. Victor, Y. Vaish, P. Tripathi, Blockchain in the healthcare domain and performing various security analysis, in *Blockchain and IoT Based Smart Healthcare Systems* (2024), pp. 114–136. https://doi.org/10.2174/9789815196290124010010
32. O. Arshi, A. Chaudhary, R. Singh, Navigating the future of healthcare: AI-powered solutions, personalized treatment plans, and emerging trends in 2023, in *2023 International Conference on Artificial Intelligence for Innovations in Healthcare Industries (ICAIIHI)*, Raipur, India (2023), pp. 1–6. https://doi.org/10.1109/ICAIIHI57871.2023.10489554

33. S. Wang et al., Blockchain-powered parallel healthcare systems based on the ACP approach. IEEE Trans. Comput. Social Syst. **5**(4), 942–950 (2018). https://doi.org/10.1109/TCSS.2018.2865526
34. E.C. Schiza, T.C. Kyprianou, N. Petkov, C.N. Schizas, Proposal for an eHealth based ecosystem serving national healthcare. IEEE J. Biomed. Health Inform. **23**(3), 1346–1357 (2019). https://doi.org/10.1109/JBHI.2018.2834230
35. W.A.N.A. Al-Nbhany, A.T. Zahary, A.A. Al-Shargabi, Blockchain-IoT healthcare applications and trends: a review. IEEE Access **12**, 4178–4212 (2024). https://doi.org/10.1109/ACCESS.2023.3349187
36. H.-S. Nguyen, M. Voznak, A bibliometric analysis of technology in digital health: exploring health metaverse and visualizing emerging healthcare management trends. IEEE Access **12**, 23887–23913 (2024). https://doi.org/10.1109/ACCESS.2024.3363165
37. R. Ranchal et al., Disrupting healthcare silos: addressing data volume, velocity and variety with a cloud-native healthcare data ingestion service. IEEE J. Biomed. Health Inform. **24**(11), 3182–3188 (2020). https://doi.org/10.1109/JBHI.2020.3001518
38. P. Dhawan, et al., Current and future challenges in point-of-care technologies: a paradigm-shift in affordable global healthcare with personalized and preventive medicine. IEEE J. Transl. Eng. Health Med. **3**, 1–10, Art no. 2800110 (2015). https://doi.org/10.1109/JTEHM.2015.2400919
39. M. Attaran, Blockchain technology in healthcare: challenges and opportunities. Int. J. Healthcare Manage. **15**(1), 70–83 (2020). https://doi.org/10.1080/20479700.2020.1843887
40. S. Dhingra, R. Raut, K. Naik, K. Muduli, Blockchain technology applications in healthcare supply chains—a review. IEEE Access **12**, 11230–11257 (2024). https://doi.org/10.1109/ACCESS.2023.3348813
41. N. Mohamed, J. Al-Jaroodi, I. Jawhar, N. Kesserwan, Leveraging digital twins for healthcare systems engineering. IEEE Access **11**, 69841–69853 (2023). https://doi.org/10.1109/ACCESS.2023.3292119
42. M.S. Arbabi, C. Lal, N.R. Veeraragavan, D. Marijan, J.F. Nygård, R. Vitenberg, A survey on blockchain for healthcare: challenges, benefits, and future directions. IEEE Commun. Surv. Tutor. **25**(1), 386–424 (2023). https://doi.org/10.1109/COMST.2022.3224644
43. K. Katsaliaki, S. Kumar, The past, present, and future of the healthcare delivery system through digitalization. IEEE Eng. Manage. Rev. **50**(4), 21–33 (2022). https://doi.org/10.1109/EMR.2022.3223112
44. A. Singh, K.K. Singh, Blockchain with federated learning for secure healthcare applications, in *Blockchain and Deep Learning for Smart Healthcare* (2023), pp. 35–44. https://doi.org/10.1002/9781119792406.ch2
45. S. Ramzan, A. Aqdus, V. Ravi, D. Koundal, R. Amin, M.A. Al Ghamdi, Healthcare applications using blockchain technology: motivations and challenges. IEEE Trans. Eng. Manage. **70**(8):2874–2890 (2023). https://doi.org/10.1109/TEM.2022.3189734
46. M.A. Islam et al., Distributed ledger technology based integrated healthcare solution for Bangladesh. IEEE Access **11**, 51527–51556 (2023). https://doi.org/10.1109/ACCESS.2023.3279724
47. M. Hiwale, R. Walambe, V. Potdar, K. Kotecha, A systematic review of privacy-preserving methods deployed with blockchain and federated learning for the telemedicine. Healthcare Anal. **3**, 100192 (2023). https://doi.org/10.1016/j.health.2023.100192
48. Y. Zhuang, L.R. Sheets, Y.-W. Chen, Z.-Y. Shae, J.J.P. Tsai, C.-R. Shyu, A patient-centric health information exchange framework using blockchain technology. IEEE J. Biomed. Health Inform. **24**(8), 2169–2176 (2020). https://doi.org/10.1109/JBHI.2020.2993072
49. M. Kassab, J. DeFranco, T. Malas, P. Laplante, G. Destefanis, V.V.G. Neto, Exploring research in blockchain for healthcare and a roadmap for the future. IEEE Trans. Emerg. Top. Comput. **9**(4), 1835–1852 (2021). https://doi.org/10.1109/TETC.2019.2936881
50. V. Louis, P. Maheshwari, Blockchain technology in healthcare, in *Current and Future Trends—A Systematic Review* (2023). https://doi.org/10.21203/rs.3.rs-2613305/v1

51. S.T. Manion, Y. Bizouati-Kennedy, The roadmap, in *Blockchain for Medical Research* (2020), pp. 123–140. https://doi.org/10.4324/9780429327735-18
52. L.D. Costa, B. Pinheiro, W. Cordeiro, R. Araújo, A. Abelém, Sec-health: a blockchain-based protocol for securing health records. IEEE Access **11**, 16605–16620 (2023). https://doi.org/10.1109/ACCESS.2023.3245046
53. S.A. Alzakari, A. Sarkar, M.Z. Khan, A.A. Alhussan, Converging technologies for health prediction and intrusion detection in internet of healthcare things with matrix-valued neural coordinated federated intelligence. IEEE Access **12**, 99469–99498 (2024). https://doi.org/10.1109/ACCESS.2024.3420078
54. M. Zarour et al., Evaluating the impact of blockchain models for secure and trustworthy electronic healthcare records. IEEE Access **8**, 157959–157973 (2020). https://doi.org/10.1109/ACCESS.2020.3019829
55. N. Pradhan, S. Sagar, K. Singh, K. Saini, Blockchain technology for IOT-based healthcare applications, in *Blockchain and IoT integration*, pp. 27–42 (2021). https://doi.org/10.1201/9781003094210-2
56. B. Ren, L.T. Yang, Q. Zhang, J. Feng, X. Nie, Blockchain-powered tensor meta-learning-driven intelligent healthcare system with IOT Assistance. IEEE Trans. Network Sci. Eng. **10**(5), 2503–2513 (2023). https://doi.org/10.1109/tnse.2022.3227317
57. D. Palanisamy, S. Rajendran, P. Venkatesan, Blockchain powered medical sector—application, challenges and future research scope, in *Blockchain and IoT Based Smart Healthcare Systems* (2024), pp. 91–113. https://doi.org/10.2174/9789815196290124010009
58. G. Subramanian, A. Sreekantan Thampy, Implementation of blockchain consortium to prioritize diabetes patients' healthcare in pandemic situations. IEEE Access **9**, 162459–162475 (2021). https://doi.org/10.1109/ACCESS.2021.3132302
59. G. Al-Sumaidaee, R. Alkhudary, Z. Zilic, A. Swidan, An evaluation framework for assessing IPFS performance within a blockchain-based healthcare system, in *2023 IEEE International Conference on Blockchain (Blockchain)*, Danzhou, China (2023), pp. 100–104. https://doi.org/10.1109/Blockchain60715.2023.00025
60. N. Badr, Blockchain or distributed ledger technology what is in it for the healthcare industry?, in *Proceedings of the 11th International Joint Conference on Knowledge Discovery, Knowledge Engineering and Knowledge Management*, pp. 277–284 (2019). https://doi.org/10.5220/0008348902770284
61. O. Ajayi, M. Abouali, T. Saadawi, Secured inter-healthcare patient health records exchange architecture, in *2020 IEEE International Conference on Blockchain (Blockchain)*, Rhodes, Greece (2020), pp. 456–461. https://doi.org/10.1109/Blockchain50366.2020.00066
62. Q. Geng, Z. Chuai, J. Jin, An integrated healthcare service system based on blockchain technologies. IEEE Trans. Comput. Social Syst. **11**(5), 6278–6295 (2024). https://doi.org/10.1109/TCSS.2024.3392591
63. R. Myrzashova, S.H. Alsamhi, A.V. Shvetsov, A. Hawbani, X. Wei, Blockchain meets federated learning in healthcare: a systematic review with challenges and opportunities. in IEEE Internet Things J **10**(16), 14418–14437 (2023). https://doi.org/10.1109/JIOT.2023.3263598
64. S. Panda, A. Mukherjee, R. Halder, S. Mondal, Blockchain-enabled emergency detection and response in mobile healthcare system, in *2022 IEEE International Conference on Blockchain and Cryptocurrency (ICBC)*, Shanghai, China (2022), pp. 1–5. https://doi.org/10.1109/ICBC54727.2022.9805544
65. A. Haddad, M.H. Habaebi, M.R. Islam, S.A. Zabidi, Blockchain for healthcare medical records management system with sharing control, in *2021 IEEE 7th International Conference on Smart Instrumentation, Measurement and Applications (ICSIMA)*, Bandung, Indonesia (2021), pp. 30–34. https://doi.org/10.1109/ICSIMA50015.2021.9526301
66. P. Vidap, A. Bhargav, R. Paswan, A. Jewalikar, blockchain solution to electronic healthcare records, in *2023 International Conference on Intelligent and Innovative Technologies in Computing, Electrical and Electronics (IITCEE)*, Bengaluru, India (2023), pp. 659–664. https://doi.org/10.1109/IITCEE57236.2023.10090970

67. J. Mišeikis et al., Lio-A personal robot assistant for human-robot interaction and care applications. IEEE Robot. Autom. Lett. **5**(4), 5339–5346 (2020). https://doi.org/10.1109/LRA.2020.3007462
68. G. Yang et al., Homecare robotic systems for Healthcare 4.0: visions and enabling technologies. IEEE J. Biomed. Health Inform. **24**(9), 2535–2549 (2020). https://doi.org/10.1109/JBHI.2020.2990529
69. M. Sahu et al., ENTRI: enhanced navigational toolkit for robotic interventions. IEEE Trans. Med. Robot. Bionics **6**(4), 1405–1408 (2024). https://doi.org/10.1109/TMRB.2024.3475827
70. O. Maki, M. Alshaikhli, M. Gunduz, K.K. Naji, M. Abdulwahed, Development of digitalization road map for healthcare facility management. IEEE Access **10**, 14450–14462 (2022). https://doi.org/10.1109/ACCESS.2022.3146341
71. M. Sharma, S. Singh, A. Deep, D. Garg, A. Kumar, Blockchain's frontier: enhancing data security and collaboration for healthcare, in *2024 11th International Conference on Reliability, Infocom Technologies and Optimization (Trends and Future Directions) (ICRITO)*, Noida, India (2024), pp. 1–6. https://doi.org/10.1109/ICRITO61523.2024.10522385
72. A. Al-Marridi, A. Mohamed, A. Erbad, Optimized blockchain-based healthcare framework empowered by mixed multi-agent reinforcement learning. J. Network Comput. Appl. **224**, 103834 (2024). https://doi.org/10.1016/j.jnca.2024.103834
73. S.R. Mallick, R.K. Lenka, V. Goswami, S. Sharma, A.K. Dalai, H. Das, R.K. Barik, BCGeo: blockchain-assisted geospatial web service for smart healthcare system. IEEE Access **11**, 58610–58623 (2023). https://doi.org/10.1109/access.2023.3283776

Face Mask Detection Using Multimedia Data and Deep Learning Techniques

Ahmed J. Obaid, Muthmainnah Muthmainnah, and Christina Innocenti Tumiar Panggabean

Abstract The data from COVID-19 has led us to develop a deep learning system for detecting face masks with high efficiency and accuracy. Our model detects whether a person is wearing a mask based on convolutional neural networks (CNNs) and VGG-16 architecture. In order to enhance the model's accuracy and computational efficiency, we used transfer learning with the VGG-16, which took advantage of pre-trained layers to classify images with various characteristics. Various datasets containing images of people wearing masks and those without were used to fine-tune the model, resulting in reliable mask and face detection. The methods were evaluated extensively using benchmark datasets, ensuring reproducibility and comparison with existing ones. Based on the proposed method, accurate results can be achieved in complex, real-world scenarios that may include occlusions and diverse lighting conditions with a short inference time and efficient memory usage.

Keywords Facemask detection · Deep learning · COVID-19 · Machine learning · People safety

1 Introduction

When people are wearing masks, face mask detection can be used to determine whether they are wearing them. It is actually reverse engineering the face detection algorithm in order to use it for security, authentication, and surveillance purposes.

A. J. Obaid (✉)
Faculty of Computer Science and Mathematics, University of Kufa, Najaf, Iraq
e-mail: ahmedj.aljanaby@uokufa.edu.iq

College of Technical Engineering, National University of Science and Technology, Dhi Qar, Iraq

M. Muthmainnah
Universitas Al Asyariah Mandar, Polewali Mandar, Indonesia

C. I. T. Panggabean
Universitas PGRI Ronggolawe Tuban, Tuban, Jawa Timur, Indonesia

R. Sharma and G. Jeon (eds.), *Data Analytics for Smart Robotics and Its Applications*, Intelligent Systems Reference Library 272,
https://doi.org/10.1007/978-3-031-87697-4_2

Computer vision and pattern recognition are both concerned with the detection of faces. Researchers have studied face detection algorithms extensively over the past few decades. A major breakthrough in face recognition was made in 2001 when traditional machine learning algorithms were combined with handcrafted features to create effective classifiers [1–3]. Because of this approach, features are highly complex and detection accuracy is low. Deep convolutional neural networks (CNNs) have recently been developed to detect faces in recent years [4–7]. As a result of their study, it is strongly recommended that general population use face masks in order to prevent Coronavirus spreading. Furthermore, reopening countries from COVID-19 lockdown have recommended masks as a precaution against infection. Some mechanisms are required to enforce facemask wear in public places in order to mandate their use.

Adding computer vision to deep learning advancements leads to new developments in a wide range of fields of technology [8]. An essential component of deep learning methods are deep neural networks (DNNs), which can be used to detect objects, classify images, and segment images [8, 9]. There are many problems related to computer vision that can be solved with DNNs, such as convolutional neural networks [10]. As well as extracting features, CNNs are capable of identifying and identifying minor differences between facial images, as well as classifying them. With images from a database, a deep learning model is used to distinguish between facemask wearers and non-facemask wearers [11]. Experimental studies have shown that artificial neural networks are a robust method for identifying features from unprocessed data [12]. Facemask classifiers are constructed using convolutional neural networks, and their prediction accuracy is affected by how many layers they contain. Raspberry Pi is used to implement this project, which uses OpenCV and TensorFlow [13].

The spread of COVID-19 has led to an increase in face coverings, which has made it more difficult for facial recognition systems to recognize and classify facial expressions. Despite the best facial recognition algorithm, there can be up to 50% error when comparing face-masked and unmasked images [14, 15]. A facial recognition algorithm rate error occurs when people's faces are sufficiently covered by masks. Face recognition systems are also less likely to be able to identify a face if the mask covers much of the nose. Face recognition algorithms are more accurate when the features of an individual are placed at a distance from each other. Masks reduce accuracy by eliminating or hiding most of the key features of an individual's face.

2 Related Work

Using the YOLO version 2 model, we were able to detect medical face masks using three elements: feature extraction from ResNet-50 and mask detection from YOLO-50. Based on the mean IoU, the authors improved detection performance. Anchor boxes were estimated based on this information. This result led the authors to conclude that Adam optimizer achieved an accuracy of 81% [16]. An algorithm has been developed that distinguishes between mask-wearers and mask-less people using machine learning techniques. PCA was used to analyze the data. In the paper, mask-less people were identified through PCA. Face mask wearers have less difficulty extracting features than people without masks, according to researchers. In addition, accuracy dropped significantly after people wearing face masks were classified [17].

Research in the area has previously focused on gray values and edges of face images based on pattern recognition and face model knowledge. It was an excellent training classifier to use Adaboost [18]. A breakthrough in facial detection technology was achieved with the Viola Jones Detector [19], which enhances the capability to detect facial expressions in real time. As a result of the Viola Jones Detector, Haar [1] was refined, but it was unable to solve real-world challenges and was affected by many factors, including facial brightness and orientation. While Viola Jones would work in well-lit environments, it would not work in dark environments or non-frontal shots. Research by independent researchers has led to the creation of new deep-learning models to detect various facial conditions better. A new generation of deep learning-based detectors has recently demonstrated high extraction abilities and robustness, which is much better than hand-carved features. A two-stage object detector is available as well as a one-stage object detector. A two-stage object detector is available as well as a one-stage object detector. A two-stage detector produces a high detection output at a low speed as well. This seminal work proposes some feature vectors that might contain objects using selective search in the R-CNN model [2]. Typically, CNN models extract feature information from proposal transmissions, and support vector machines (SVMs) identify object classes. However, the second stage of R-CNN [4] requires another-by-one proposal detection, as well as a separate SVM for classification. Stacks of convolutional layers are found in fundamental architectures, such as AlexNet [6] and VGGNet [7]. The ImageNet LSVRC-2012 competition was won by AlexNet, which replaced large kernels with 3×3 multiple kernels in a row to replace the large kernels within AlexNet. In the winning GoogleNet [11] design of ILSVRC-2014, parallel convolution kernels are concatenated together with function maps. For example, convolutions of 1×1, 3×3 and 5×5 have been performed, as well as maxpooling of 3×3. A small convolution extracts features from feature maps, while a larger convolution extracts features at a higher level. By creating skip connections, we are able to prevent deep neural networks from exhausting themselves during training. Facial recognition systems are also based on these architectures for extracting initial features. There are also adaptive K-Nearest Neighbours, adaptive weighted averages, reverse weighted averages, and exponential weighted averages that can be used for face recognition [20, 21].

In addition to analyzing big data, deep learning can recognize patterns and speech, and improve computer vision. The focus of this work is on a few of the most commonly implemented deep learning architectures as well as their applications. The section discusses a number of networks in depth, including autoencoders, CNNs, Boltzmann machines, and deep belief networks. Data that is unlabeled can be processed using unsupervised learning algorithms using deep learning. In [2] the author introduces a CNN model to detect faces correctly, discard non-face aspects of the input image, and process higher resolution regions precisely. The standardisation net stimulates the detection process. With standard VGA images, this model can achieve 14 frames per second on CPU, and can be increased to 100 frames per second on GPU. The Deep Dense Face Detector (DDFD) was used in our research to detect faces with multiple views. An application of the proposed method to recognize faces from different angles does not require segmentation, bounding box regression, or SVM classifiers. The idea of using data augmentation to detect masks in speech was presented by [19], which can be used for communication among surgeons, in forensics, or even in infectious diseases such as Coronavirus. Several ResNet models have been used in their project, as well as GANs with series constancy [22]. A multiclass problem would be the focus of their future research. As a result of the author's [6] provision of three masked face datasets, including MFDD, RMFRD, and SMFRD, it is possible to easily complete projects related to face masks.

3 Proposed Methodology

In the first step, Convolutional Neural Networks are used to layer the input pictures and classify them by using the outputs of the CNNs. Build CNN models from scratch and fine-tune them with image augmentation techniques. To categorize images and evaluate training and validation data accuracy, use VGG-16 [15].

3.1 Convolutional Neural Network

Artificial neural networks are categorized according to their types, but a convolutional neural network uses multiple perceptrons to evaluate input pictures and can learn weights and bases based on several segments from those pictures. Convolutional Neural Networks can make use of local spatial coherence in input pictures to share certain parameters, thus reducing the number of weights. There is no doubt that this procedure is efficient when it comes to memory and complexity. Generally speaking, a convolutional neural network includes the following components:

Convolution layer: Convolutional layers create feature maps by passing kernel matrices over input matrices. Kernel matrices are slid over input matrices during convolution. A matrix is used to multiply each point and display the results on the map. There are a number of disciplines that rely on convolution, such as image processing, statistics, and physics, that utilize this linear operation. There is no limit to how many axes can be convolutioned. It can be computed as follows: If I is a two-dimensional image input, then K is a two-dimensional kernel filter:

$$S(i, j) = \Sigma\Sigma I)m, n)k(i - m, j - n)$$

Non-linear Activation Function: Input signals are modified in a nonlinear way by the activation function. An after-convolutional layer node. In the rectified linear unit active function (ReLU), if the input is 1, the output will be 1, otherwise it will be zero.

Pooling layer: It does not store precise position information about input features in its feature map output, which is a limitation of this layer. The input picture will therefore be transformed into a completely new feature map when cropped, rotated, or adjusted in any way. Our proposal involves down sampling of convolutional layers to solve this problem. Applied after the nonlinearity layer, a pooling layer can be used to sample down. By pooling, the representation is made robust to small input translations as well as invariant to small input translations. According to translation invariance, the values of most pooled outputs stay the same when we translate the input a little bit.

Fully-connected layer: As a conclusion, the final pooling layer of an integrated convolutional neural network feeds the fully connected layer with its output. It is possible for there to be more than one of these levels. Fully linked nodes are those that are linked to one another in the first and second layers as well.

3.2 VGG-16 Architecture

Convolutional neural networks (CNNs) make up VGG16. VGG1-16 is a vision model architecture that has had tremendous success. 92.7% top-5 test precision was achieved on a dataset containing 15 million images, 14 million of which are categorized into 1000 categories (the ImageNet dataset)". A Softmax classifier, 13 convolutional layers, and two fully linked layers are used in this system. To simplify the construction, only 3 × 3 convolutions were layered on top of each other to create the VGG-16 network shown in Fig. 1.

Different layers of convolutions and poolings make up the VGG-16 architecture. An RGB image of 224 × 224 pixels is provided as input to convolution-1 (convo) layers. Convo 1–1 has 64 channels, and convo 1–2 has 64 channels. Both layers

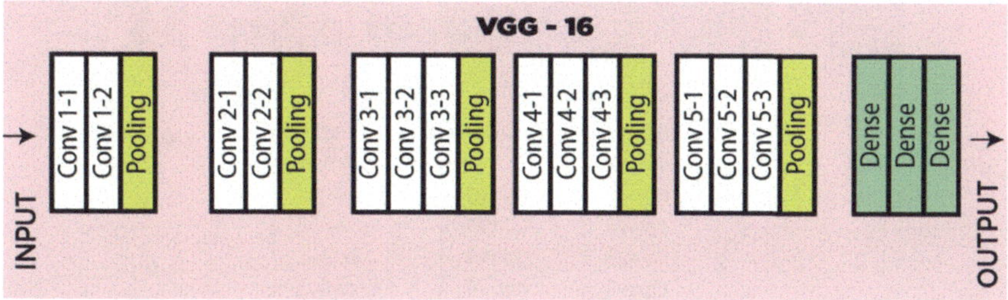

Fig. 1 The architecture of VGG-16

have the same padding. A very small field of receptivity is provided by these filters. In addition to convo1, convo2, and convo3 layers, there are two maximum pooling layers of stride 2 by 2. The following layers consist of two convolutional layers (Con2-1 and Convo2-2), followed by a pooling layer. According to Convolution-2, the filter size is 256 and the maximum pooling layer size is 2 × 2 stades, as with the prior layer. In the following set of filtering stages, there are two sets of three convolution (convo) layers followed by a max pool layer whose filter size is similar to 512 (3 × 3). An initial portion of the convolution is followed by five maxpooling layers that perform spatial pooling. A 2 × 2 pixel window is used for max-pooling, according to the second step. In order for the convolution layers to be dense, there must be three fully connected (connected layers) layers. It depends on the architecture of the system which layer has what depth. '4096' channels are available in the first two layers, and '1000' channels are found in the third layer. In all networks, fully connected layers are configured the same way (Fig. 2).

An optimization method is proposed using VGG16 architectures and transfer learning, in which weights and layers are transferred from a trained model to a untrained model. Besides VGG16, there is also the Oxford-Net, a convolutional neural network architecture named after the Visual Geometry Group at Oxford. During the 2014 ILSVR (ImageNet) competition, a model similar to this one was used. This model combines existing VGG16 with transfer learning for face recognition and detection, with custom data sets with varying backgrounds and lighting conditions for training and validation. Thousands of images were used to compile the data set, both with and without masking. We also used weights from Imagenet to improve performance. A convolutional neural network model, VGG16, is used in this analysis. Convolutional networks are used widely in computer vision. The VGG16 is one of these networks. The proposed model of better classification removes the top layer of the existing model that is fully connected, and replaces it with the flatten,

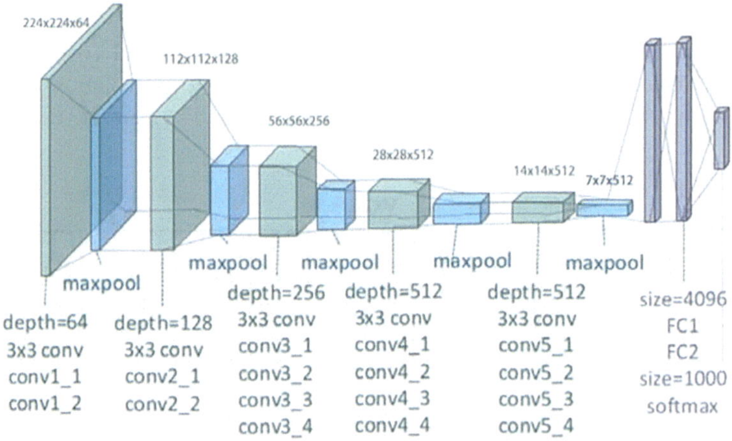

Fig. 2 The layered architecture of VGG-16

dense, and dense softmax layers. In order to prevent overfitting, certain values are dropped at random using the dropout. There is a softmax layer in an image that can be used to multiclassify facial features. All layers but the last have been activated through the ReLu mechanism. 80 percent of the images in this dataset are used during the training phase, and the remaining 20 percent are used during confirmation. As part of our preprocessing steps, all RGB images are resized 128×128 pixels.

4 Result Analysis and Discussion

Our experiments are based on benchmark datasets, and we evaluate performance using widely recognized benchmark metrics, as described in the previous section. Detailed descriptions of both aspects will be provided in this section to assist replication of the method presented here. Experiments were conducted using a benchmark dataset as a starting point. This dataset was selected because it is widely adopted and recognized by researchers. An objective evaluation of the proposed method based on comparisons with other existing methods is conducted here using a benchmark dataset. Moreover, the method was evaluated using popular standard metrics. The metrics serve as quantifiable indicators of how effective and efficient the proposed approach is. Results are credible and reliable because they are based on well-established metrics. Additionally, these metrics allow for straightforward comparisons with earlier studies, which allows a comprehensive assessment of how far the methodology has progressed and how much more work needs to be done. In order to recreate the proposed method, it is necessary to explicitly specify the benchmark dataset and the standard metrics, which are used in the experimentation procedure. Researchers and practitioners can replicate and validate the study's findings more easily if these foundational elements are clearly defined. Thus, the proposed method will be more reproducible as a result of the experimentation process. Through this transparency, the announced results can be verified and further advancements in the field can be made based on the methodology proposed.

An image classification model and a segmentation model are proposed. Python 3.7 was used to develop these models, along with popular machine learning libraries, Keras and Tensorflow. Jupiter Notebook online experiments were conducted on Anaconda using an HP GeForce(R) i7 12th Gen Intel(R) Core(TM) i7-12700H with 16.0 GB of RAM, all in conjunction with an HP GeForce(R) i7 12th Gen Intel(R) Core(TM) i7-12700H processor. As part of this study, the models were trained and tested using CovidxCT-2A dataset, which was centered around a particular problem domain. Face masks can be detected in real-time by collecting data which is used to train a model. As well as images of mask-wearing and non-masking people from Google and Bing, we also used Kaggle datasets, such as those from the RMFD (Fig. 3).

Furthermore, as shown in Fig. 4, the loss function versus learning rate during training is plotted, and as shown in Fig. 5, the same loss batches processed in a similar manner are plotted. In addition, the model accuracy continues to rise in

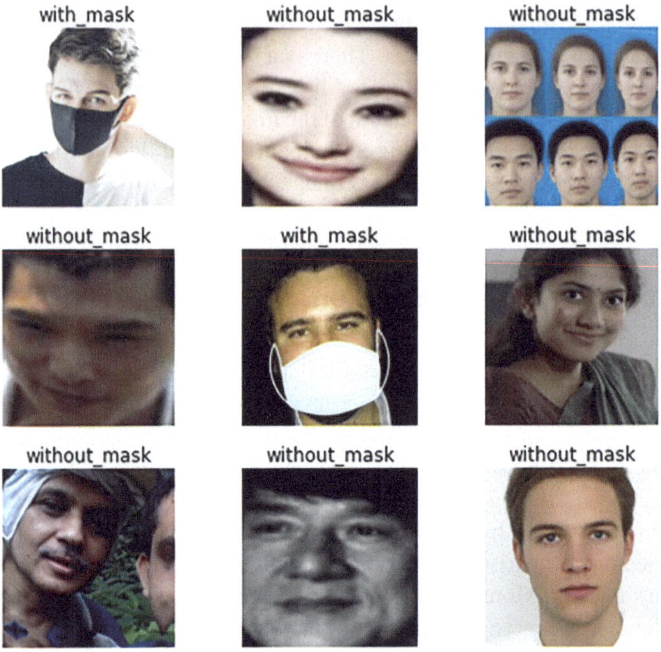

Fig. 3 Sample of the images for with_mask and without_mask

different epochs before stabilizing as shown graphically. Based on a comparison with recent methods, the proposed model shows high accuracy and low memory consumption in face and mask detection. We have had to put in considerable effort to resolve the imbalance problem in the existing dataset, which has resulted in highly suitable unbiased datasets for mask detection. (Fig. 5).

Datasets containing masks and non-masks are the primary focus of our research. There are, however, many kinds of masks, in addition to surgical/cloth masks, in the

Fig. 4 Loss versus learning rate

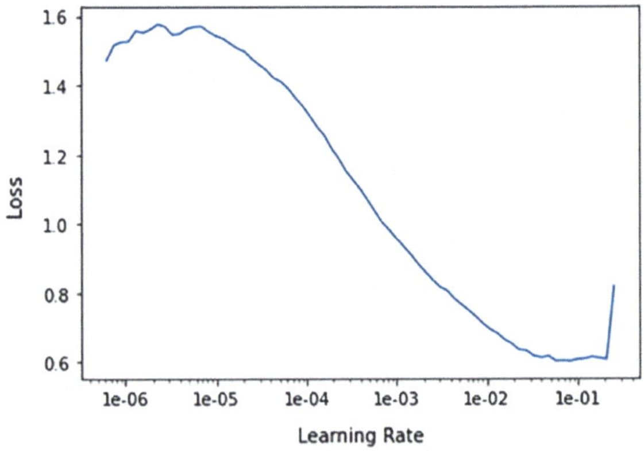

Fig. 5 Training and validation loss for batches processing

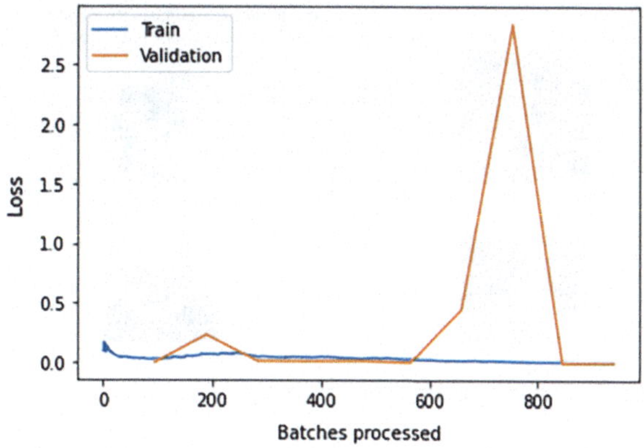

mask class. As an example, Fig. 6 illustrates how objects like people, hands, hair, and food can obstruct ROI. This occlusion correlates with the detection of faces and masks. It is difficult to balance accuracy and computation time when it comes to face detection. We propose a predictor for image complexity in this paper. It categorizes data into soft and hard images at the outset and then distinguishes masks from no masks at the end using a facemask classifier. We are primarily concerned with the hardness or softness of an image. What is the difference between a hard and a soft image (Fig. 7).

Fig. 6 Confusion matrix for with_mask and without_mask

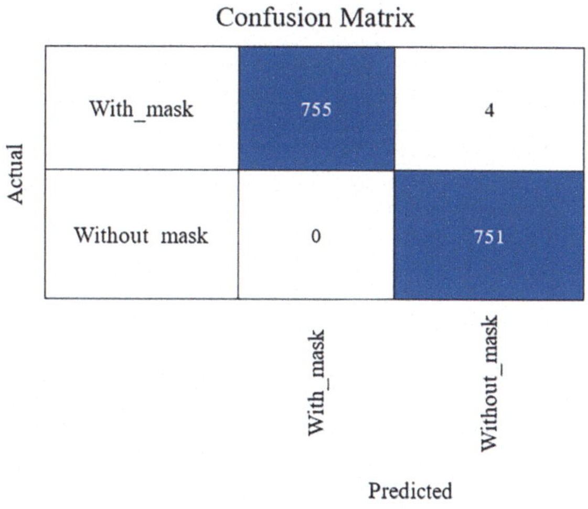

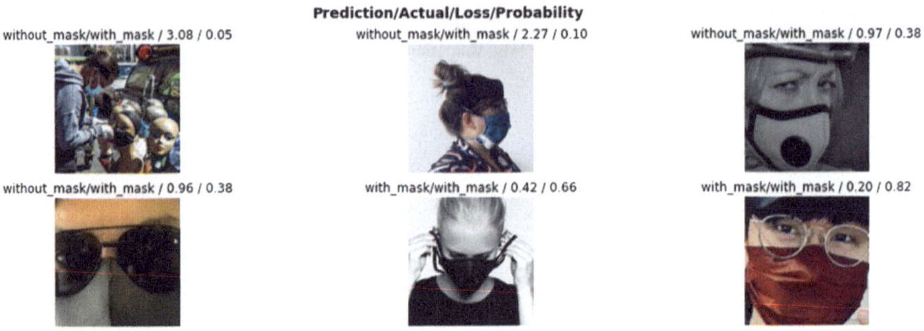

Fig. 7 Resultant outcome of mark detection

5 Conclusion

A CNN-based and transfer-learning method is used to detect mask wearers and non-maskers in this proposal. Public health monitoring is ideal for the solution due to its accuracy, speed, and compatibility with various lighting conditions. There is no doubt about the robustness and efficiency of this technique when compared to others. Further research will be focused on improving robustness against mask types and occlusions to ensure that deep learning-based models can enhance health and safety measures in complex environments.

References

1. L. Nanni, S. Ghidoni, S. Brahnam, Handcrafted vs. non-handcrafted features for computer vision classification. Pattern Recognit. **71**, 158–172 (2017). https://doi.org/10.1016/j.patcog.2017.05.025
2. Y. Jia, et al., Caffe: Convolutional architecture for fast feature embedding, in *Proceedings of the 22nd ACM international conference on Multimedia*. Orlando, Florida, USA: ACM (2014), pp. 675–678. https://doi.org/10.1145/2647868.2654889
3. S.P. Yadav, M. Jindal, P. Rani, V.H.C. de Albuquerque, C. dos Santos Nascimento, M. Kumar, An improved deep learning-based optimal object detection system from images. Multimed. Tools Appl. **83**(10), 30045–30072 (2024).
4. P. Sermanet, D. Eigen, X. Zhang, M. Mathieu, R. Fergus, Y. LeCun, OverFeat: integrated recognition, localization and detection using convolutional networks (2013). arXiv. https://doi.org/10.48550/ARXIV.1312.6229.
5. X. Fan, M. Jiang, RetinaFaceMask: a single stage face mask detector for assisting control of the COVID-19 pandemic, in *2021 IEEE International Conference on Systems, Man, and Cybernetics (SMC)*, Melbourne, Australia. IEEE (2021), pp. 832–837. https://doi.org/10.1109/SMC52423.2021.9659271
6. D. Erhan, C. Szegedy, A. Toshev, D. Anguelov, Scalable object detection using deep neural networks, in *Proceedings of the IEEE Conference on Computer Vision and Pattern Recognition* (2014), pp. 2147–2154. Accessed 28 Oct 2024. [Online]. Available: http://openaccess.thecvf.com/content_cvpr_2014/html/Erhan_Scalable_Object_Detection_2014_CVPR_paper.html

7. J. Redmon, You only look once: unified, real-time object detection, in *Proceedings of the IEEE Conference on Computer Vision and Pattern Recognition* (2016). Accessed 28 Oct 2024. [Online]. Available: http://175.27.250.89:10000/media/attachment/2024/08/YOLOV1.pdf
8. Y. LeCun, K. Kavukcuoglu, C. Farabet, Convolutional networks and applications in vision, in *Proceedings of 2010 IEEE International Symposium on Circuits and Systems*, Paris, France. IEEE (2010), pp. 253–256. https://doi.org/10.1109/ISCAS.2010.5537907
9. A. Krizhevsky, I. Sutskever, G.E. Hinton, ImageNet classification with deep convolutional neural networks. Commun. ACM **60**(6), 84–90 (2017). https://doi.org/10.1145/3065386
10. G. Cifuentes-Alcobendas, M. Domínguez-Rodrigo, Deep learning and taphonomy: high accuracy in the classification of cut marks made on fleshed and defleshed bones using convolutional neural networks. Sci. Rep. **9**(1), 18933 (2019). https://doi.org/10.1038/s41598-019-55439-6
11. B. Qin, D. Li, Identifying Facemask-wearing condition using image super-resolution with classification network to prevent COVID-19, 14 May 2020. https://doi.org/10.21203/rs.3.rs-28668/v1
12. P. Rani, S. Verma, S.P. Yadav, B.K. Rai, M.S. Naruka, D. Kumar, Simulation of the lightweight blockchain technique based on privacy and security for healthcare data for the cloud system. Int. J. E-Health Med. Commun. IJEHMC **13**(4), 1–15 (2022)
13. K. Zhang, Z. Zhang, Z. Li, Y. Qiao, Joint face detection and alignment using multitask cascaded convolutional networks. IEEE Signal Process. Lett. **23**(10), 1499–1503 (2016). https://doi.org/10.1109/LSP.2016.2603342
14. R. Materese, NIST launches studies into masks' effect on face recognition software." NIST (2020)
15. S. Verma, et al., An automated face mask detection system using transfer learning based neural network to preventing viral infection, Expert Syst., p e13507, 2024.
16. M. Loey, G. Manogaran, M.H.N. Taha, N.E.M. Khalifa, A hybrid deep transfer learning model with machine learning methods for face mask detection in the era of the COVID-19 pandemic. Measurement **167**, 108288 (2021). https://doi.org/10.1016/j.measurement.2020.108288
17. A. Singh, et al., Smart Traffic monitoring through real-time moving vehicle detection using deep learning via aerial images for consumer application. IEEE Trans. Consum. Electron. 1–1 (2024). https://doi.org/10.1109/TCE.2024.3445728
18. S.E. Eikenberry et al., To mask or not to mask: Modeling the potential for face mask use by the general public to curtail the COVID-19 pandemic. Infect. Dis. Model. **5**, 293–308 (2020). https://doi.org/10.1016/j.idm.2020.04.001
19. S. Sethi, M. Kathuria, T. Kaushik, A real-time integrated face mask detector to curtail spread of coronavirus. Comput. Model. Eng. Sci. **127**(2), 389–409 (2021). https://doi.org/10.32604/cmes.2021.014478
20. C.-Y. Fu, W. Liu, A. Ranga, A. Tyagi, A.C. Berg, DSSD : Deconvolutional single shot detector. *arXiv*. https://doi.org/10.48550/ARXIV.1701.06659
21. P. Rani, P.N. Singh, S. Verma, N. Ali, P.K. Shukla, M. Alhassan, An implementation of modified blowfish technique with honey bee behavior optimization for load balancing in cloud system environment. Wirel. Commun. Mob. Comput.Comput. **2022**, 1–14 (2022)
22. G. Ansari, P. Rani, V. Kumar, A novel technique of mixed gas identification based on the group method of data handling (GMDH) on time-dependent MOX gas sensor data, in *Proceedings of International Conference on Recent Trends in Computing: ICRTC 2022* (Springer, Berlin, 2023), pp. 641–654

Digital Twin Robots for Securing IoT Based Healthcare 4.0 Application: Key Technologies, Integration Trends and Recent Advancement

Vats Raj Kashyap, Bharat Bhushan, Yousef Farhaoui, and Ahmed Alkhayyat

Abstract Healthcare 4.0 involves integrating robotics into settings where humans predominate. Developing human-centred robotics solutions that provide safe and dependable human–robot interaction toward human–robot symbiosis is one of the primary problems in such a setting. Even though there have been significant achievements, healthcare still faces problems like data blockage and threats to information privacy. Healthcare 4.0 emerged which sees common implementation of technologies such as Cyber-Physical Systems (CPS), IoT, and machine learning to transform the healthcare landscape from being reactive towards predictive care models. Green anticipates sophisticated robotics and artificial intelligence improving healthcare by reading emotions, providing compassionate care, empowering caregivers, minimizing mistakes, and enabling real-time health monitoring. This paper analyses the changing face of healthcare as it relates to robotics, including its origins and growth through different departments such as hospitals and rehabilitation settings, surgical procedures with related practices. It also shows the effect of teleoperation in enabling distance care. Finally, the paper concludes with a review of recent accomplishments in robotics demonstrating their significant potential to drive future healthcare delivery.

Keywords Healthcare · Industry 4.0 · Digital twins · Cyber physical systems · Robotics · Internet of things

---

V. R. Kashyap (✉) · B. Bhushan
Department of Computer Science and Engineering, School of Engineering and Technology, Sharda University, Greater Noida, India
e-mail: vatsraj966@gmail.com

Y. Farhaoui
IMAI Laboratory, T-IDMS Faculty of Sciences and Techniques of Errachidia, Moulay Ismail University of Meknès, Errachidia, Morocco
e-mail: y.farhaoui@fste.umi.ac.ma

A. Alkhayyat
College of Technical Engineering, The Islamic University, Najaf, Iraq
e-mail: ahmedalkhayyat85@iunajaf.edu.iq

R. Sharma and G. Jeon (eds.), *Data Analytics for Smart Robotics and Its Applications*, Intelligent Systems Reference Library 272,
https://doi.org/10.1007/978-3-031-87697-4_3

1 Introduction

Robotics with its highly advanced technologies is very useful to do work which is dangerous, dirty, dull or even impossible for humans. Healthcare, manufacturing and space exploration are among the industries that have been most affected by robotics [1]. Robots in hospitals—medical, rehabilitation, mobile and humanoid robots are re-defining healthcare services. Surgical robots, exemplified by da Vinci type knee systems perfect accuracy in the operating room and postoperative recovery, while rehabilitation kit return motion); transport (mobile sys-to economy logistics), and care for sick individuals [2]. The human-like behaviour of humanoid robots can help in the implementation of healthcare interactions. The robots for healthcare end up saving time and reducing errors in it helps in diagnosis, rehabilitation and also as the labour they work can easily be replaced hence the senior care will take a lot less time from the medical professionals which can now assist to another type of intensive care. They also reduce risk by replacing human labour with machines useful for pandemics [3].

In this approach, Healthcare 4.0 modernize healthcare by blending cyber-physical systems (CPS) with the Internet of Things (IoT), as well as leading-edge technologies such as machine learning, cloud computing, blockchain and fog computing to enable healthcare from reactionary to predictive, preventive and person-centred processes [4]. This model drives increased efficiency, economic benefits and optimizes patient care by leveraging real-time data from centralized devices to serve personalized treatment and diagnostics. Combining CPS and IoT for continuous health monitoring, cloud computing for cost-effective data storage Secure Immutable Health Records via Blockchain While several obstacles such as those related to data management, scalability and security issues exist in Healthcare 4.0, it is still a significant move towards more efficient technology-rich healthcare solutions [5, 6].

The transition from Healthcare3.0 to Healthcare4.0 places a great deal of emphasis on Human–Robot Interaction that referred to as (HRI), which has been developed in such a way robots are intended to not just perform tasks, but also understand, interpret and respond to human emotions [7]. This makes it possible for robots to easily communicate with the infrastructure of healthcare and thus, exploit CPS, IoT and intelligent systems. Human–Robot Symbiosis advancements are enabling more direct, physical collaboration by removing the barrier between human and robot, enhancing safety and efficiency [8]. Caregiving efforts are also reinforced by teleoperation, or the ability for healthcare workers or even patients to remotely control robots. Using the term CPS-Human–Robot System (CPS-HRS) as a guiding principle, near future generation of AI enabled human–robot interacting in cyber physical spaces will augment increased demands for caregiver support, where robots will now play an essential role to deliver healthcare [9, 10].

Despite the continuous advancements in robotics in healthcare over recent years, it is crucial to focus on developing advanced techniques and security frameworks tailored for the next generation of healthcare robotics. Robotics in healthcare enhances precision, efficiency, and patient care through automation, surgical assistance, and personalized rehabilitation. The major contribution of this paper is the overview of integrating robotics in the field of healthcare4.0. A summary of the contribution of the paper is enumerated below.

- This paper presents an overview of robotics and its types and also discussed the two most prominent robotic-assisted platform—Da Vinci Surgical System and Mako Surgical System with application of robotics.
- This paper gives an overview of evolution of healthcare from Healthcare1.0 to Healthcare4.0 highlighting key technologies used in Healthcare4.0 specifically Cyber Physical System, Internet of Things, Cloud Computing, Machine Learning and Fog Computing and discussed the challenges in healthcare.
- This paper contributes towards highlighting the integration of robotics in healthcare for improved patient outcomes, enhanced Patient Experience, reduced costs, better communication, comprehensive care for complex needs and increased accessibility.
- This paper throws light on the major recent advancements in the use of robotics for healthcare for enhancement in the field of healthcare including patient management, hospital management, surgery, medicine etc.

The remainder of the paper proceeds as follows. Section 2 presents Robotics is an important aspect of the health sector as it provides high-functionality in terms of undertaking the most complex tasks, alongside increased efficiency and reduced risk for the experts. Section 3 highlights Healthcare 4.0 is an advanced technology-related approach that offers a move from reactive to pro-active patient care, in which healthcare systems will enable well-being by predicting your health in pre-symptomatic early stages. Section 4 emphasized the role of HRI more effectively in caregiving by perceiving emotions and cooperating with interactions. Finally, the paper concludes in Sect. 5 with role of Robots in Healthcare4.0 and recent advancements of robotics in healthcare.

2 Robotic

Robotics is a complex and interdisciplinary field that integrates computer science, engineering and technology to produce, build up function robots [11]. These items were intended to perform tasks that are typically repetitive, dangerous or beyond human scale. Robotics are widely utilized in many industries such as manufacturing, healthcare or space exploration and for domestic services. Robotics is the study and development of robots, a field that includes mechanical engineering, electrical engineering, computer science and others. Robots word is derived from the Czech word "robota" which basically means labour by force and was very first seen in Karel Capek's play back 1920 [12]. Henceforth, the age of robotics was initiated in the early twentieth century due to technological breakthroughs and needs from sectors such as manufacturing and aircraft. The first industrial robots came from these two sources: the digitally operated ones introduced by Joe Engelberger, along with his partner George Devol (in the 1950s), which was in fact “the very first commercially applied robot”, and also ultimate of UNIVAC [13]. Since that time, wide range of applications have emerged from robotics in health care to defence work force and entertainment and home use. At the moment, there are roughly 8 million robots in use globally. While home health robots are still in research and testing stages, it has been predicted that healthcare robot sales will grow as the world’s population ages, there is a shortage of health care providers available to care for them, and pressure increases on quality improvement. Return it and affordable rates [14].

2.1 Types of Robots

The multidisciplinary area of robotics focusses on creating, constructing, and managing robots to carry out tasks-often in an autonomous or semi-autonomous manner. Robotics, which combines AI, computer science, and engineering, improves automation, productivity, and creativity in a variety of sectors [15]. The types of robots have been explained further and also summarized in Table1.

2.1.1 Medical Robots

It includes Precision operations, rehabilitation assistance, and even patient involvement in the form of medicine dispensing robots are all examples of how medical robots have grown to become game changers. Examples include the da Vinci surgical system, which is meant to facilitate minimally invasive procedures, and robotic prosthesis for mobility aid. These robots aid in accurate surgery, shorten healing times, and support healthcare workers; this provides enhanced patient experiences, which benefit both patients who receive timely treatment when needed and health care professionals who deliver more efficiently [16].

2.1.2 Rehabilitation Robots

It can help people restore motor movement after an accident or neurological condition. Providing repeated, exact motions recuperation assistance. Exoskeleton derivation assists therapists in daily training by motivating them for therapy and strengthening their joints through robotics, which augment efficiency and productivity in rehabilitative exercise processes with many variables to maintain diurnal workout programs improving patient outcomes. These robots also assist patients by ensuring that they perform exactly what is required to promote quicker than expected improvement. Their inclusion into rehabilitation programs has altered the way physical therapy is delivered [17].

Table 1 Summary of various types of robots

Robot type	Primary use	Key characteristics
Industrial robots [16]	Production and installation	Articulated arms, SCARA, and Cartesian types; accurate, repeatable jobs in manufacturing
Agricultural robots [16]	Farming and crop management	Includes harvesting robots, drones, and weeding robots; automates labour-intensive farm activities
Exploration robots [17]	Exploration in space, the seas, and hazardous zones	Built to survive harsh environments; examples include space rovers and underwater vehicles
Military robots [17]	Surveillance, logistics, and combat support	UAVs (drones) and UGVs are utilized for reconnaissance, search and rescue, and remote operations
Mobile robots [18]	Navigation and transportation	Autonomous mobile robots (AMRs), AGVs, and drones use sensors and AI for mobility
Telepresence robots [18]	Remote Presence and Interaction	equipped with cameras and speakers; utilized in healthcare, education, and corporate environments
Humanoid Robots [19]	Research, caring, customer service	Human-like look; can communicate with others and replicate human duties
Swarm robots [19]	Coordinated group tasks	Work in big groups with decentralized control; utilized for environmental monitoring and search missions
Service robots [20]	Customer service, healthcare, and household responsibilities	include household (cleaning), medicinal (surgical), and instructional robots
Collaborative robots [20]	Assisting workers in industrial environments	Safe for intimate human collaboration; utilized for jobs like as packing, inspection, and material handling

2.1.3 Mobile Robots

A mobile robot is a self-contained machine that can determine its location in its surroundings and carry out activities without human involvement. The Creator AGV robotic platform is employed directly here by utilizing unambiguous references. They contribute to the safer and more efficient operations of logistics, healthcare, and manufacturing. These robots can perceive the surroundings and function alone [18].

2.1.4 Humanoid Robots

According to leading experts, humanoid robots are designed to mimic human movements and communicate with one another. They may provide customer service, education, and health care services with a single tele-operational camera and sensors powered by artificial intelligence. Sophia and Pepper, both humanoid robots that converse with you, recognize who is speaking to them, and otherwise assist you, provide some examples: These robots communicate with humans, and they are increasingly being used in industries that need a sense of human presence [19].

2.1.5 Service Robots

Service robots are made to help people by carrying out practical activities in a variety of settings, either fully or partially on their own. Service robots work in homes, workplaces, healthcare facilities, and public areas, in contrast to industrial robots, which are usually limited to manufacturing. Cleaning, delivery, security, surgical assistance, and even friendship is among the duties they provide. Service robots that are outfitted with sensors, artificial intelligence, and machine learning are able to naturally interact with people, recognise items, and adjust to changing circumstances. Service robots are becoming more and more essential as technology develops to boost output, improve people's quality of life, and solve labour shortages in numerous industries [20]. These various types of robots are summarised in Table 1.

2.2 Types of Robotics-Assisted Platform

A robotics-assisted platform combines specialised software with robotic technology to improve accuracy, control, and productivity in a range of jobs. These platforms, which are frequently used in industries like manufacturing, logistics, and surgery, offer improved accuracy and support, enabling people to carry out difficult, high-stakes activities more safely and successfully with robot assistance [21].

2.2.1 Da Vinci Surgical System

The System is a robotic-assisted platform that allows surgeons to do minimally invasive procedures with excellent precision, control, and flexibility. The method is extensively used in operations affecting the heart, prostate, urology, gynaecology, and general surgery, and it is intended to enhance patient outcomes by decreasing trauma, lowering infection risk, and allowing for faster recovery [22]. The Da Vinci Surgical System working have been explained below.

- The Surgeon Console, Patient Cart, and Vision System are the three primary parts of the Da Vinci Surgical System. The vision system offers 3D, high-definition visualisation for improved surgical precision, the patient cart has robotic arms and surgical instruments, and the console enables precise control [23].
- From Surgeon console, surgeon control robotic arms using foot pedals and hand controllers. The surgical field can be magnified 10–12 times with this setup's 3D HD vision, enabling more accurate and detailed movements than are feasible with a manual tool [23].
- Robotic arms are scaled down for greater accuracy, but they nevertheless imitate the surgeon's movements in real time. They can do complex rotations and pivots for intricate surgical procedures because Endo-Wrist technology allows them to have a wider range of motion than human wrists [24].
- By providing a sharp, enlarged picture of the surgical site, a 3D high-definition camera improves spatial awareness and depth perception. Even in cramped areas, surgeons can do delicate procedures with extreme precision thanks to this visual aid [24].
- By enabling surgery through tiny incisions (1–2 cm), the technique lessens the amount of bodily harm. Tremor-filtration technology, which eliminates accidental hand motions for increased accuracy, helps surgeons work with precise control in confined places [25].
- Automatic arm locking in the event that the surgeon's attention wanders and real-time error detection are two examples of safety features. By facilitating a comfortable working position, this device enhances surgical results, shortens recovery periods, minimises scarring, and decreases operator fatigue [25].

2.2.2 Mako Surgical System

The system is a robotic arm-assisted surgical platform intended primarily for orthopaedic surgeries, notably complete hip and knee replacements. It improves the surgeon's capacity to do accurate, minimally invasive procedures, resulting in better patient outcomes [26]. The Mako Surgical System explained below.

- Designed mainly for joint replacement procedures, the Mako Surgical System combines cutting-edge robotic technology with 3D imaging software. It helps surgeons plan and carry out exact incisions and placements for knee, hip, and other joint implants by giving them tactile and visual feedback in real time [27].
- The Mako system generates a customised 3D model of the patient's anatomy from a CT scan. With the use of this model, surgeons can create an exact surgical plan in advance of the procedure, increasing precision and aligning the body for the best possible implant fit [27].
- Robotic Arm Assistance: The Mako system's robotic arm precisely directs the surgeon's hand during surgery. It lowers the possibility of mistakes and permits accurate bone excision and implant placement based on the patient's anatomy by limiting movements to the predetermined parameters [28].

- Real-Time Adjustments: During surgery, the Mako system gives surgeons real-time input, enabling them to modify bone incisions in response to current circumstances. This flexibility guarantees that each patient's implant placement is unique, fostering joint stability and optimum performance [28].
- Robotic Arm Assistance: The Mako system's robotic arm precisely directs the surgeon's hand during surgery. It lowers the possibility of mistakes and permits accurate bone excision and implant placement based on the patient's anatomy by limiting movements to the predetermined parameters [29].
- Real-Time Adjustments: During surgery, the Mako system gives surgeons real-time input, enabling them to modify bone incisions in response to current circumstances. This flexibility guarantees that each patient's implant placement is unique, fostering joint stability and optimum performance [29].

2.3 Need of Robots in Healthcare

According to NASA, a robot is a machine that can perform a complicated series of operations automatically or under the control of an operator [30]. A healthcare robot is a gadget that can do certain health-related duties. Medical facilities have lately been developed in large hospitals such as Siriraj Hospital in Bangkok. Healthcare in Thailand includes recognizing symptoms and disorders, diagnosing and treating ailments, and preventing them from repeating. Healthcare robots can not only increase system accuracy through screening and auditing, but they also aid to reduce the high cost of medical mistakes [31]. Medical robots are widely recognized as a valuable asset in the health of healthcare personnel, aided by diagnostic research support and physiosystem augmentation training [32]. This paper focuses on medical robots such as surgical robots, telepresence robots, rehabilitation robots, hospital automation, and pharmacy automation, among others. The instruments improve surgical accuracy and efficiency, lowering human error and enhancing patient outcomes. Robots can undertake redundant activities like prescription administration and patient observation, giving healthcare professionals more time to provide direct care [33]. They also lower the risk of job injuries from dangerous duties like heavy lifting and other menial labour. In the event of a pandemic, robots will eliminate human contact, drastically limiting the spread [34]. Furthermore, robots give continuous 24-h help to ensure that patients are treated swiftly and efficiently. They improve healthcare delivery by assisting with diagnosis and treatment planning. Overall, robots in healthcare improve patient safety as well as efficiency and quality of care [35, 36].

3 Healthcare 4.0

The introduction of Heath 4.0 competencies in healthcare is aimed at superior service delivery but doing so more efficiently and effectively; using less cost to achieve maximum resource utilization [37]. This paper explores on how Health 4.0 characteristics align with several suggested solutions to meeting these two aims. High quality medical care must be provided by improving the efficiency of systems between increased use of resources, patient satisfaction and hospital staff [38]. Enhances patient experience such as Appointments booking, Therapies, Health monitoring and Patient engagement with doctors this makes them extremely comfortable for the patient. Improve the Preparedness of Health Workers and Resource Allocation After Historical Operations and Needs [39]. It will optimize resource and people allocation, ultimately improving employee satisfaction and patient experience. Health 4.0 strives for improved and quality resources in healthcare services World Science-tech Org. Decreasing the price of upkeeps and rising gear Performance, reducing disruptions created attributable to improper in solution repairs will conjointly trigger beside well grooming [40].

3.1 *Evolution of Healthcare*

From traditional herbal cures to sophisticated robotic surgery and AI-driven diagnostics, healthcare has undergone substantial change. Medical science advancements including vaccines, antibiotics, and imaging technology have improved life expectancy and transformed therapy [6]. Modern healthcare emphasizes individualised medication, minimally invasive procedures, and telemedicine, making care more accessible and efficient. This continuous development keeps changing the way healthcare is delivered, emphasising patient-centred strategies and better health outcomes everywhere [41]. The evolution of healthcare is illustrated in Fig. 1.

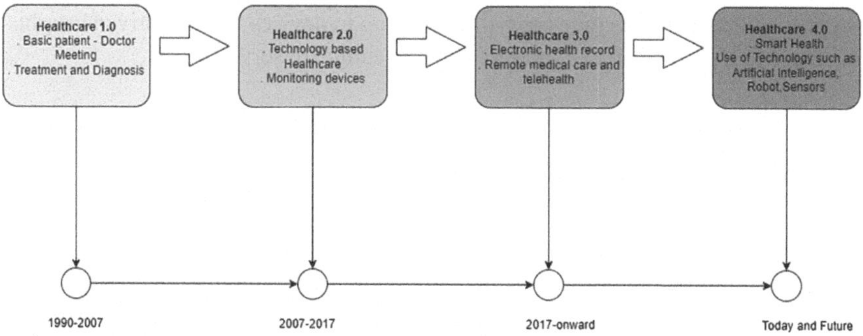

Fig. 1 Expansion in healthcare

3.1.1 Healthcare 1.0

During this time, patients only went to doctors for diagnosis and treatment after they had been ill, and the focus was on conventional, reactive healthcare. With minimal use of technology for patient management or diagnostics, the focus was on in-person consultations and physical examinations. The majority of healthcare was hospital-focused, with no programs for preventive care [42].

3.1.2 Healthcare 2.0

With the advent of electronic health records (EHRs), which allowed for improved data management and exchange, the digital era got underway. Basic telemedicine services improved remote patients' access to healthcare by enabling video calls for consultations. Despite growing interconnectedness, healthcare systems remained mostly predicated on conventional modes of service delivery [43].

3.1.3 Healthcare 3.0

With an emphasis on personalised medicine, healthcare moved towards patient-centred care. Patients now have more control over their health thanks to wearable technology, data analytics, and mobile health apps. Care became more proactive with the adoption of technological equipment to identify problems and monitor health, enabling early intervention and customised therapies [44].

3.1.4 Healthcare 4.0

This age creates a highly linked, predictive, and preventative healthcare system by combining cutting-edge technology like AI, robotics, IoT, and machine learning. In order to deliver individualised treatment, enhance patient experiences, and improve results, healthcare 4.0 leverages real-time data, smart devices, and AI-driven insights. This leads to an ecosystem of interconnected healthcare services [45].

3.2 Key Technologies of Healthcare4.0

Healthcare 4.0 aims at achieving healthcare digitization using essence of latest technological novelties. These technologies offer increased reliability, convenience and patient satisfaction in addition to enhanced transparency for the patients [46]. This section provides an overview of some key technologies used in Healthcare4.0 to enhance the healthcare experience as highlighted in Fig. 2.

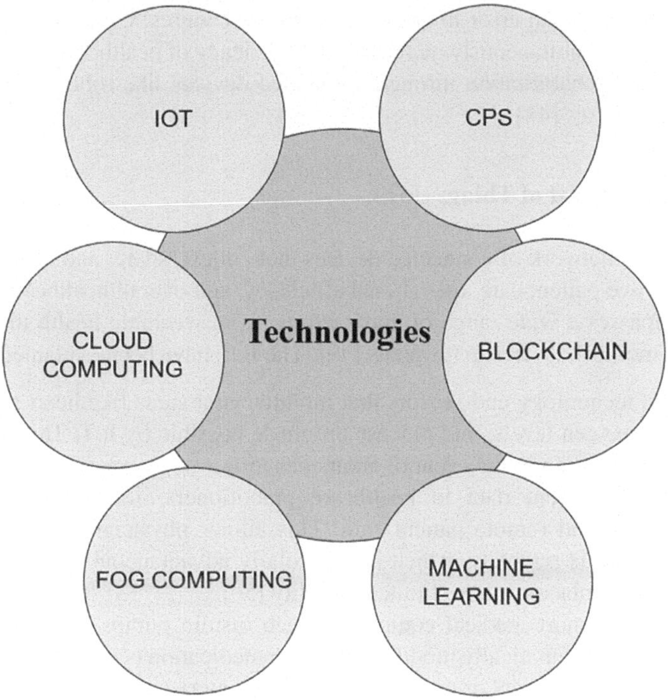

Fig. 2 Various technologies of healthcare

3.2.1 Cyber-Physical Systems (CPS)

It is defined as the integration of computer, networking, and physical processes to improve medical practices, patient care, and operational efficiency. These systems connect physical items, such as medical equipment and wearable sensors, to cyber components, such as software and data analytics platforms, allowing for real-time monitoring, data analysis, and automated decision-making [47]. The CPS functionality have been explained below.

- To monitor, diagnose, and treat patients in real-time, CPS in healthcare combines digital and physical systems, such as data processing and cloud computing, with medical sensors, wearable technology, and robotic systems [48].
- By using IoT devices to provide continuous remote monitoring of patient health indicators (such as blood pressure, heart rate, and glucose levels), CPS makes it possible to identify medical problems early and eliminates the need for in-person visits [48].
- CPS can offer individualised healthcare solutions, such modifying drug dosages or creating customised treatment plans based on patient data, by gathering and evaluating data from several sensors [48].

- By lowering human error in crucial medical procedures, CPS improves patient safety while simultaneously increasing the efficiency of healthcare processes like surgery and rehabilitation through automated devices like robotic surgery and smart prostheses [48].

3.2.2 The Internet of Things (IoT)

It refers to the network of connected devices that collect, share, and analyse health data to improve patient care, operational efficiency, and overall healthcare delivery. IoT encompasses a wide range of applications, from wearable health monitors to connected medical devices in hospitals [49]. The IOT have been explained below.

- Wearable technology and sensors that monitor vital signs like heart rate, blood pressure, oxygen levels, and glucose are made possible by IoT. This allows for timely intervention in the event of anomalies [50].
- By sending patient data to healthcare practitioners, IoT devices facilitate telemedicine and remote patient care. This allows physicians to monitor and manage patients remotely, which is particularly advantageous for patients who live in distant places or have limited mobility [50].
- IoT-powered smart medical equipment, such insulin pumps, pacemakers, and inhalers, can automatically modify how much medication is administered, record usage, and notify carers or medical experts if an intervention is required [50].
- By monitoring the whereabouts and state of medical equipment, controlling inventories, and enhancing workflows, IoT simplifies hospital operations, resulting in lower expenses, better resource management, and better patient care outcomes [50].

3.2.3 Cloud Computing

It refers to the use of cloud-based services and solutions to store, manage, and analyse health data, as well as to improve cooperation and patient outcomes. Cloud computing has altered healthcare by providing scalable resources, greater data accessibility, and more cooperation among providers [51]. Here is a comprehensive review of cloud computing in healthcare.

- Cloud computing improves data availability for more informed decision-making and continuity of treatment by allowing healthcare providers to safely store large volumes of patient data and make it instantly accessible from any authorised device [52].
- Scalable Infrastructure: Without requiring expensive physical gear, cloud solutions enable healthcare facilities to extend their infrastructure as needed to handle massive volumes of data from telemedicine apps, medical imaging, and electronic health records (EHRs) [52].
- Data Security and Compliance: To safeguard sensitive health data and adhere to laws like HIPAA, cloud providers frequently use strong security features like

encryption and multi-factor authentication, guaranteeing patient privacy and data protection [52].
- Collaboration and Telemedicine: Cloud computing facilitates smooth collaboration by allowing real-time data sharing and treatment plan cooperation between healthcare practitioners, supporting telemedicine platforms, and facilitating distant patient care [52].

3.2.4 Machine Learning (ML)

The use of algorithms and statistical models to allow computers to learn from data and make predictions or judgments based on it. Machine learning has the potential to change several aspects of healthcare by analysing massive volumes of health data, including diagnosis and treatment, operational efficiency, and patient involvement [53]. A comprehensive summary of machine learning in healthcare is discussed below.

- Early and more precise diagnosis is made possible by machine learning algorithms that examine enormous volumes of patient data to find trends and forecast the beginning of diseases including cancer, diabetes, and cardiovascular disorders [54].
- By evaluating genetic, lifestyle, and medical history data, ML assists in customising treatment plans for each patient, guaranteeing more efficient therapies and enhancing patient outcomes through precision medicine [54].
- By accurately identifying abnormalities like tumours or fractures, ML improves the interpretation of medical images (such as X-rays, MRIs, and CT scans) and helps radiologists make quicker and more precise diagnoses [54].
- By using ML models to forecast patient deterioration, readmission risk, or recovery timeframes, healthcare practitioners can proactively allocate resources, provide priority to high-risk patients, and enhance the quality of care provided [54].

3.2.5 Fog Computing

It is also known as edge computing, extends cloud computing capabilities to the network's edge, where data is created. In healthcare, fog computing is crucial for processing and analysing data from numerous medical equipment, sensors, and apps in real time. This method improves the efficiency and efficacy of healthcare delivery, especially in remote monitoring, telemedicine, and smart health systems [55]. The fog computing in healthcare is explained below.

- Fog computing reduces latency and makes it possible to monitor and respond to patients in real-time, which is essential for time-sensitive applications like emergency care. It does this by processing data closer to the source (such as wearable technology or medical sensors) [56].

- Fog computing helps healthcare providers stay in compliance with data protection laws by processing data locally, which minimises the quantity of sensitive health data sent to the cloud [56].
- By processing and analysing data locally, fog computing reduces network congestion and eliminates the need to send massive amounts of data to central cloud servers. This is particularly advantageous in healthcare settings with limited bandwidth [56].
- By allowing healthcare equipment to analyse and respond to data at the network edge, fog computing facilitates quick decisions in urgent situations, like remote diagnostics or intensive care unit monitoring, without exclusively depending on cloud connectivity [56].

3.3 Issue and Challenges in Healthcare4.0

Different healthcare frameworks such as the use of leading to variety are produced in Healthcare 4.0 technologies, yet many research issues and challenges must be considered so that it can really implement real-time. The challenges in healthcare 4.0 is depicted in Fig. 3 and explored in the subsections below.

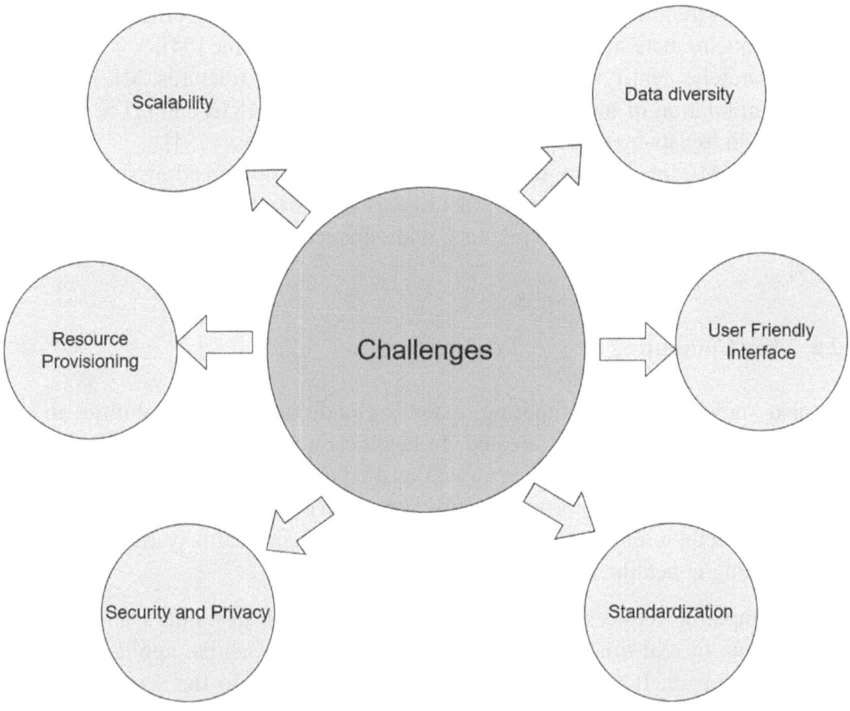

Fig. 3 Various challenges of healthcare

- The healthcare frameworks tackle info coming from an unparalleled array of inputs as EMR records, user apps, IoT devices para phrasal. Therefore, having well-defined standard practices and file formats is an absolute must for handling data which comes in so many different forms (like text pdfs jpg images etc.) from numerous databases [57].
- The most advanced solutions in Healthcare4.0 must be expanded at the level of our country or region to meet rising patient expectations and encouraging better health activities. They are based on a human-centric approach. This way it would save time of the patients while waiting for consults and reports etc. and they could be given some basic medical care right away as well [57].
- On the other hand, now a days a substantial amount of data is collected by different applications and handling it becomes cumbersome. Thus, it is necessary to store the collected data and analyse and interpret these large quantities of information in real time. In such an environment that is both very unpredictable and dynamically changing, message-passing of compute-intense resources are bound to be limited. To avoid this, an autonomous resource management has to be employed so that these are present in hand while the demand fluctuates [57].
- The possibilities are immense as healthcare 4.0 can improve real-time health outcomes and experience of the patients, through simplified services along like never before. But ensuring the privacy and security of some healthcare framework in an era when physical health data is highly appealing to hackers, remains a continuing challenge. This places the data of patients at a significant risk. Therefore, it is essential to take major strides in securing and protecting data at various phases of healthcare systems designing [58].
- A critical issue is the complete absence of healthcare regulations. Security, Privacy and Data transfer are major constraint when we want the sharing or synchronized data with layer to layer. This is where standardization activities need to come in, tele-health thought be standardized by an organization just looking at the health technologies. Enables Real-time response and resolution of data discrepancies. Also, a solid regulatory system needs to be established before even patients arrive for monitoring and control [58].
- The development team should comprise healthcare consumers to know their thoughts, choices as well as other expectations. Therefore, Healthcare 4.0 will be characterized as a very user friendly and patient centre care [58].

4 Integration of Robotics in Healthcare 4.0

Human–Robot Interaction (HRI) is a growing research area in psychology, behaviour science, and neuroscience. It includes both robot production and human contact, as well as research into human–robot interactions. Past HRI scientists are investigating how people control robots. Medical robots, on the other hand, interact with humans in order to improve their performance. As a result, the robot must be able to recognize, grasp, and respond to people's emotions, which express fundamental components of a connection, such as thoughts and sentiments [59]. The Internet of Things and intelligent facilities have helped to create single-point electronic devices and open-loop networks that allow architecture and robotics to communicate seamlessly. Using closed-loop comments, the CPS-HRS proposes to link each smart gadget in the house, allowing for seamless infrastructure-robot connectivity. Learning more about architecture and the natural systems that surround it may have an influence on human health.

Human–Robot Symbiosis a Rapid advances in robotics have helped to integrate sophisticated robots into human professions and living settings. Robots can collaborate with humans and assist them with certain activities in the absence of physical boundaries. Maintaining security and creating human–robot cooperation are upcoming research topics [60].

Teleoperation of Collaborative Robots: With the population aging, the scarcity of home caregivers will be a major worry. Teleoperation is a possible method for running automated machinery virtually depending on the operator's preferences. Robotics may do small helpful tasks autonomously. The robot may be teleoperated by the elderly who have become partially disabled, allowing them to live independently. Teleoperation enables specialists, such as nurses in medical groups, to care for elderly people in their homes. The combination of improved sensors, artificial intelligence, novel materials, and machines increases robot safety interface capabilities. This allows robots to provide healthcare in the future of HRS. Enhanced motion capture technology and specialized motion tracing addresses will result in a more natural and efficient method of capturing the operator's desire to teleoperate. The CPS-HRS will ultimately include caretakers or other skilled healthcare personnel serving as virtual tele-operators [61].

Healthcare 4.0 systems will be used in the future, and such advancements are already affecting almost every aspect of everyday life. Healthcare 4.0 is a fresh revolution, and the different uses it offers are shown in the examples below.

- Physiological and pathological signal sensing: The use of high-quality medical sensors (acceleration, temperature, humidity; ECG glucose blood pressure); and gas sensors allows for the continuous acquisition of physiological/pathological health features of patients in smart homes. These traits can assist maintain track of patients and improve care facilities [62].
- Well-being monitoring and prevention: In the Healthcare 4.0 era, self-management options are strongly encouraged. These solutions, which use algorithms, have the ability to identify and improve health habits that increase disease risk in order to avoid diseases [62].
- Personalized healthcare and recommender systems: In Healthcare 4.0, recommender systems are a groundbreaking finding for understanding user requirements by automatically forecasting users based on their interactions with a group of objects. Used to provide personalized patient-led care and planting education according on user needs. They've even gone so far as to match physicians and patients based on their medical histories [62].

- Telemedicine and disease monitoring are the use of information technology from remote places to transmit health-related data, allowing physicians and other members of the medical team to give accurate diagnosis and treatment. Moving towards hospitals is extremely risky given the present Covid 19 scenario. There is one region where waterborne infections are common, and that is Africa, where health 4.0 provides telemedicine services for a variety of serious disorders that require immediate action. Previously, diabetes and chronic respiratory failure were continuously monitored and treated [62].
- Assisted living (AL) is the intermediate step between community care and nursing homes. Various AL services, primarily for older patients, in which key parameters are continually detected and communicated to a central hub employed in healthcare 4.0. When any of these characteristics vary, they will be treated in an emergency setting [62].

5 Recent Advancement

Alotaibi et al. [63] highlighted the contributions of robots in healthcare, specifically in surgical operations, patient support, and the management of hospital environments. Martin et al. [64] proposed a solution involving a socially assistive robot for exercise recognition, showcasing a 99.87% accuracy achieved through a CNN-LSTM model, and underscores its significance for elderly care and clinical applications. Yang et al. [65] discussed innovative concepts and functionalities of Cyber-Physical Systems (CPS) in Human–Robot Systems (HRS), highlighting progress in areas such as artificial intelligence, sensing technologies, materials science, cloud computing, communication methods, motion capture, and mapping techniques. Hill et al. [66] presented the latest advancements in bone-targeting surgical robots, introduces a novel framework for autonomy, and discusses potential future developments in robotic surgery. Qian et al. [67] discussed augmented reality-integrated robotic-assisted surgery, covering aspects such as hardware, applications, and clinical significance, while emphasizing its practicality and the promise it holds for enhancing surgical results, despite its current developmental stage.

In another work, Rahmani et al. [68] highlighted the design and effectiveness of 72 robotic devices for drop foot assistance, highlighting their success in improving gait while noting limitations in combining advanced features for optimal performance. Lei et al. [69] proposed a control system for wheelchair navigation that predicts user intentions and improves task completion speed by 10%, enhancing navigation for both disabled and healthy individuals while allowing path following with similar accuracy. Fikry et al. [70] discussed the role of robots in healthcare during and after the COVID-19 pandemic, emphasizing the increasing demand for these technologies, their advantages, the challenges they face, and the implications for society and engineering management. Maki et al. [71] presented innovative digital and facility management methods for healthcare, using a causal loop diagram and a survey to explore technology adoption, while discussing AI techniques for efficient, cost-effective healthcare delivery. Kim et al. [72] proposed a motorless surgical device with a compensation mechanism that reduces end-effector backlash, improving precision and accuracy in surgeries involving unpredictable curved paths at a low cost. Thompson et al. [73] proposed a Soft Active Transseptal Puncture Simulator (SATPS) to provide low-risk training for new TP operators, accurately simulating heart dynamics and receiving positive feedback from clinicians for improving catheterization skills.

Similarly, Tang et al. [74] proposed a define design criteria that relate clinical demands to design specifics and encourage the creation of a scientific design framework in robotic tracheostomy. Cai et al. [75] proposed a motor recovery training and evaluation technique for upper limb rehabilitation robotic systems is suggested. Huang et al. [76] proposed a quick and accurate approach for determining human ankle motion sensory acuity. Dinesh et al. [77] proposed a quick overview of sensors and components used in robotic surgery that have been researched in recent literature. Dyck et al. [78] proposed robotic platform for intraoperative neurosurgery ultrasound scanning, using 3D surface reconstruction and collaborative probe guidance, with preliminary tests showing potential for soft-tissue scanning. Li et al. [79] discussed challenges in robotic systems for endovascular treatment, emphasizes future design perspectives, and advocates for collaboration between engineers and physicians to enhance clinical adoption and innovation. Jin et al. [80] presented a robotic platform utilizing magnetorheological fluids for haptic feedback, along with a force sensing method and collaborative operation techniques to aid interventionists in navigating blood vessel branches. Oquendo et al. [81] discussed training in a repelling force field would enhance performance in a virtual reality task, with 38 surgical novices evaluated on completion time and path errors after five days of training under different conditions. Jiang et al. [82] proposed a detailed overview of platforms and current advancements in robotic perception of transparent objects. The above discussed recent advances are summarized in table 2.

6 Conclusion

The transformative impact of robotics in healthcare is increasingly clear, signalling a movement toward more efficient and patient-focused care [83]. This paper outlines the historical development of healthcare robotics, highlighting how technological progress has responded to the needs of various environments, such as hospitals and rehabilitation facilities. Nevertheless, significant challenges, including data blockage and risks to patient privacy, persist. The emergence of Healthcare 4.0, which integrates CPS, IoT and Machine learning offers a pathway to tackle these issues. These technologies enable a shift from reactive to predictive care models, improving both decision-making and operational processes [84]. Green's insights into the ability of robots to understand human emotions and provide empathetic care emphasize the promise of enhanced HRI in building meaningful connections between patients and healthcare providers [85].

Table 2 Recent advancements

References	Year	Contribution
Alotaibi et al. [63]	2019	Robots that have been recognized enhance surgical operations, patient care, and hospital administration
Martinez-Martin et al. [64]	2019	Developed a robotic device for workout identification that achieved 99.87% accuracy
Yang et al. [65]	2020	Innovative CPS methods to HRS were explored, including advances in AI, sensing, and communication
Hill et al. [66]	2020	Presented changes in bone-targeting robots, a self-sufficient structure, and upcoming developments
Qian et al. [67]	2020	Described AR-integrated surgery using robots, emphasizing hardware, applications, and relevance
Rahmani et al [68]	2022	The highlighted examines 72 artificial legs that improve movement for drop leg
Lei et al. [69]	2022	The designed operating system identifies intents, which improves wheelchair movement and activity performance
Fikry et al. [70]	2023	The entire conversation focused on robotic in healthcare after COVID-19: need, rewards, and problems
Maki et al. [71]	2023	Suggested healthcare management systems that utilize AI and tests
Kim et al. [72]	2023	The suggested motorless surgical tools boosts accuracy in curved route procedures while remaining affordable
Thompson et al. [73]	2023	The idea for SATPS aims to provide minimal-risk instruction for transseptal insertion abilities
Tang et al. [74]	2023	The detailed overview of platforms and current advancements in robotic awareness of tracheostomy framework
Cai et al. [75]	2023	For efficient motor recovery in upper limb rehabilitation robotics, use adaptive, task-oriented activities that provide real-time feedback
Huang et al. [76]	2023	A rapid and precise method for assessing the sensory acuity of human ankle motion
Dinesh et al. [77]	2023	A brief summary of the sensors and parts utilized in robotic surgery that have been studied recently
Dyck et al. [78]	2024	The designed artificial platform accelerates neurosurgical diagnostic ultrasound through joint path
Li et al. [79]	2024	Teamwork improves medical creativity by addressing obstacles in robot endovascular systems
Jin et al. [80]	2024	The designed artificial platform optimizes interventional guidance through haptic input and cooperation
Oquendo et al. [81]	2024	The mentioned instruction in the force field increased the accuracy of laparoscopic beginners on VR tasks
Jiang et al. [82]	2024	AI, depth sensors, polarized light, and haptic feedback are all used in the development of robotic vision for transparent objects

Additionally, artificial intelligence is pivotal in assisting healthcare professionals by reducing errors and improving real-time health monitoring. The significance of teleoperation is also discussed, as it allows for remote healthcare delivery, enhancing accessibility [86]. The latest developments in robotics showcase their considerable potential to transform healthcare services, leading to better patient outcomes and increased efficiency. As this field advances, continuous research and innovation will be crucial in overcoming current challenges and maximizing the advantages of robotics in healthcare [87, 88].

References

1. J. Collins, S. Chand, A. Vanderkop, D. Howard, A review of physics simulators for robotic applications. IEEE Access **9**, 51416–51431 (2021). https://doi.org/10.1109/ACCESS.2021.3068769
2. P. Chen, Q. Ding, Y. Liu, Z. Deng, J. Huang, Programmable Pressure control in pneumatic soft robots with 2-way 2-State solenoid valves. IEEE Robot. Autom. Lett. **9**(7), 6448–6455 (2024). https://doi.org/10.1109/LRA.2024.3406201
3. X. Qin, Y. Yang, M. Pan, L. Cui, L. Liu, Morphobot: a platform for morphogenesis in robot swarm. IEEE Robot. Autom. Lett. **8**(11), 7440–7447 (2023). https://doi.org/10.1109/LRA.2023.3320021
4. G.K. Getele, T. Li, J. T. Arrive, The role of supply chain management in healthcare service quality. IEEE Eng. Manage. Rev. **48**(1), 145–155 (2020). https://doi.org/10.1109/EMR.2020.2968429
5. A.K. Jameil, H. Al-Raweshidy, AI-enabled healthcare and enhanced computational resource management with digital twins into task offloading strategies. IEEE Access **12**, 90353–90370 (2024). https://doi.org/10.1109/ACCESS.2024.3420741
6. K. Katsaliaki, S. Kumar, The past, present, and future of the healthcare delivery system through digitalization. IEEE Eng. Manage. Rev. **50**(4), 21–33 (2022). https://doi.org/10.1109/EMR.2022.3223112
7. J. Mišeikis et al., Lio-A personal robot assistant for human-robot interaction and care applications. IEEE Robot. Autom. Lett. **5**(4), 5339–5346 (2020). https://doi.org/10.1109/LRA.2020.3007462
8. A.N. Navaz, M.A. Serhani, H.T. El Kassabi, N. Al-Qirim, H. Ismail, Trends, technologies, and key challenges in smart and connected healthcare. IEEE Access **9**, 74044–74067 (2021). https://doi.org/10.1109/ACCESS.2021.3079217
9. A. Luschi, C. Petraccone, G. Fico, L. Pecchia, E. Iadanza, Semantic ontologies for complex healthcare structures: a scoping review. IEEE Access **11**, 19228–19246 (2023). https://doi.org/10.1109/ACCESS.2023.3248969
10. S. Gandah, et al., An Integrated sensorized platform for environmental monitoring in healthcare. IEEE Sens. Lett. **7**(9), 1–4, Art no. 5502204 (2023). https://doi.org/10.1109/LSENS.2023.3301836
11. D. Stogl, O. Armbruster, M. Mende, B. Hein, X. Wang, P. Meyer, Robot-based training for people with mild cognitive impairment. IEEE Robot. Autom. Lett. **4**(2), 1916–1923 (2019). https://doi.org/10.1109/LRA.2019.2898470
12. S. Zhang et al., Design and characteristics of 3D magnetically steerable guidewire system for minimally invasive surgery. IEEE Robot. Autom. Lett. **7**(2), 4040–4046 (2022). https://doi.org/10.1109/LRA.2022.3146909
13. M. Zhu, S. Biswas, S.I. Dinulescu, N. Kastor, E.W. Hawkes, Y. Visell, Soft, wearable robotics and haptics: technologies, trends, and emerging applications. Proc. IEEE **110**(2), 246–272 (2022). https://doi.org/10.1109/JPROC.2021.3140049

14. N. Shvalb, S. Hacohen, O. Medina, Statistical Robotics: Controlling multi robot systems using statistical-physics. IEEE Access **12**, 134739–134753 (2024). https://doi.org/10.1109/ACCESS.2024.3406599
15. S. Aoki, T. Yonezawa, N. Kawaguchi, Robot NEST: toward a viable testbed for IoT-enabled environments and connected and autonomous robots. IEEE Sens. Lett. **6**(2), 1–4, Art no. 6000304 (2022). https://doi.org/10.1109/LSENS.2021.3139624.
16. G. Kosa, O. Morozov, A. Lehmann, H. Pargger, S. Marsch, P. Hunziker, Robots and intelligent medical devices in the intensive care unit: vision, state of the art, and economic analysis. IEEE Trans. Med. Robot. Bionics **5**(1), 2–17 (2023). https://doi.org/10.1109/TMRB.2023.3240537
17. J. Zhou, Z. Li, X. Li, X. Wang, R. Song, Human–robot cooperation control based on trajectory deformation algorithm for a lower limb rehabilitation robot. IEEE/ASME Trans. Mechatron. **26**(6), 3128–3138 (2021). https://doi.org/10.1109/TMECH.2021.3053562
18. A. Stefek, T.V. Pham, V. Krivanek, K.L. Pham, Energy comparison of controllers used for a differential drive wheeled mobile robot. IEEE Access **8**, 170915–170927 (2020). https://doi.org/10.1109/ACCESS.2020.3023345
19. J.-H. Kim, Multi-axis force-torque sensors for measuring zero-moment point in humanoid robots: a review. IEEE Sens. J. **20**(3), 1126–1141 (2020). https://doi.org/10.1109/JSEN.2019.2947719.
20. X. Zhu, B. Zhang, Y. Qiu, S.A. Chepinskiy, An interaction behaviour decision-making model of service robots for the disabled based on human-robot empathy. IEEE Access **12**, 15778–15790 (2024). https://doi.org/10.1109/ACCESS.2024.3358614
21. G. Caccianiga, A. Mariani, E. De Momi, G. Cantarero, J.D. Brown, An evaluation of inanimate and virtual reality training for psychomotor skill development in robot-assisted minimally invasive surgery. IEEE Trans. Med. Robot. Bionics **2**(2), 118–129 (2020). https://doi.org/10.1109/TMRB.2020.2990692
22. H. Song et al., Real-time intraoperative surgical guidance system in the da Vinci surgical robot based on transrectal ultrasound/photoacoustic imaging with photoacoustic markers: an ex vivo demonstration. IEEE Robot. Autom. Lett. **8**(3), 1287–1294 (2023). https://doi.org/10.1109/LRA.2022.3191788
23. D.G. Black, A.H.H. Hosseinabadi, S.E. Salcudean, 6-DOF force sensing for the master tool manipulator of the da Vinci surgical system. IEEE Robot. Autom. Lett. **5**(2), 2264–2271 (2020). https://doi.org/10.1109/LRA.2020.2970944
24. O. Özgüner et al., Camera-robot calibration for the Da Vinci robotic surgery system. IEEE Trans. Autom. Sci. Eng. **17**(4), 2154–2161 (2020). https://doi.org/10.1109/TASE.2020.2986503
25. N. Yilmaz, B. Burkhart, A. Deguet, P. Kazanzides, U. Tumerdem, Sensorless transparency optimized haptic teleoperation on the da Vinci research kit. IEEE Robot. Autom. Lett. **9**(2), 971–978 (2024). https://doi.org/10.1109/LRA.2023.3335779
26. E.C. Ng, S. Xu, X.E. Liu, J.B. Lim, M.H. Liow, H.N. Pang, D.K. Tay, S.J. Yeo, J.Y. Chen, Enhanced recovery after surgery day surgery for Mako® Robotic-arm assisted TKA; better outcome for patients, improved efficiency for hospitals. J. Orthop. **56**, 77–81 (2024). https://doi.org/10.1016/j.jor.2024.05.013
27. Y. Zhu, T. Mo, J. Zhang, C. Jiang, The use of Mako Robot-assisted total hip arthroplasty in the treatment of Crowe IV developmental dysplasia of the hip. Asian J. Surg. **45**(6), 1333–1335 (2022). https://doi.org/10.1016/j.asjsur.2022.02.008
28. M. Roche, Robotic-assisted unicompartmental knee arthroplasty. Clin. Sports Med. **33**(1), 123–132 (2014). https://doi.org/10.1016/j.csm.2013.08.007
29. O. Shalash, P. Rowe, Computer-assisted robotic system for autonomous UNICOMPARTMENTAL Knee arthroplasty. Alex. Eng. J. **70**, 441–451 (2023). https://doi.org/10.1016/j.aej.2023.03.005
30. P.E. Dupont, N. Simaan, H. Choset, C. Rucker, Continuum robots for medical interventions. Proc. IEEE **110**(7), 847–870 (2022). https://doi.org/10.1109/JPROC.2022.3141338
31. R. Raina, R.K. Jha, Intelligent and interactive healthcare system (I2HS) using machine learning. IEEE Access **10**, 116402–116424 (2022). https://doi.org/10.1109/ACCESS.2022.3197878

32. M. Nasr, M.M. Islam, S. Shehata, F. Karray, Y. Quintana, Smart healthcare in the age of AI: recent advances, challenges, and future prospects. IEEE Access **9**, 145248–145270 (2021). https://doi.org/10.1109/ACCESS.2021.3118960
33. X. Li, Q. Lu, J. Chen, N. Jiang, K. Li, A robust region control approach for simultaneous trajectory tracking and compliant physical human-robot interaction. IEEE Trans. Syst. Man Cybern. Syst. **53**(10), 6388–6400 (2023). https://doi.org/10.1109/TSMC.2023.3285603
34. M.M.R. Khan et al., Development of a robot-assisted telerehabilitation system with integrated IIoT and digital Twin. IEEE Access **11**, 70174–70189 (2023). https://doi.org/10.1109/ACCESS.2023.3291803
35. H. Manh Do, W. Sheng, E.E. Harrington, A.J. Bishop, Clinical screening interview using a social robot for geriatric care. IEEE Trans. Autom. Sci. Eng. **18**(3), 1229–1242 (2021). https://doi.org/10.1109/TASE.2020.2999203
36. J. Li, H. Shi, K.-S. Hwang, Using goal-conditioned reinforcement learning with deep imitation to control robot arm in flexible flat cable assembly task. IEEE Trans. Autom. Sci. Eng. **21**(4), 6217–6228 (2024). https://doi.org/10.1109/TASE.2023.3323307
37. A. Kumar, R. Krishnamurthi, A. Nayyar, K. Sharma, V. Grover, E. Hossain, A novel smart healthcare design, simulation, and implementation using healthcare 4.0 processes. IEEE Access **8**, 118433–118471 (2020). https://doi.org/10.1109/ACCESS.2020.3004790
38. S. Abbate, P. Centobelli, R. Cerchione, E. Oropallo, E. Riccio, Investigating healthcare 4.0 Transition through a knowledge management perspective. IEEE Trans. Eng. Manage. **70**(9), 3297–3310 (2023). https://doi.org/10.1109/TEM.2022.3200889
39. J. Al-Jaroodi, N. Mohamed, E. Abukhousa, Health 4.0: On the way to realizing the healthcare of the future. IEEE Access **8**, 211189–211210 (2020). https://doi.org/10.1109/ACCESS.2020.3038858
40. N. Taimoor, S. Rehman, Reliable and resilient AI and IoT-based personalised healthcare services: a survey. IEEE Access **10**, 535–563 (2022). https://doi.org/10.1109/ACCESS.2021.3137364
41. E. Cano-Marin, S. Sánchez-Alonso, M. Mora-Cantallops, Unleashing competitive intelligence: news mining analysis on technology trends and digital health driving healthcare innovation. IEEE Trans. Eng. Manage. **71**, 12311–12325 (2024). https://doi.org/10.1109/TEM.2023.3326233
42. Y. Su, Y. Zhang, L. Weigang, Multi-stage information spreading model in simplicial complexes driven by spatiotemporal evolution of public health emergency. IEEE Access **11**, 128316–128336 (2023). https://doi.org/10.1109/ACCESS.2023.3332656
43. S. Thapliyal, M. Wazid, D.P. Singh, A.K. Das, S. Shetty, A. Alqahtani, Design of robust blockchain-envisioned authenticated key management mechanism for smart healthcare applications. IEEE Access **11**, 93032–93047 (2023). https://doi.org/10.1109/ACCESS.2023.3310264
44. I.S. Khayal, A.M. Farid, A dynamic system model for personalized healthcare delivery and managed individual health outcomes. IEEE Access **9**, 138267–138282 (2021). https://doi.org/10.1109/ACCESS.2021.3118010
45. H.R. Chi, M. de Fátima Domingues, H. Zhu, C. Li, K. Kojima, A. Radwan, Healthcare 5.0: in the perspective of consumer internet-of-things-based fog/cloud computing. IEEE Trans. Consum. Electron. **69**(4), 745–755 (2023). https://doi.org/10.1109/TCE.2023.3293993
46. M. Wazid, A.K. Das, N. Mohd, Y. Park, Healthcare 5.0 security framework: applications, issues and future research directions. IEEE Access **10**, 129429–129442 (2022). https://doi.org/10.1109/ACCESS.2022.3228505
47. F. Farivar, M.S. Haghighi, A. Jolfaei, M. Alazab, Artificial intelligence for detection, estimation, and compensation of malicious attacks in nonlinear cyber-physical systems and industrial IoT. IEEE Trans. Industr. Inf. **16**(4), 2716–2725 (2020). https://doi.org/10.1109/TII.2019.2956474
48. A. Fu, J.A. McCann, Dynamic decentralized periodic event-triggered control for wireless cyber-physical systems. IEEE Trans. Control Syst. Technol. **29**(4), 1783–1790 (2021). https://doi.org/10.1109/TCST.2020.3016131

49. H. Tran-Dang, N. Krommenacker, P. Charpentier, D.-S. Kim, Toward the internet of things for physical internet: perspectives and challenges. IEEE Internet Things J. **7**(6), 4711–4736 (2020). https://doi.org/10.1109/JIOT.2020.2971736
50. C.W. Chen, Internet of video things: next-generation IoT with visual sensors. IEEE Internet Things J. **7**(8), 6676–6685 (2020). https://doi.org/10.1109/JIOT.2020.3005727
51. Y. Motai, E. Henderson, N.A. Siddique, H. Yoshida, Cloud colonography: distributed medical testbed over cloud. IEEE Trans. Cloud Comput. **8**(2), 495–507 (2020). https://doi.org/10.1109/TCC.2015.2481414
52. B. Wang, C. Wang, W. Huang, Y. Song, X. Qin, A survey and taxonomy on task offloading for edge-cloud computing. IEEE Access **8**, 186080–186101 (2020). https://doi.org/10.1109/ACCESS.2020.3029649
53. K.S. Bohnsack, M. Kaden, J. Abel, T. Villmann, Alignment-free sequence comparison: a systematic survey from a machine learning perspective. IEEE/ACM Trans. Comput. Biol. Bioinform. **20**(1), 119–135 (2023). https://doi.org/10.1109/TCBB.2022.3140873
54. R.G.L. Koh et al., A scoping review of machine learning applied to peripheral nerve interfaces. IEEE Trans. Neural Syst. Rehabil. Eng. **32**, 3689–3698 (2024). https://doi.org/10.1109/TNSRE.2024.3468995
55. T.-A.N. Abdali, R. Hassan, A.H.M. Aman, Q.N. Nguyen, Fog computing advancement: concept, architecture, applications, advantages, and open issues. IEEE Access **9**, 75961–75980 (2021). https://doi.org/10.1109/ACCESS.2021.3081770
56. B. Ali, M. Adeel Pasha, S. Islam, H. Song, R. Buyya, A volunteer-supported fog computing environment for delay-sensitive IoT applications. IEEE Internet Things J. **8**(5), 3822–3830 (2021). https://doi.org/10.1109/JIOT.2020.3024823
57. M.C. Chibuike, S.S. Grobbelaar, A. Botha, Overcoming challenges for improved patient-centric care: a scoping review of platform ecosystems in healthcare. IEEE Access **12**, 14298–14313 (2024). https://doi.org/10.1109/ACCESS.2024.3356860
58. A.K. Pandey et al., Key issues in healthcare data integrity: analysis and recommendations. IEEE Access **8**, 40612–40628 (2020). https://doi.org/10.1109/ACCESS.2020.2976687
59. S. Jayaraman, E.K. Phillips, D. Church, L.D. Riek, Privacy and utility perceptions of social robots in Healthcare. Comput. Human Behav. Artif. Humans **2**(1), 100039 (2024). https://doi.org/10.1016/j.chbah.2023.100039
60. L. Patrício, C.R. de Sousa Costa, L. Varela, M.M. Cruz-Cunha, Literature review on the sustainable implementation of robotic process automation (RPA) in medical and Healthcare Administrative Services. Procedia Comput. Sci. **239**, 166–176 (2024). https://doi.org/10.1016/j.procs.2024.06.159
61. L. Aymerich-Franch, E. Gómez, Public perception of socially assistive robots for Healthcare in the EU: a large-scale survey. Comput. Human Behav. Rep. **15**, 100465 (2024). https://doi.org/10.1016/j.chbr.2024.100465
62. A. Ahmed, R. Xi, M. Hou, S.A. Shah, S. Hameed, Harnessing big data analytics for healthcare: a comprehensive review of frameworks, implications, applications, and impacts. IEEE Access **11**, 112891–112928 (2023). https://doi.org/10.1109/ACCESS.2023.3323574
63. M. Alotaibi, M. Yamin, Role of robots in healthcare management, *in 2019 6th International Conference on Computing for Sustainable Global Development (INDIA.Com)*, New Delhi, India (2019), pp. 1311–1314
64. E. Martinez-Martin, M. Cazorla, A socially assistive robot for elderly exercise promotion. IEEE Access **7**, 75515–75529 (2019). https://doi.org/10.1109/ACCESS.2019.2921257
65. G. Yang et al., Homecare robotic systems for healthcare 4.0: visions and enabling technologies. IEEE J. Biomed. Health Inform. **24**(9), 2535–2549 (2020). https://doi.org/10.1109/JBHI.2020.2990529
66. D. Hill, T. Williamson, C.Y. Lai, M. Leary, M. Brandt, Robots and tools for remodeling bone. IEEE Rev. Biomed. Eng. **13**, 184–198 (2020). https://doi.org/10.1109/RBME.2019.2949749
67. L. Qian, J.Y. Wu, S.P. DiMaio, N. Navab, P. Kazanzides, A Review of augmented reality in robotic-assisted surgery. IEEE Trans. Med. Robot. Bionics **2**(1), 1–16 (2020). https://doi.org/10.1109/TMRB.2019.2957061

68. N. Al-Rahmani, D.M. Mohan, M.I. Awad, S.A. Wasti, I. Hussain, K. Khalaf, Lower-limb robotic assistance devices for drop foot: a review. IEEE Access **10**, 51964–51994 (2022). https://doi.org/10.1109/ACCESS.2022.3174686
69. Z. Lei, B.Y. Tan, N.P. Garg, L. Li, A. Sidarta, W.T. Ang, An intention prediction based shared control system for point-to-point navigation of a robotic wheelchair. IEEE Robot. Autom. Lett. **7**(4), 8893–8900 (2022). https://doi.org/10.1109/LRA.2022.3189151
70. A. Fikry, I.S.M. Shafie, Y.L.M. Yusof, S.F. Mahamood, N. Jamil, When contactless service matters: the use of robotic services in the healthcare sector. IEEE Eng. Manage. Rev. **51**(2), 26–34 (2023). https://doi.org/10.1109/EMR.2023.3266604
71. O. Maki, M. Alshaikhli, M. Gunduz, K.K. Naji, M. Abdulwahed, Development of digitalization road map for healthcare facility management. IEEE Access **10**, 14450–14462 (2022). https://doi.org/10.1109/ACCESS.2022.3146341
72. J. Kim, S.-I. Kwon, Y. Moon, K. Kim, A hand-held non-robotic surgical device to compensate for wire length in unpredicted paths. IEEE Access **9**, 60629–60639 (2021). https://doi.org/10.1109/ACCESS.2021.3073139
73. N.A. Thompson, S. Shin, A.G. Kocheril, E.T. Hsiao-Wecksler, G. Krishnan, Design and validation of a soft robotic simulator for transseptal puncture training. IEEE Trans. Biomed. Eng. **70**(10), 3003–3014 (2023). https://doi.org/10.1109/TBME.2023.3278651
74. Y. Tang et al., Design requirements of robotic systems for assisting percutaneous tracheostomy: a scoping review and development framework. IEEE Trans. Med. Robot. Bionics **5**(4), 754–767 (2023). https://doi.org/10.1109/TMRB.2023.3309946
75. H. Cai, S. Guo, Z. Yang, J. Guo, A motor recovery training and evaluation method for the upper limb rehabilitation robotic system. IEEE Sensors J. **23**(9), 9871–9879 (2023). https://doi.org/10.1109/JSEN.2023.3258980
76. Q. Huang, B. Zhong, N. Elangovan, M. Zhang, J. Konczak, A robotic device for measuring human ankle motion sense. IEEE Trans. Neural Syst. Rehabil. Eng. **31**, 2822–2830 (2023). https://doi.org/10.1109/TNSRE.2023.3288550
77. S. Dinesh, U.K. Sahu, D. Sahu, S.K. Dash, U.K. Yadav, Review on sensors and components used in robotic surgery: recent advances and new challenges. IEEE Access **11**, 140722–140739 (2023). https://doi.org/10.1109/ACCESS.2023.3339555
78. M. Dyck, Toward safe and collaborative robotic ultrasound tissue scanning in neurosurgery. IEEE Trans. Med. Robot. Bionics **6**(1), 64–67 (2024). https://doi.org/10.1109/TMRB.2024.3349626
79. N. Li, Y. Wang, H. Zhao, H. Ding, Robotic systems design in endovascular treatment. IEEE Trans. Med. Robot. Bionics **6**(2), 367–383 (2024). https://doi.org/10.1109/TMRB.2024.3377374
80. X. Jin, S. Guo, A. Song, P. Shi, X. Li, M. Kawanishi, A novel robotic platform for endovascular surgery: human–robot interaction studies. IEEE Trans. Instrument. Measure. **73**, 1–9, Art no. 4000509 (2024). https://doi.org/10.1109/TIM.2023.3338682
81. Y.A. Oquendo et al., Haptic Guidance and haptic error amplification in a virtual surgical robotic training environment. IEEE Trans. Haptics **17**(3), 417–428 (2024). https://doi.org/10.1109/TOH.2024.3350128
82. J. Jiang, G. Cao, J. Deng, T.-T. Do, S. Luo, Robotic perception of transparent objects: a review. IEEE Trans. Artif. Intell. **5**(6), 2547–2567 (2024). https://doi.org/10.1109/TAI.2023.3326120
83. C. Rauch, R. Long, V. Ivan, S. Vijayakumar, Sparse-dense motion modelling and tracking for manipulation without prior object models. IEEE Robot. Autom. Lett. **7**(4), 11394–11401 (2022). https://doi.org/10.1109/LRA.2022.3200177
84. C. Liu, P.-C. Lin, Development of a 2-DOF singularity-free spherical parallel remote center of motion mechanism with extensive range of motion. IEEE Robot. Autom. Lett. **9**(6), 5663–5670 (2024). https://doi.org/10.1109/LRA.2024.3396113
85. J. Park, B. Lim, J. Yoon, Virtual frame rotation: a novel two-stage pose estimation scheme of permanent magnet marker for medical applications. IEEE Robot. Autom. Lett. **9**(11), 10575–10582 (2024). https://doi.org/10.1109/LRA.2024.3474480

86. J. Wang et al., Online multi contact receding horizon planning via value function approximation. IEEE Trans. Rob. **40**, 2791–2810 (2024). https://doi.org/10.1109/TRO.2024.3392154
87. L. Wang, Y. Sun, Q. Li, T. Liu, J. Yi, Two shank-mounted IMUs-based gait analysis and classification for neurological disease patients. IEEE Robot. Autom. Lett. **5**(2), 1970–1976 (2020). https://doi.org/10.1109/LRA.2020.2970656
88. H. Eskandari, H.K. Lee, A decision support system on postoperative care management of total joint replacement patients using clinical outcome measurement. IEEE Robot. Autom. Lett. **7**(3), 8558–8565 (2022). https://doi.org/10.1109/LRA.2022.3187618

Evolution of FinFET SRAM Cells for Smart Robotics: A Comprehensive Review

Deepak Garg and Devendra Kumar Sharma

Abstract These days, the majority of embedded systems contain microprocessors with volatile cache memory based on SRAM technology. The performance of SRAM is essential and demands greater attention since it is a fundamental element of computing. While logic and arithmetic processes utilize less than 25% of the worldwide integrated circuits, cache memory occupy approximately 75%, highlighting the need for more performant systems-on-chips (SoCs). Moreover, AI-based implantable electronic components and robotics applications require efficient and reliable SRAM circuits with faster compute-in-memory reaction times. To meet these demands, the three-dimensional gate structure of nanoscale transistors, specifically FinFET, has been introduced for SRAM construction. FinFET has enhanced key performance metrics, including efficiency, power consumption, and space utilization. Additionally, because FinFET is resistant to short-channel effects (SCEs), it has emerged as the preferred transistor for these applications. The performance characteristics of cutting-edge FinFET SRAM cell designs for computing and robotics applications are compared across multiple technologies in this review study.

Keywords Short channel effect · FinFET · SRAM · Leakage current · MOSFET · Leakage power

---

D. Garg (✉)
Department of Electronics and Communication Engineering, SRM Institute of Science and Technology, Ghaziabad, India
e-mail: deepakgarg1985@gmail.com

D. K. Sharma
Department of Electronics and Communication Engineering, ABES Engineering College, Ghaziabad, India

R. Sharma and G. Jeon (eds.), *Data Analytics for Smart Robotics and Its Applications*, Intelligent Systems Reference Library 272,
https://doi.org/10.1007/978-3-031-87697-4_4

1 Introduction

In recent developments, memory now occupies more than 80–85% of the chip's area [1]. Both Static Random Access Memory (SRAM) and Dynamic Random Access Memory (DRAM), the two primary types of memory, significantly enhance the performance of SSDs. These days, memory cells come in a variety of forms, including as DRAM, ferroelectric RAM (F-RAM), SRAM, and phase-change RAM [2]. Power consumption and speed scaling are now crucial considerations for their use in digital circuits due to continuous technological improvements.

For VLSI applications, SRAM [3, 4] is an essential element that provides excellent performance and low power consumption. However, rapid scaling in CMOS [5] design introduces challenges such as leakage problems, operational variances, and short-channel effects (SCEs), which is compounded by dependability concerns. While dimensional scaling is enabled by CMOS dimension reduction, it also results in power and stability issues. The key problem with CMOS is that supply voltage (V_{DD}) scaling leads to threshold voltage (V_{TH}) scaling. According to Moore's Law, CMOS scaling has progressed into the nanoscale regime [6]. As a result, CMOS scalability now faces limits, prompting the exploration of alternative technologies, such as FinFETs [7], tunnel FETs (TFETs) [8], and carbon nano-tube transistors (CNTs) [9]. Among these options, FinFET technology [10, 11] has emerged as the most suitable alternative to CMOS.

Because of their superior transistor scaling, enhanced mobility, reduced leakage, lower power consumption, improved gate control over the channel, and lack of random dopant fluctuation, FinFET has been proposed as a potential replacement for upcoming technologies. With n being the number of fins and h being the fin height, the effective gate width is 2nh. Unlike other double-gate (DG) devices, FinFET devices have a quasi-planar construction, meaning that the device width can only be increased in quanta that are dictated by the fin height. FinFETs are dual-gate devices, with the second gate working in opposition to the first. FinFETs optimize several performance parameters, including sub-threshold leakage, short-channel effects, low-voltage operation, and variability in device parameters induced by doping [12–14].

Traditional planar MOSFET scaling's declining returns are the driving force behind the creation of FinFET technology [15]. Problems include gate leakage, subthreshold slope degradation, and short-channel effects worsen as transistor sizes continue to decrease, making further downsizing and performance enhancements extremely difficult. Due to their integration into advanced process nodes by key semiconductor manufacturers and foundries, FinFETs have emerged as the de facto standard for cutting-edge semiconductor technologies in recent years. The significance of FinFET technology in maintaining semiconductor innovation cannot be emphasized, as Moore's Law continues to propel the quest for increased transistor densities and enhanced performance.

With advancement in technology, the need for low-power devices has become increasingly important. Low power consumption is now prioritized over speed in

VLSI designs, particularly for embedded SRAMs and latches [16]. In an SRAM cell, storing the same value over time causes unbalanced stress, which has a major effect on the cell's dependability. Since an SRAM cell contains a latch, it can retain data as long as it receives power, eliminating the need for periodic refresh operations. In an SRAM, each bit is stored using two cross-coupled inverters, with each inverter comprising four transistors. High-performance on-chip caches are a critical component of the memory architecture in modern computing systems.

Several low-power methods, including as VTCMOS, MTCMOS, SVL, stacking, and power gating, can lower leakage current and power dissipation [17, 18]. Reducing power dissipation, achieving faster access times, and minimizing leakage current are key needs for memory design. Therefore, SRAM cells using FinFETs are recommended over conventional SRAM cells [19]. Compared to traditional SRAM design architectures, FinFET technology minimizes SCEs [20]. Additionally, reducing the leakage parameters of SRAM cells is essential for improving cell stability [21]. Controlling the V_{TH} and current of the device is made possible by the ability to differentially bias the independent front and back gates, which is one of the most beneficial characteristics of FinFETs. FinFETs can further decrease the power consumption of SRAM cells by modifying the back gate [22, 23].

The different design problems in SRAM cell and their solutions are illustrated in Fig. 1 [16]. This work provides an extensive review of numerous modern FinFET-based SRAM architectures, offering an evaluation of the key performance factors to consider for different applications. The survey highlights the technology used in SRAM design, different out-come metrics and the advantages and disadvantages of each architecture. This section also addresses the some research gaps in current methods for improving the design of FinFET based SRAM cells.

- **Less Extent of Novelty**: An obstacle to future progress is the fact that most research on enhancing the performance of SRAM cells focuses on either changing the number of transistors or using FinFET devices of varying sizes.
- **Less Studies toward Optimization**: An optimization model must be developed in order to address the fault tolerance concerns and enhance the SRAM FinFET design performance. Mathematical optimization techniques backed by probability theory are needed to enhance the design concepts of SRAM-based FinFETs.
- **Less number of computational Modeling**: The majority of current methods for FinFET-based SRAM rely on hardware-based modeling environments or experimental methods. Although experimental approaches provide highly reliable results, their application for large-scale, real-time commercial use is limited due to the lack of thorough consideration of computational complexity. On the other hand, hardware-based methods rely on specific simulation environments, which limit the amount of computational power that can be utilized.

The main goal of study is to review various SRAM memories for robotics and other electronic applications that utilize FinFET technology. FinFETs offer distinctive features such as volume inversion, the corner effect, and enhanced mobility. FinFET-based design structures offer superior performance, reduced costs, and

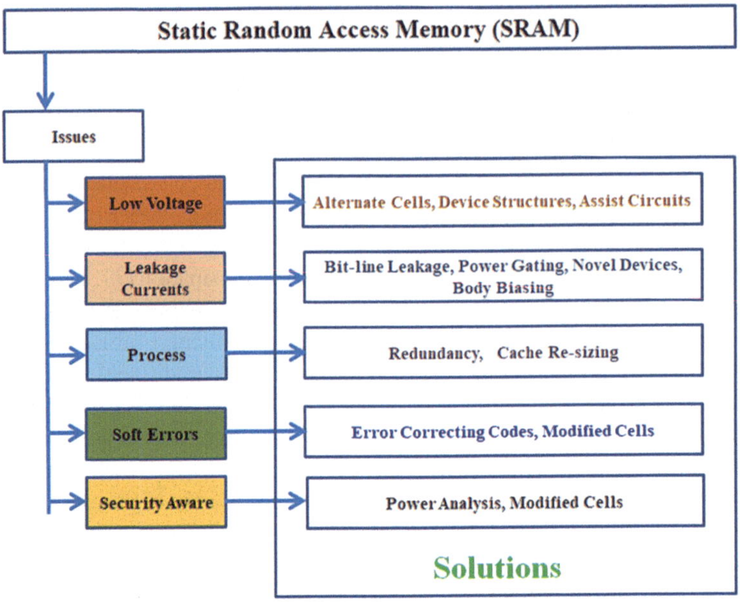

Fig. 1 An outline of the SRAM design problems and their solutions

enhanced circuit functionality. This survey includes only English-language publications, reports, review articles, and editorial remarks. The key criteria for selecting sources are the accessibility of studies, the techniques used, and the visibility of the indexed articles.

2 FinFET Technology

Modern semiconductor technology is based on FETs, which have made it possible to create increasingly sophisticated and potent electronic devices. Out of all the different kinds of FETs, the FinFET has become a ground-breaking development, providing better scalability and performance than conventional planar MOSFETs [24]. In order to overcome the limits caused by scaling constraints, the FinFET introduces a three-dimensional (3D) architecture, marking a substantial shift from the planar MOSFET structure. More control over current flow and improved electrostatic integrity are two benefits of the FinFETs unique fin-shaped channel that extends over the silicon substrate.

Figure 2 illustrates the construction of a FinFET. The FinFETs structure includes multiple 90° channels resembling the 'fin' of a fish, hence its name. This is also recognized as a Multi Gate Device (MGD) and is constructed on a silicon substrate. Placing the gate on two, three, or even four sides of the channel creates a double-gate design. In between the drain and source areas, a 'fin' forms on the silicon surface.

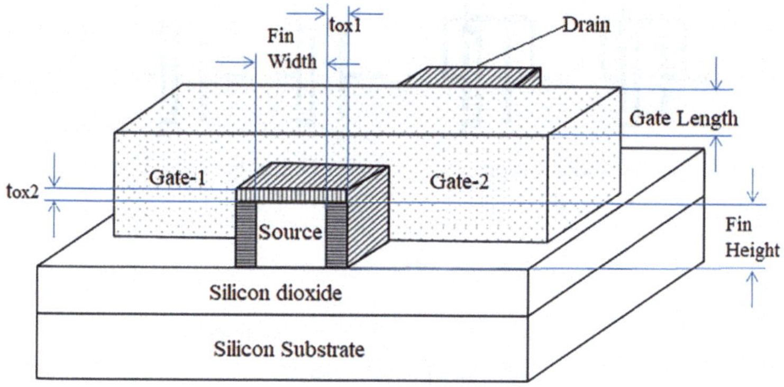

Fig. 2 FinFET device structure

FinFETs operate similarly to conventional MOSFETs, using gate, drain, and source terminals to regulate the motion of current. However, the channel among the drain and source in a FinFET is the key feature that distinguishes it from a MOSFET.

The FinFET gate is surrounded by the 3D bar design, which produces numerous gate electrodes on each side of the channel, as seen in Fig. 2. These electrodes raise the drive current and lessen the effects of leakage [25]. By switching to a 3D architecture, FinFETs provide a number of significant benefits while overcoming several scaling issues [26]. The fin structures enhanced electrostatic control, which enhances gate control and lowers leakage current, is one of the most obvious advantages. Furthermore, the channel's vertical orientation improves gate coupling, which raises the device's speed, power consumption, and reliability.

The given areas are existing in the FinFET model:

- **Gate-Oxide Region (silicon dioxide)**: high doping poly-silicon region.
- **Silicon Fin with Low Doping**: high doping contact region among drain and source.

FinFET transistors operate in two different modes: Independent-Gate (IG) and Shorted-Gate (SG).

Figure 3a shows the symbol of FinFET- IG and SG mode in 3 dimensions [27] and Fig. 3b represents the structure of two-dimensional. IG and SG FinFETs are also recognized as 3-terminal (3 T) FinFETs and 4-terminal (4 T) FinFETs, respectively. In reality, both the front and rear gates of SG FinFETs are on same potential (shorted). Using these two gates to regulate the electrostatics of the channel is standard procedure. So, SG FinFETs have higher on-currents (I_{on}) and subthreshold currents (I_{on} and I_{off}) compared to IG FinFETs [28]. Because of its higher gate-to-channel coupling, the SG mode is more effective at preventing SCEs.

The front gates and rear gates of IG FinFETs are physically separated. The versatility of feeding different signals or voltages at the terminals of two gates is provided by this type of FinFET. This increases the back-gate bias's application to linearly modify the forward-facing gate's V_{TH}. To provide comparable rising and falling

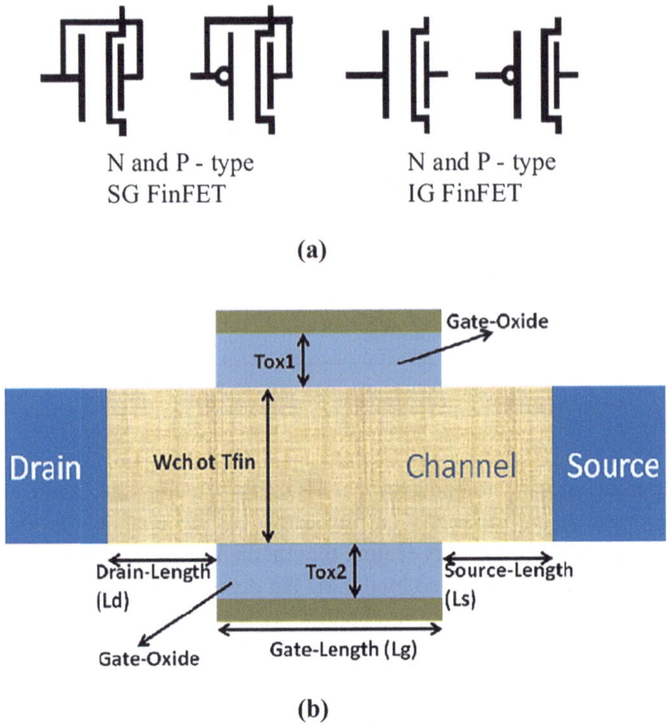

Fig. 3 (**a**) FinFET device symbol. (**b**) Two dimensional structure of FinFET

delays, the IG-FinFET syndicates the P-type parallel FinFET and combines with the type-N's back gate to ground [29]. In addition to mitigating short channel effects, the back gate biasing technique allows for dynamic regulation of a FinFETs V_{TH} in the IG mode. In the IG mode, two separate channels form beneath every gate oxide. The two channels are connected by a capacitively coupled connection. The back gate Vgb (bias voltage) can be suitably varied to find the necessary V_{TH} for a FinFET. However, IG FinFETs incur a significant area consequence due to the necessity of placing two different gate contacts.

FinFET structures give a range of design features. It has different modes of operation, such as hybrid, SG, IG, and low-power [30]. Better performance advantages at extremely low power are made possible by FinFETs' increased gate control. SCEs are suppressed using FinFET devices with gate-di-electric leakage currents. FinFET is a helpful technique for overcoming the scientific divide between other state-of-the-art devices such as CNTFETs, graphene FETs, and bulk CMOS.

FinFETs (Fin Field-Effect Transistors) provide numerous advantages over traditional planar MOSFETs, making them the preferred choice for modern semiconductor technologies, particularly at advanced process nodes. These advantages include:

i. **Improved Electrostatic Control**
- **Advantage**: In contrast to planar MOSFETs, FinFETs have a three-dimensional construction in which the gate encircles the fin (channel), offering better electrostatic control over the channel.
- **Benefit**: Enables better scalability to smaller dimensions and reduce SCEs.

ii. **Reduced Leakage Current**
- **Advantage**: Enhanced gate control minimizes subthreshold leakage, a critical issue in low-power applications.
- **Benefit**: Lower static power consumption and Better performance in energy-efficient designs.

iii. **Higher Drive Current**
- **Advantage**: The 3D structure increases the effective channel width (Weff), allowing higher current flow for the same footprint.
- **Benefit**: Improves device speed and performance and Suitable for high-performance computing and mobile devices.

iv. **Enhanced Performance at Lower Supply Voltages**
- **Advantage**: FinFETs maintain excellent performance even at reduced supply voltages (V_{DD}) due to their superior gate control and reduced variability.
- **Benefit**: Enables aggressive power scaling and Reduces dynamic power consumption, a critical factor in battery-operated devices.

v. **Better Scalability**
- **Advantage**: FinFETs are well-suited for scaling down to sub-5 nm nodes, addressing the limitations of planar MOSFETs at smaller dimensions.
- **Benefit**: Prolongs Moore's Law by supporting further miniaturization and Facilitates higher transistor density in integrated circuits.

vi. **Reduced Variability**
- **Advantage**: FinFETs exhibit less variability in device characteristics, such as threshold voltage (V_{TH}), due to reduced dependence on random dopant fluctuations (RDF).
- **Benefit**: Enhances reliability and uniformity across devices and Improves yield in manufacturing.

vii. **Improved Thermal Performance**
- **Advantage**: The 3D structure allows for better heat dissipation compared to planar designs.
- **Benefit**: Reduces thermal issues in densely packed circuits and enhances the reliability of high-performance applications.

viii. **Increased Energy Efficiency**
- **Advantage**: Lower leakage and reduced power consumption per transistor contribute to better overall energy efficiency [31].
- **Benefit**: Essential for applications requiring long battery life or high computational efficiency and Supports sustainable semiconductor design by reducing power demands.

ix. **Compatibility with Advanced Design Techniques**
- **Advantage**: FinFET technology is compatible with techniques like multi-gate architectures and gate-all-around (GAA) FETs, paving the way for next-generation devices [32].
- **Benefit**: Enables innovative design solutions for future nodes and offers a pathway to integrate emerging materials and structures.

The changeover from planar MOSFETs to FinFETs marks a significant technological leap, addressing the challenges of traditional planar designs while meeting the demands of modern applications in performance, efficiency, and scalability.

3 SRAM Memory

SRAM is a kind of semiconductor memory that save the data using bistable flip-flops. Unlike DRAM, which requires periodic refreshing to retain data, SRAM retains its contents as long as power is supplied, making it "static." SRAM is known for its low latency and high speed making it best for applications require fast access, such as processor cache memory, networking devices, and high-performance computing systems. It consumes more power and space per bit compared to DRAM but offers better reliability and simpler control logic, as it does not rely on capacitors for data storage. Its scalability and efficiency have made it a cornerstone of modern computing systems, especially in scenarios where performance outweighs density and cost concerns.

An SRAM cell's latch is made by connecting two cross-coupled inverters, which store a single bit of data. This latch is linked to two bit lines through transistors M1 and M2. The word line controls these transistors, determining whether they are open or closed. In read mode, the word line signal closes both M1 and M2, allowing data to be accessed. As shown in Fig. 4, the values at Q and $\overline{Q}$ are always complementary. Table 1 outlines the detailed operation of the SRAM cell [33].

As seen in Fig. 4, the architecture of SRAM is made up of many memory cells arranged in an array. Usually, a flip-flop configuration of several transistors makes up each memory cell. Here's a simplified overview of the structure:

- **Memory Cell**: The memory cell is a small, bistable circuit used to save a single bit of data (0 or 1). It is designed to maintain its state as long as power is supplied, without requiring periodic refreshing. A cross coupled pair of inverters connected in a feedback loop usually makes up memory cell, and the fundamental component of SRAM [34]. With this setup, a bi-stable circuit is produced that, with enough power, can maintain a stable state indefinitely.
- **Word Lines and Bit Lines**: Bit lines and Word lines are essential components of the memory array structure. They facilitate addressing, reading, and writing data to the individual memory cells. The word line is a horizontal line in the memory arrays that connects to the gate terminals of the access transistors in a row of

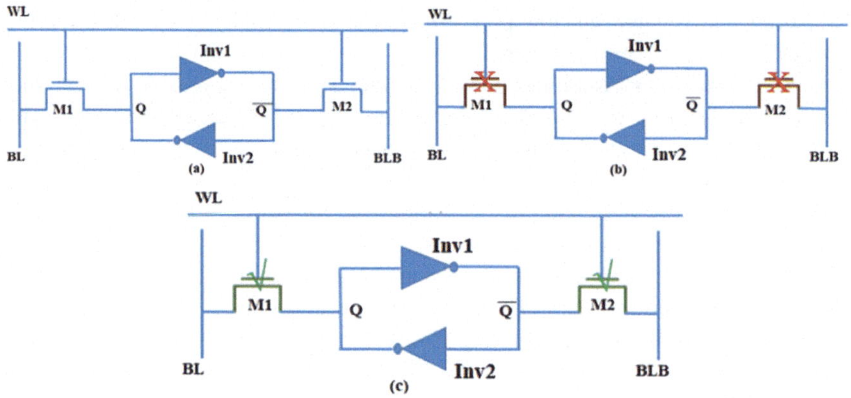

Fig. 4 Models of SRAM cell (**a**) General, (**b**) Hold (M2 and M1- both OFF), (**c**) Write and Read (M2 and M1- both ON) [33]

Table 1 Basic SRAM cell operations [33]

Operation	WL	Q	QB	M1	M2	BL (V_{DD})	BLB	Sense amplifier output
Hold	0	0	1	OFF	OFF	–	–	–
	0	1	0					
Write	1	0	1	ON	ON	0.5	GND	1
	1	1	0					0
Read	1	0	1	ON	ON	0.5	0.5 V_{DD}	0 (BL < BLB)
	1	1	0					1 (BL > BLB)

SRAM cells. It is used to enable or disable access to the SRAM cells in a specific row. When the word line (WL) is **high (active)**, the access transistors turn on, connecting the storage node of the cell to the bit lines for reading or writing data. Similarly when the WL is Low **(inactive)**, the access transistors are off, and the SRAM cell is isolated from the bit lines, maintaining its stored data. The bit lines are vertical lines in the memory array that connect to the source/drain terminals of the access transistors of all cells in a column. Bit lines are used to read data from or write data to the SRAM cell. At the time of write operation, the bit lines are driven with the desired data (BL = 1, $B\overline{L}$ = 0 for writing a 1, and BL = 0, $B\overline{L}$ = 1 for writing a 0). During a read operation, the bit lines sense the state of the SRAM cell by detecting a small voltage difference between $B\overline{L}$ and BL, which is then amplified by a sense amplifier.

- **Sense Amplifiers**: Sense amplifiers are critical components in SRAM (Static Random-Access Memory) used at the time of read operation to detect and amplify the few voltage difference on the bit lines (B $\overline{L}$ and BL). They play a key role in ensuring fast, reliable, and low-power operation of SRAM by quickly determining whether the stored bit is a 0 or a 1.

- **Address Decoders**: Address decoder is a crucial circuit responsible for selecting a specific memory cell based on the address provided during a read or write operation. It interprets the binary address input and activates the corresponding WL or BL in the memory array.
- **Write Circuitry**: Write circuitry is responsible for storing a bit of data (0 or 1) into a specific memory cell. This process involves driving the appropriate voltage levels onto the bit lines to overwrite the existing data in the selected SRAM cell.
- **Power and Ground Lines**: The power and ground lines are vital components that provide the necessary operating voltages for the SRAM cells and peripheral circuitry. These lines ensure that the SRAM operates reliably, efficiently, and within design specifications.
- **Control Logic**: Control logic is responsible for managing and coordinating various operations such as read, write, and standby. It ensures that the correct memory cells are accessed, data is written or read properly, and the SRAM functions efficiently in response to external commands.

Figure 5 displays a simplified block diagram of an SRAM cell. The behavior of SRAM during various stages of operation, such as read, write, and standby modes, is referred to as its operation state. Here's an overview of each operation state:

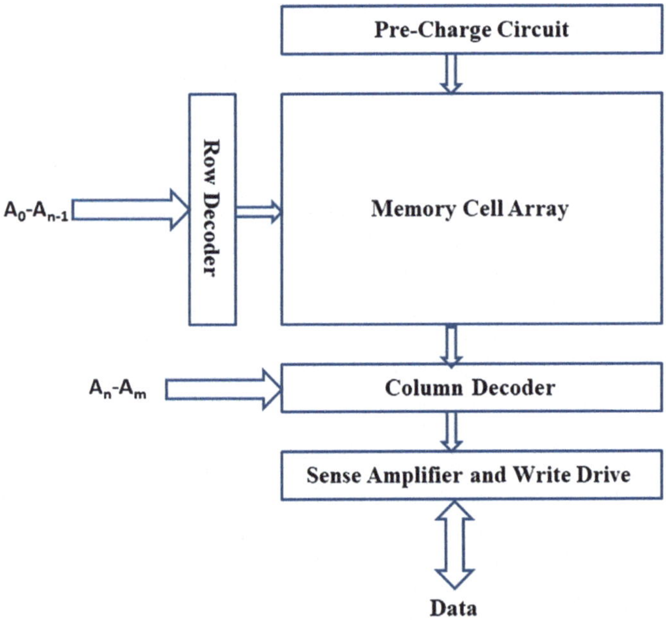

Fig. 5 Simplified block diagram of a SRAM

Read Operation:

- It is the process of retrieving the stored data (1 or 0) from a specific memory cell. This operation relies on the interaction between the SRAM cell, word lines, bit lines, sense amplifiers, and control logic.

Write Operation:

- It is the process of storing new data (0 or 1) into a specific memory cell. This operation modifies the bistable latch within the SRAM cell by forcing it to transition to the desired state. The process involves the coordinated action of bit lines, word lines, write drivers, and control logic.
- The write operation is a crucial process that involves overwriting the data stored in a memory cell by forcing its bistable latch to transition to the desired state. Its efficiency and reliability are critical for the performance of SRAM in high-speed, low-latency applications such as processor caches, embedded systems, and networking hardware.

Standby Mode:

- It refers to the state when the memory is powered on but not actively involved in read or writes operations. In this mode, the SRAM retains the data stored in its memory cells while consuming minimal power.
- The standby mode operation of SRAM is designed to retain stored data while minimizing power consumption when the memory is not actively accessed. This mode is crucial for energy efficiency in modern computing systems, especially in applications requiring high-speed, low-latency memory with minimal downtime.

To guarantee data retention over long stretches of time without active operation, SRAM designs may include refresh processes in certain low-power or standby modes.

Comparing SRAM cells to other memory types reveals a number of benefits. Here are a few of the main benefits:

i. **High Speed**: SRAM is substantially faster than DRAM. SRAM can offer faster access times because it doesn't require refreshing cycles to preserve data, which makes it perfect for applications like CPU cache memory where speed is crucial.
ii. **Low Latency**: Low latency in SRAM operations is a result of direct access to stored data and the absence of refresh cycles. This quick response time is advantageous for tasks that need immediate access to data, such as high-performance computing and real-time processing.
iii. **Low Power Consumption**: Generally speaking, SRAM uses less power than DRAM when operating in both standby and active modes. In many applications, particularly those with frequent access patterns, SRAM cells use less power because they are bi-stable and do not need to be refreshed constantly [35].
iv. **Ease of Use**: SRAM is easier to integrate into digital systems because of its straightforward operation and interface. Because it doesn't need complicated refresh control hardware, it's easier to develop and use in a variety of applications.

v. **Predictable Timing**: SRAM's deterministic and predictable timing waveforms make it perfect for uses where timing constraint is serious. This predictability timing is mainly beneficial in high-speed communication interface systems and real-time systems and.
vi. **High Reliability**: Different from DRAM, which can suffer bit flips because of charge leakage in memory cells, SRAM is less vulnerable to soft mistakes created by cosmic rays or alpha particles [34]. SRAM is favored for mission-critical applications and aerospace systems due to its increased reliability.

Low power consumption,Fast speed, low latency and dependability make SRAM a good choice for a variety of applications like high-performance computing, embedded systems, CPU cache memory and networking devices [35].

4 SRAM Cell Using FinFET

High-tech FinFETs are a type of three-dimensional transistor structure that, in contrast to traditional planar CMOS transistors provide better control over current flow. When compared to prior SRAM cell designs, this enhanced control allows SRAM cells with FinFETs to achieve reduced power consumption and higher performance. FinFETs improve transistor performance by enabling better electrostatic control of the transistor channel, which reduces leakage current. For contemporary high-performance computing applications, SRAM cells with FinFETs are preferred due to their ability to achieve increased density and lower power consumption.

4.1 FinFET-6 T SRAM Cell

In comparison to conventional CMOS-based SRAM cells, a 6T FinFET SRAM cell uses FinFET technology for its device structure, offering enhanced performance, power efficiency, and density. A 6T SRAM cell using FinFET uses two pairs of inverters linked in a cross-coupled topology to create a bi-stable latch for data storage, much like SRAM cells utilizing CMOS. The four FinFETs that make up the two cross-coupled inverters are referred to as M1, M2, M3, and M4. Two FinFETs (of both types) are placed vertically in a fin configuration to form each inverter. When compared to planar CMOS transistors, the fins of FinFETs offer superior electrostatic control and lower leakage, which enhances the latch's stability and dependability [36]. The SRAM cell can read and write data thanks to the access transistor function of two more FinFETs (M5, M6). For read and write operations, the BL is linked to the access transistor's source terminal (M5), and the bit-line bar (BLB) to the other terminal (M6). Figure 6 displays the schematic of an SRAM 6T cell based on FinFETs.

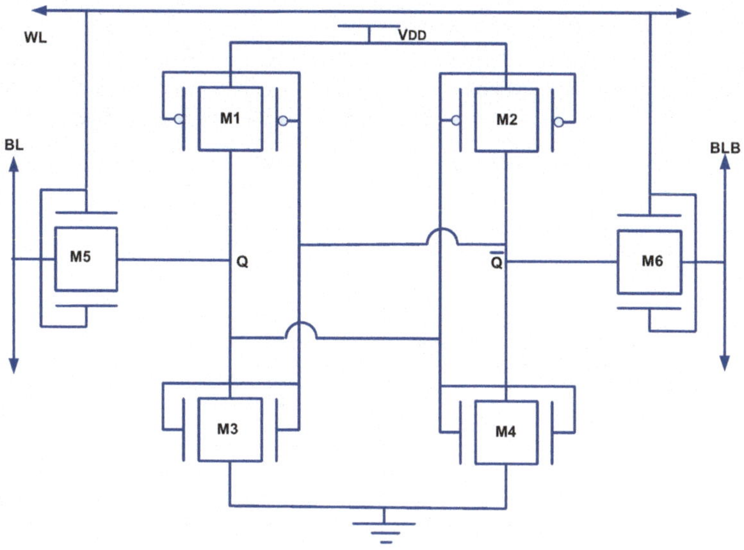

Fig. 6 6 T FinFET based SRAM Cell

Data can be read from or written to the SRAM cell thanks to the WL, which regulates the activation of the access transistors (M5, M6). Access to the stored data is made possible by the activation of the access transistors by the WL. The three basic operations are hold, write and read [37, 38].

Write Operation: The word-line is turned on ($WL = V_{DD}$) during the write process, which activates the access transistors (M5, M6). The voltage level at which $\overline{Q}$ is sampled to turn on the transistors M3 and M6 and Q lowers is then reached by the node voltage level rising.

Read Operation: The transistors M2 and M4 turn on during the read process, and the WL is activated ($WL = V_{DD}$). The access transistors (M5, M6) are turned on by the activated WL, enabling the bit lines to receive the stored data. If $Q = 1$ and $\overline{Q} = 0$ and M1 and M6 turn ON, then the transistors M5 and M3 are in the OFF state. As a result, BLB voltage level discharges and BL voltage level maintains at V_{DD}.

Hold Operation: The word line remains fixed to the ground ($WL = 0$) throughout the hold operation. Transistors M2 and M4 switch off to maintain the separation of the latching circuit and bit lines. The remaining transistors (M1, M5, M6 and M3) form a latch structure that stores data until it is detached from the bit lines.

Tables 2 [38] and 3 [3] present the performance analysis of traditional and FinFET-based SRAM at temperature of 25 °C.

Table 2 6T SRAM cell: performance analysis

S. No.	Device type	Technology (nm)	Delay		Power		SNM (mV)
			Write (ps)	Read (ps)	Write (nW)	Read (nW)	
1	SRAM using FinFET	22	98	610	1.6	1.98	280
2	Conventional SRAM		112	947	243	746	240

Table 3 6T SRAM cell using MOSFET and FinFET: performance analysis

S. No.	Device type	Temp (°C)	SNM (mV)	Write margin (mV)	Cell standby leakage current (mA)	Read current (mA)	Leakage current (nA)
1	MOSFET	40	156.07	265.11	6.99	0.0223	2.07
			165.17	274.43	0.0155	0.0232	0.0446
2	FinFET		184.2	292.02	4	0.117	0.359
			192.74	302.43	0.0149	0.125	0.0304
3	MOSFET	27	156.12	266.21	4.41	0.0224	1.15
4	FinFET		184.23	292.31	1.02	0.118	0.235
5	MOSFET	125	129.15	242.43	9.81	0.19	4.37
6	FinFET		175.57	276.63	6.22	0.11	0.702

4.2 *FinFET-7 T SRAM Cell*

A bi-stable latch for data storage is formed by the core of the SRAM cell, which contains two sets of cross-coupled inverters, designated M3, M4, M1, and M2. Both an n-type and a p-type FinFET make up each inverter. Read and write operations in the SRAM cell are made possible by two extra FinFETs (M5, M6) acting as access transistors. For read and write operations, the BL is connected to the access transistor's source terminal (M5), and the BLB to the other (M6). Figure 7 displays the schematic of an SRAM 6T cell based on FinFETs.

A 7T SRAM cell has an additional FinFET transistor (M7), usually of the n-type, called the write enable (WE) transistor, in addition to the 6T SRAM structure [39, 40]. The 6T cell's leakage problem is fixed with the 7T Fin-FET construction. To perform the read and write functions, WL is connected to both the M5 and M6 transistors.

Write Operation: Both the write enable signal and the word line are turned on during a write operation. The bit lines are able to drive the cross-coupled inverters to the required state by turning on the access transistors. The write-enable transistor controls the ability to modify the data stored in the cell.

Read Operation: At the time of read operation, the WL is triggered and the access transistors (M5, M6) are activated. By detecting the voltage differential between the

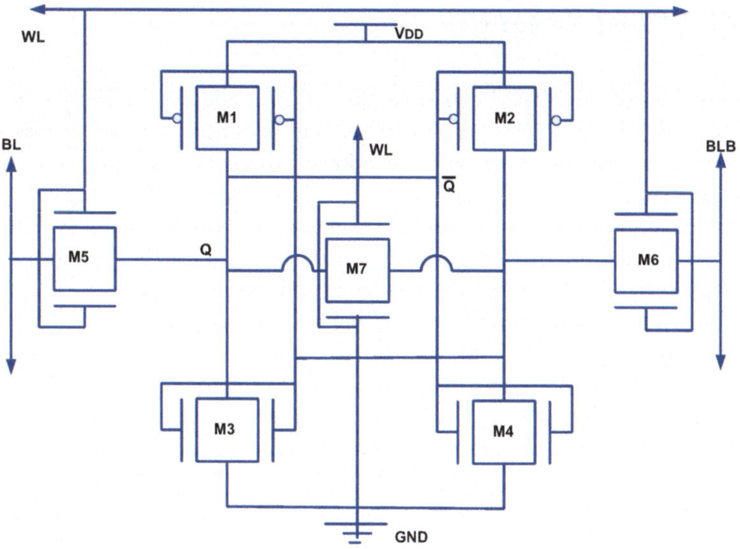

Fig. 7 7T FinFET based SRAM Cell

bit lines, the cell's condition is known. The data is then recovered without changing the previously preserved state.

Hold Operation: The WL is fixed to the ground during the hold action. The M1 and M2 transistors are therefore shut off. Logic 0 in the SRAM cell causes a sub-threshold leakage current to flow through the transistors when they are in the OFF state. Furthermore, the SRAM cell only requires the BLB to perform write operations, and the extra transistor M7 connects and disconnects feedback [41, 42].

The performance analyses of 7T SRAM cells are shown in Tables 4 [43] and 5 [5] respectively.

Table 4 7T SRAM using FinFET and CMOS technology: performance analysis

S. No.	Parameters	Technology (nm)	Average power (μW)	Delay (ns)	PDP (fj)
1	CMOS based	22	0.169	0.129	0.0219
		16	0.139	0.223	0.0311
2	FinFET based	22	0.023	0.09	0.0021
		16	0.015	0.002	0.0028

Table 5 7T FinFET-based and conventional SRAM cell: leakage current and power calculation

S. No.	Device type	Leakage power	Leakage current
1	FinFET	38.10 nW	3.094 nA
2	Traditional	26.34 μW	14.12 μA

4.3 FinFET-8T SRAM Cell

The FinFET based 8T SRAM cell was introduced to address the shortcomings of the 6T SRAM cell [36]. The main trouble is that the writes and read processes are not separated. In read mode, the cell with a lower SNM might be able to write more efficiently. Therefore, if the read and write operations are properly segregated, circuit designers can optimize the write and read functions with more flexibility. The write and read processes are separated using 8T FinFET-based SRAM architecture. SRAM offers improved stability during low-voltage operations as a result of this separation [44, 45]. The leakage power and current of an 8T SRAM cell based on a conventional and FinFET are displayed in Table 6 [46].

The FinFET-8T SRAM cell's circuit diagram is shown in Fig. 8. Two additional n-type FinFETs are used in this SRAM to enable a read procedure without interfering with internal nodes. As a result, this technique requires a separate Read Word Line (RWL) and Write Word Line (WWL). SRAM functions in read mode when RBL is pre-charged and RWL is engaged. The sensing amplifier recognizes that when one is stored at Q, the transistor M4 turns on and offers a minimum resistance channel for the cell current to pass via RBL to Ground [47, 48].

Table 6 Traditional and FinFET based 8T SRAM cell: leakage power and current analysis [46]

S. No.	Device type	Leakage power (nW)	Leakage current (nA)
1	FinFET	60.90	55.56
2	Traditional	35.87	34.67

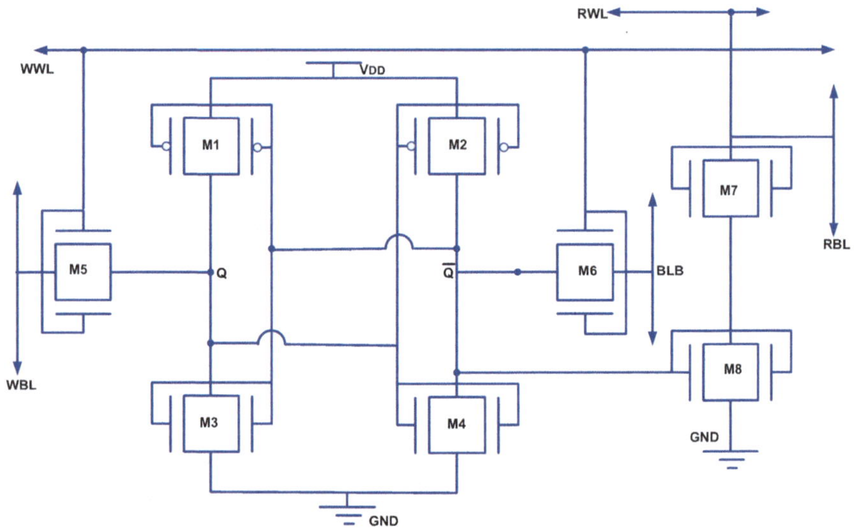

Fig. 8 FinFET based 8T SRAM Cell

4.4 FinFET-9 T SRAM Cell

With the development of the 9T SRAM cell design with FinFET, RBL leakage problem with 8T SRAM cells has been fixed [49]. An 8T cell can only be used for low -density applications that can be handled with a 9T structure, such as putting an M7 transistor between an M9 and an M8. Consequently, leakage in BL is significantly decreased by the stack effect phenomenon. The FinFET-9T SRAM cell is schematically represented in Fig. 9. The SRAM cell with 9T FinFET has two main sub-sections [49–52]. The components of the 9T SRAM's top section—M1, M5, M3, M6, and Q—are the same as those of the 6T cell design. Data storage is carried out in this primary subsection.

The upper portion of the 9T SRAM is made by the combination of two bit-line access transistors (M2 and M4) and one read access transistor (M7). Transistors (M8 and M9) rely on the information in the cell to function, but M7 requires an independent RD. WBL or WBLB and RWL (read word lines), respectively, measure the read and write access that are carried out by the read and write access transistors.

When OFF-state transistors are coupled in series, stacking takes place in this SRAM cell. This means that the source voltage of the upper transistor in the stack will be somewhat higher than that of the lower transistor. The VTH rises as a result of the top transistor's increased voltage. The improvement in the threshold voltage results in a reduction in leakage. The stack effect is the name given to this reduction. The examination of multi parameters for a 9T SRAM cell using FinFET at various technologies is displayed in Table 7 [53].

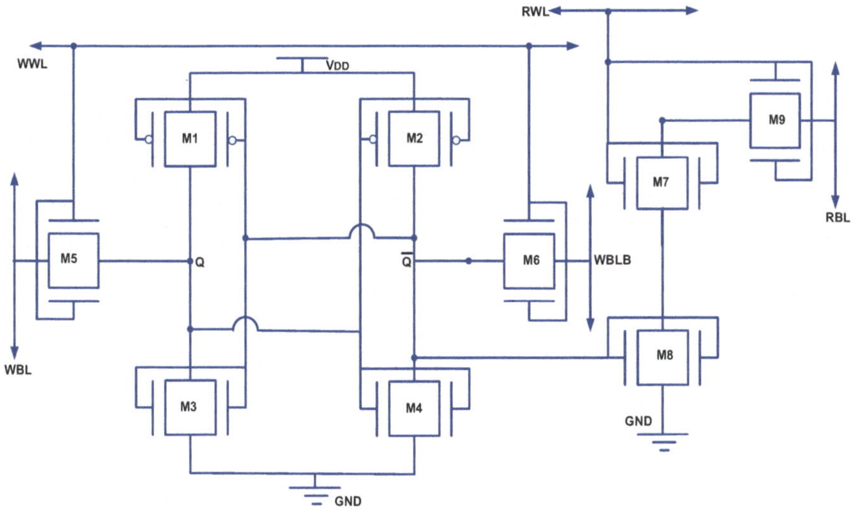

Fig. 9 FinFET based 9T SRAM Cell

Table 7 FinFET based 9 T SRAM cell: analysis of several parameters [53]

Technology (nm)	Leakage power (μW)	SPNM (μW)	Read power (nW)		Write power (nW)		Read access (ps)		Write access (ps)	
			1	0	0	1	0	1	1	0
180	3.26	7.55	64	162.3	110.2	89.5	20.4	15.4	120	82.5
7	0.804	7.53	54	48	15.41	56.3	15.8	7.2	78.5	14.7

4.5 FinFET-10 T SRAM Cell

10T SRAM cell using FinFET achieves reduced leakage current as compared to 9T SRAM cells by incorporating transistors M7, M8, M9, and M10. The diagram of the 10T FinFET SRAM cell is shown in Fig. 10. In the read path, a dual V_{TH} method is applied using specific transistors, enhancing the ON/OFF current ratio. The access transistors M5 and M6 are connected to the bit lines and the write word line (WWL). The read word line (RWL) is linked to the gate of transistor M9 and the source of transistor M7. Transistors M8 and M10 are included to improve the write margin of the cell. To minimize static current, the access transistors are designed to be twice the size of the pull-up transistors. All other transistors in the 10T design, except the access transistors, are implemented with the minimum achievable gate length [54].

Read Operation: The two access transistors are used to link the pseudo nodes. The read stability is effectively conserved since there is no read current flow by storage nodes.

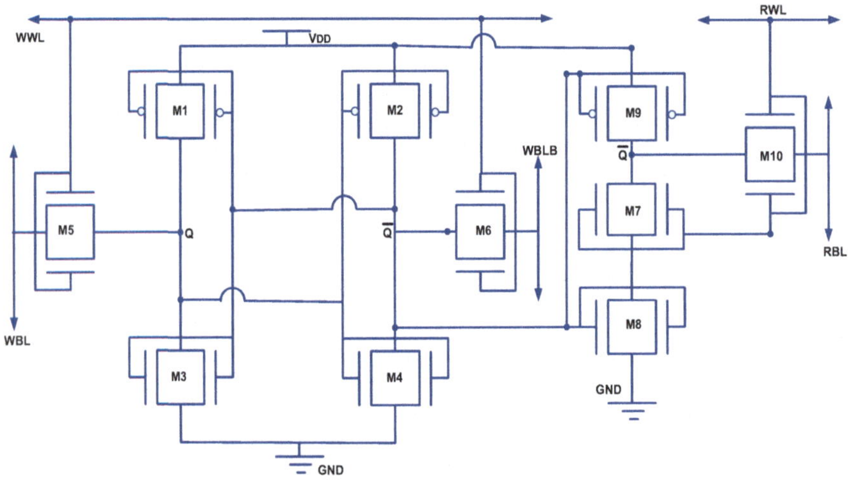

Fig. 10 10 T FinFET based SRAM cell

Table 8 Analysis of Write and Read operations of 10T INDEP FinFET based SRAM cell [55]

S. No.	Parameter	10T INDEP FinFET Operations	
		Write	Read
1	Leakage power (μW)	26.5	7.26
2	SNM (mV)	106.06	309.7
3	PDP (aJ)	129.6	426.61
4	Propagation Delay (ps)	16.1	18.01

Write Operation: The node $\overline{Q}$ and Q store the values 0 and 1, respectively, for writing operations. Due to discharge via the pull-up transistors and access, a high supply voltage is fed into the node Q, dragging it down to "0".

A number of factors for a 10T SRAM cell with an INDEP FinFET are analyzed in Table 8 [55].

4.6 FinFET-11 T SRAM Cell

Static power consumption poses a significant challenge in the design of SRAMs. The 11T SRAM cell is specifically designed to minimize energy usage, decrease static power consumption, and improve performance, particularly in the sub-threshold region [56]. The diagram of the 11T FinFET-based SRAM cell is shown in Fig. 11. The transistors M2, M4, M3, and M6 in this design share similarities with those in a standard 6T SRAM cell. Additionally, the sizes of M1 and M6 transistors are optimized to align with the dimensions of the p-MOS transistors. According to [57], this SRAM cell incorporates word lines (WL), bit lines (BL), a read word line (RWL), and dedicated read/write ports. Its low power dissipation is attributed to the use of series-connected drivers powered by BL, BLB, and read buffers. The analysis of the 11T SRAM design against other SRAM designs with different transistor configurations at 10 nm technology is provided in Table 9 [58].

4.7 FinFET-12 T SRAM Cell

Figure 12 depicts the architecture of a 12T FinFET-based SRAM cell. Transistors (M1, M5, M3, M2, M9, M4, M8, M6, M7, M12, M10, and M12) and differential read/write ports are features of this SRAM cell. The primary objective of the 12T bit-cell design is to enable efficient low-voltage functionality. During the writing process, the supply voltage to either the right or left half of the cell is turned off to weaken the pull-up network, simplifying the process and removing the requirement of additional timing control or peripheral write-assist circuits. This design improves the speed of write operations. The signal configuration includes row-based signals

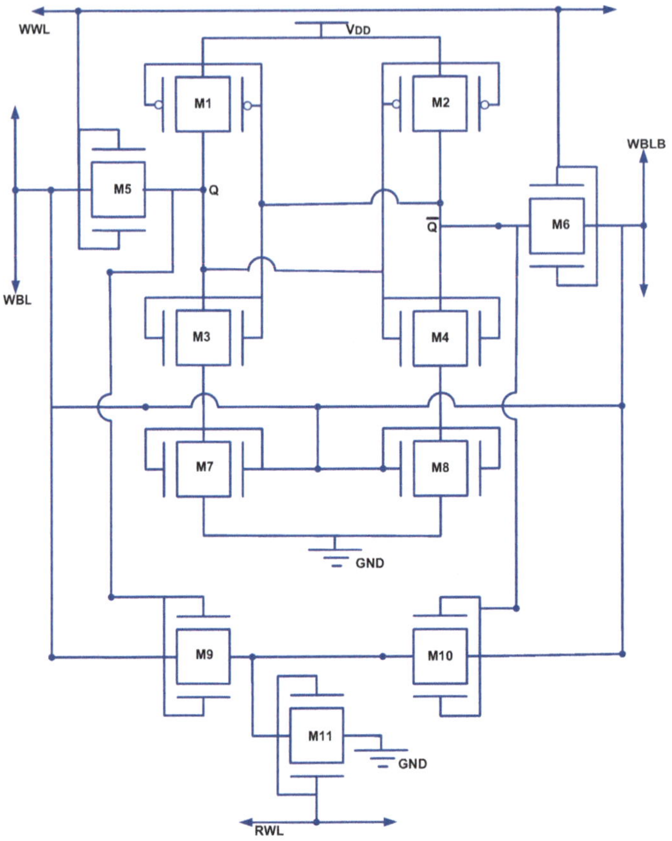

Fig. 11 11 T FinFET based SRAM cell

Table 9 Analysis of various SRAM cells with different transistor count [58]

Type/Parameters	Delay (ps)		SNM (mV)		Power (nW)		PDP (aJ)
	Read	Write	Read	Write	Average	Static	
11T	676.17	218.27	175	225	137.66	5.17	168.57
9T	676.61	275.09	183	225	631.17	4.37	445.54
8T	640.26	1749.2	183	80	226.72	6.85	701.38
7T	491.8	230.46	70	198	194.95	4.76	128.28

such as RWL and Virtual Ground (VGND) and column-based signals like WWL1, WWL2 (Write Word Lines), BL, and BLB. Studies [59, 60] highlight the benefits of the 12T design in terms of reduced power consumption and current. Table 10 [60] provides a comparison of the average power consumption between 12T SRAM cells implemented with FinFET and CMOS technologies.

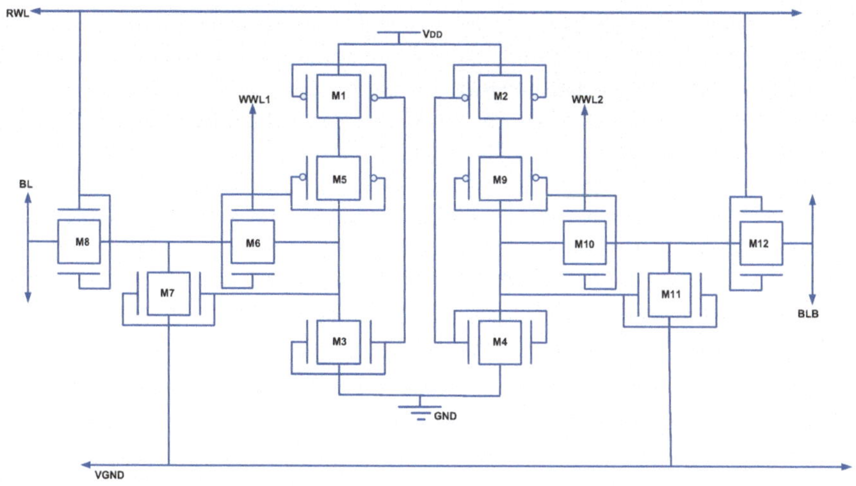

Fig. 12 12T FinFET based SRAM cell

Table 10 12T SRAM cell using FinFET and CMOS: average power analysis [60]

Device type	Average power		
	During write operation (nW)	During read operation (nW)	During hold operation (nW)
FinFET	60.06	45.64	14.06
CMOS	201.61	161.31	180.79

5 Performance Metrics of FinFET Based SRAM Cell Structure

This section discusses the performance metrics of SRAM cells with FinFET. Some common metrics used to evaluate the performance of FinFET architectures are as follows:

- **Static Noise Margin (SNM)**: An SRAM cell with FinFET technology's SNM determines the threshold voltage difference required to reverse the cell's state in relation to the noise in the system [61]. This parameter is critical for ensuring the reliable operation of SRAM cells, as it safeguards stored data from being corrupted by random voltage fluctuations. Several factors influence SNM, including device geometry, process variations, temperature, and supply voltage. It depends on parameters such as the cell ratio (CR), Pull-up ration (PUR) and V_{DD}. CR is the ratio of the driver transistor's size to the load transistor's size during a read operation, while the PUR is defined as the size of the load transistor relative to the access transistor during a write operation [62]. SNM is the minimum voltage difference between write and read access levels required to reliably differentiate between logic '0' and logic '1,' the two stable states of the SRAM cell. This margin ensures the cell can tolerate voltage noise without causing unintended state flips, maintaining data integrity.

- **Read Noise Margin (RNM)**: The RNM represents the difference between the voltage levels corresponding to the logic '0' and logic '1' states during a read operation. It is calculated by the voltage difference between the stable logic states (commonly represented as V_{DD} for logic '1' and V_{SS} for logic '0') and the noise voltage present during the read process. RNM has a direct relationship with the CR. To achieve a higher RNM, the pull-down FinFET must be stronger than the access transistors. The transistor sizing influences the pull-up ratio, and an increase in the pull-up ratio results in an improved RNM.

The RNM of SRAM is a critical parameter that defines the robustness of a SRAM cell during a read operation. It measures the ability of the cell to maintain its stored data while being accessed by the read circuitry. RNM is influenced by factors such as transistor sizing, supply voltage, and process variations. A higher RNM indicates better stability and resistance to noise, ensuring reliable operation. Optimization of RNM is essential in SRAM design, particularly for modern low-power and high-density memory applications, where process scaling and voltage reductions pose challenges to cell stability. RNM of an SRAM cell represents the maximum noise voltage that can be applied to the cell during a read operation without causing a state flip or data corruption. It is a vital metric for assessing the stability of the memory cell under read conditions, where the internal storage nodes are exposed to potential disturbances due to the interaction between the bit-lines and the access transistor. RNM is typically visualized using a butterfly curve, which is a plot of the input-output voltage characteristics of the cross-coupled inverters in the SRAM cell [63]. The width of the lobes in the butterfly curve quantifies the noise margin; a wider lobe indicates a more stable cell. Factors such as process variations, supply voltage scaling, temperature, and transistor mismatch can significantly affect the RNM. For example, reduced supply voltage in low-power SRAM designs can lead to a narrower butterfly curve, thereby reducing the RNM and increasing the risk of read instability.

To enhance RNM, designers often focus on optimizing the transistor sizing within the SRAM cell. For instance, rising the strength of the pull-down transistors relative to the pass transistors can improve RNM but may adversely impact write margin and area efficiency. Thus, achieving a balance between read stability, write performance, and area considerations is a key challenge in SRAM design.

- **Write Noise Margin (WNM)**: The WNM represents the voltage margin required to write a '0' or a '1' into an SRAM cell without being disrupted by noise during the write operation. It is defined as the voltage difference between the levels representing logic states and the noise voltage present during the write process. WNM is influenced by the pull-up ratio, with stability improving as WNM increases. WNM is the maximum noise voltage tolerated at the BLs during a successful write operation. A write failure occurs only if the noise voltage exceeds the WNM threshold. Stability and write performance are enhanced by using strong access FinFETs and weaker pull-up transistors, enabling faster discharge during a write

'0' operation while facilitating the storage of a '1.' Consequently, WNM improves near the cell's read margin with a combination of strong access transistors and weak pull-up FinFETs. WNM is a critical parameter for reliable and accurate data writing in SRAM cells, ensuring that stored data is not compromised by noise or unintended bit flips during the write operation.

- **Temperature**: Temperature plays a major role in the design and performance of FinFET-based SRAM, as it directly affects reliability, device characteristics, speed and leakage currents. An increase in temperature typically reduces the threshold voltage of transistors, altering their operational behavior. At elevated temperatures, reduced carrier mobility can lead to longer access times and increased read/write latency in SRAM cells. Additionally, extreme temperatures may compromise the reliability of SRAM, making it more prone to soft errors and accelerating device degradation over time. Higher temperatures also contribute to increased sub-threshold leakage currents in FinFET devices, resulting in higher static power consumption. This rise in leakage negatively affects the power efficiency of SRAM cells, posing challenges for maintaining performance and reliability in high-temperature environments.
- **Propagation Delay**: In FinFET SRAM architecture, propagation delay is a crucial component since it directly affects how quickly data is retrieved and processed with in the memory array. Longer propagation delays lead to increased access times, slowing down the overall performance of the memory. This delay impacts the critical path in SRAM circuitry, which defines the highest operating frequency the memory can achieve. Increased propagation delays also contribute to higher dynamic power consumption in FinFET SRAM because circuit elements remain active for a longer duration during data access. Additionally, propagation delay affects the scalability of operating voltage and frequency, as longer delays can hinder frequency scaling and limit performance improvements through voltage and frequency adjustments. Specifically, the delay in sense amplifiers during read operations directly impacts the speed and precision of data retrieval, making it a crucial parameter for optimizing high-performance SRAM systems.
- **Power Delay Product (PDP)**: The PDP is used to evaluate the energy efficiency of a circuit by combining power consumption and propagation delay. In the context of SRAM cells, PDP serves as an indicator of energy performance. A lower PDP value suggests that the circuit consumes less energy per operation while maintaining the required performance. In FinFET-based SRAM, there is a direct correlation between PDP and dynamic power dissipation. Greater dynamic power consumption during read and write operations is the result of higher PDP values, which raises the memory array's overall power consumption. PDP is therefore essential to the design of energy-efficient SRAM systems.

6 Relative Study of FinFET Based SRAM Cells

The relative study of various SRAM cells with FinFET is conducted based on their design methods, features and technology. Table 11 [64] presents a comparison of different SRAM cells, while Table 12 gives the summary of FinFET based SRAM cells. Additionally, Table 13 provides an overview of SRAM cells employing various power reduction techniques [33].

Figure 13 shows a comparative chart for SRAM cell designs with varying transistor counts. It illustrates power dissipation using dynamic threshold power reduction and power gating strategies.

7 Conclusion and Future Scope

This survey provides an in-depth analysis of FinFET-based SRAM designs, focusing on various optimization methods and performance metrics. Compared to conventional SRAM, FinFET-based SRAM exhibits superior performance in areas such as sub-threshold leakage, reduced short-channel effects, lower doping variability, operation at lower voltages, and increased versatility in device parameters. It has been observed that these architectures offer improved read, write, and hold noise margins, as well as reduced power consumption and delays. As the industry moves toward lesser device sizes and higher integration densities, the insights from this review offer valuable guidance for circuit designers and researchers seeking to fully leverage FinFET-based memory architectures.

To meet the growing demands of microprocessors, future SRAM cells will need to improve both performance and durability. In the realm of in-memory digital domain computing, especially for AI-driven miniature electronic devices, computing-in-memory solutions require further exploration. Given the power constraints of these devices, energy-efficient and reliable computing processes are essential. This work sets the stage for identifying key challenges in SRAM, laying a foundation for exploring SRAM-based CIM architectures for machine learning. Future efforts may focus on reducing power dissipation, addressing variability, and enhancing speed, power efficiency and cost-effectiveness.

Table 11 Comparison of different FinFET-equipped SRAM cells [64]

S. No.	Pub. year [Ref. No.]	FinFETs used	Technology (nm)	Power reduction technique	Observations
1	2008 [22]	8	32	Back gate voltage	During Write operation, discharging activity reduced
2	2017 [40]	7	45	Supply voltage-scaling	RBL charging is not necessary, and the voltage supply is getting lower
3	2016 [41]	7	14	Near V_{TH} operation	High performance and low energy consumption are combined in this efficient design solution for the NTV region
4	2015 [42]	7	20	NTV operation	Improved read and write margins and lower write times
5	2017 [48]	8	16	supply voltage-floating	Improved read and write ability
6	2014 [50]	9	22	Near V_{TH} operation and yield estimation technique	Minimum operating voltage is 0.3 V
7	2015 [51]	9	22	Multi-V_{TH}	Reduce the power consumption by 3 times
8	2016 [52]	9	22	Power gated 9T SRAM	For each read and write operation: lower energy consumption and smaller bit cell area
9	2021 [65]	6	16	Biasing scheme	At the body terminal, a dynamic voltage is applied
10	2017 [66]	8	22	Power gating	Two sleep transistors are loaded into standard SRAM cells
11	2016 [67]		28	Reverse body bias	Complimentary voltage is fed at body terminal
12	2016 [68]	10	16	V_{TH} technique	Memory cell using FinFET will be utilized without any SCE under 32 nm technology
13	2014 [69]	8	20	Supply voltage	FinFET design has lower Sub threshold swing

(continued)

Table 11 (continued)

S. No.	Pub. year [Ref. No.]	FinFETs used	Technology (nm)	Power reduction technique	Observations
14	2014 [70]	8 CMOS and FinFET	90	Bit-line floating	Bit-line floating
15	2013 [71]	10	32	Back-gate biasing and Built-in feedback mechanism	Stability will also rise with the application of back gate biasing and positive feedback

Table 12 Comparison of FinFET based SRAM Design

Pub. year	Ref. No.	Technology (nm)	Transistors count	Power consumption	Propagatio delay	Power delay product
2021	[72]	22	12	99% more efficient than conventional SRAM	79.2% more efficient as compare to conventional SRAM	99.7% more efficient as compare to traditional SRAM
		22	14	99% more efficient than conventional SRAM	76.5% more efficient as compare to conventional SRAM	99.4% more efficient as compare to traditional SRAM
2020	[73]	32	6	Least read power consumption	NA	NA
2020	[55]	7		For read and write operations, the leakage power is 10.68 μW and 31.01 μW, respectively	NA	136.8 aJ and 19.5 aJ for read and write operation respectively
				For read and write operations, the leakage power is 7.26 μW and 26.50 μW, respectively	18.01 ps and 16.1 ps, respectively, for a read and write operation	129.6 aJ and 426.6 aJ, respectively. for read and write operation
2020	[74]	20	8	At 0.5 V, the SG and IG SRAMs had leakage powers of 12.61 nW and 8.92 nW, respectively	NA	NA

(continued)

Table 12 (continued)

Pub. year	Ref. No.	Technology (nm)	Transistors count	Power consumption	Propagatio delay	Power delay product
2020	[75]	22	8	At 0.5 V, the traditional CMOS and FinFET leakage powers are 5.63 nW and 52.3 pW, respectively At 0.9 V, the traditional CMOS and FinFET leakage powers are 62.2 nW and 572 pW, respectively	NA	NA
2020	[76]	18	7,8,9,10	At $V_{DD} = 0.1$ v, 0.001224 mW, 0.001239 mW, 6.99 mW for write operation of 7T, 8T and 9T respectively	At $V_{DD} = 0.1$ v, the write operations for 7T, 8T, and 9T take 0.2943 ns, 0.915 ns, and 0.2255 ns, respectively	At $V_{DD} = 0.1$ v, the write operations for 7T, 8T, and 9T take 0.003602 pJ, 0.01135 pJ, and 0.01 pJ, respectively
2019	[77]	32	6,8,10	0.0236 μW, 0.0268 μW, 0.02696 μW for writing operation of 6T, 8T and 10T respectively	0.0219 ns, 0.0383 ns, 0.0385 ns for writing operation of 6T, 8T and 10T respectively	0.000516 fJ, 0.001026 fJ, 0.001037 fJ for writing operation of 6T, 8T and 10T respectively
2018	[58]	10	11	5.17 nW and 137.66 nW, respectively, represent the static and average power	Delay for Read and write is 491.8 ps and 230.46 ps respectively	168.57 aJ
2017	[78]	22	13	0.096 nW leakage power	7.8 ns	0.7488 fJ

(continued)

Table 12 (continued)

Pub. year	Ref. No.	Technology (nm)	Transistors count	Power consumption	Propagatio delay	Power delay product
2016	[68]	16	10	138.82 nW and 2.4538 nW for writing operation of 10T and 6T respectively 24.52 nW and 62.249 nW for reading operation of 10T and 6T respectively	0.010403 ns and 60.069 ns, for writing operation of 6 T and 10 T respectively 0.00024056 ns and 0.000 49619 ns for reading operation of 10T and 6T respectively	1.4442 aJ and 147.40 aJ for writing operation of 10 T and 6 T respectively 0.0058986 aJ and 0.030887 aJ for reading operation of 10T and 6T respectively
2015	[79]	45	6	9.45 nW for reading and 0.010 μW for writing at 7 nm technology	1.51 ps for reading and 2.64 ps for writing operation at 7 nm technology	0.0143 aJ for reading and 0.028 aJ for writing operation at 7 nm technology
2015	[80]	45	6	1.709 μW for reading and 7.561 nW for writing operation	20.55 ns for writing and 21.44 ns for reading operation	NA

Table 13 SRAM cells using various power-saving methods: performance comparison [33]

S. No.	Transistors count	Power dissipation (nW)			Power delay product		
		Along with power gatting	Along with dynamic threshold	Normal cells	Along with power gatting (aws)	Along with dynamic threshold (aws)	Normal cells (fws)
1	6	72.21	23.51	17,050	690.33	225.68	163.68
2	7	43.99	20.20	15,220	420.53	193.58	145.86
3	8	55.98	33.22	20,880	532.15	318.23	244.96
4	9	62.78	34.89	25,570	600.16	333.53	199.61

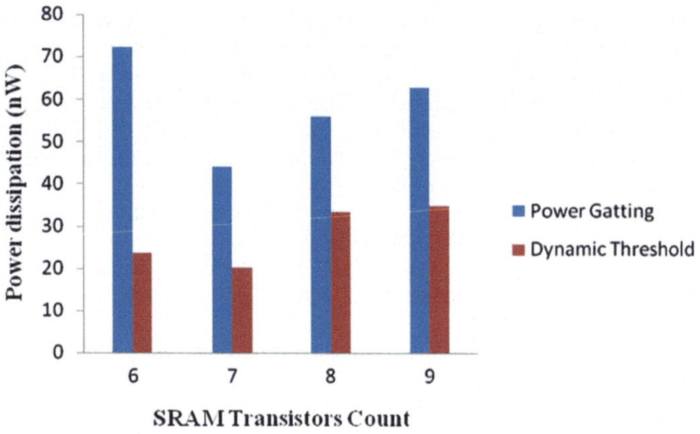

Fig. 13 Comparison of power gating and dynamic threshold power reduction techniques for SRAM cell design

References

1. S. Yu, H. Jiang, S. Huang, X. Peng, A. Lu, Compute-in-memory chips for deep learning: recent trends and prospects. IEEE Circuits Syst. Mag. **21**(3), 31–56 (2021)
2. V. Kumbar, M. Waje, A comparative analysis of FinFET based SRAM design, in IJEER, vol. 10, no. 4 (2022), pp. 1191–1198
3. R. Agrawal, V.K. Tomar, Analysis of low power reduction techniques on cache (SRAM) memory, in *2018 9th International Conference on Computing*. (IEEE, Communication and Networking Technologies (ICCCNT), 2018), pp.1–7
4. G. Pasandi, K. Mehrabi, B. Ebrahimi, S.M. Fakhraei, A. Afzali-Kusha, M. Pedram, Low-power data encoding/decoding for energy-efficient static random access memory design. IET Circuits Devices Syst. **13**(8), 1152–1159 (2019)
5. Z. Lin, X. Wu, Z. Li, L. Guan, C. Peng, C. Liu, J. Chen, A pipeline replica bitline technique for suppressing timing variation of SRAM sense amplifiers in a 28-nm CMOS process. IEEE J. Solid-State Circuits **52**(3), 669–677 (2016)
6. K. Zhang, Circuit design in nano-scale CMOS technologies, in *2018 IEEE Asian Solid-State Circuits Conference (A-SSCC)* (IEEE, 2018), pp. 1–4
7. S.S. Ensan, M.H. Moaiyeri, M. Moghaddam, S. Hessabi, A low-power single-ended SRAM in FinFET technology. AEU-Int. J. Electron. Commun. **99**, 361–368 (2019)
8. R. Liautard, L. Trojman, A. Arevalo, L.M. Procel, TFET and FinFET hybrid technologies for SRAM Cell: performance improvement over a large VDD-range, in *2019 IEEE Fourth Ecuador Technical Chapters Meeting (ETCM)* (IEEE, 2019), pp. 1–5
9. X. Jia, F. Wei, Advances in production and applications of carbon nanotubes, in *Single-Walled Carbon Nanotubes* (Springer, Cham, 2019)
10. S. Birla, FinFET SRAM cell with improved stability and power for low power applications. J. Integr. Circuits Syst. **14**(2), 1–8 (2019)
11. S. Birla, Variability aware FinFET SRAM cell with improved stability and power for low power applications. Circuit World (2019)
12. P. Praveen, R.K. Singh, Performance analysis of 9T SRAM using 180 nm, 90 nm, 65 nm, 32 nm, 14 nm CMOS technologies. IJEER **10**(2), 381–386 (2022)
13. C. Saxena, Leakage current analysis of FinFET based 7T SRAM at 32 nm technology. IJEER **3**(2), 31–34 (2015)

14. A. Dharmireddy, S.R. Ijjada, I. Hemalatha, Performance analysis of various fin patterns of hybrid tunnel FET. 1St Int. J. Electr. Electron. Res. **10**, 806–810 (2022)
15. A. Malinowski, J. Chen, S.K. Mishra, S. Samavedam, D.K. Sohn, What is killing Moore's law? Challenges in advanced FinFET technology integration, in *2019 MIXDES-26th International Conference Mixed Design of Integrated Circuits and Systems* (IEEE, 2019), pp 46–51
16. W. Gul, M. Shams, D. Al-Khalili, SRAM cell design challenges in modern deep sub-micron technologies: an overview. Micromachines **13**(8), 1332 (2022)
17. R. Lorenzo, S. Chaudhury, Review of circuit level leakage minimization techniques in CMOS VLSI circuits. IETE Tech. Rev. **33**, 1–23 (2016)
18. A. Keshavarzi, K. Roy, S. Member, C.F. Hawkins, Intrinsic leakage in deep submicron CMOS ICs—measurement-based test solutions. IEEE Trans. Very Large Scale Integr. (VLSI) Syst. **8**, 717–723 (2000)
19. J. Panchal, R. Vishal, Design and implementation of 6t Sram using Finfet with low power application, pp. 2395–0056 (2017)
20. V. Narendar, R.A. Mishra, Analytical modeling and simulation of multigate FinFET devices and the impact of high-k dielectrics on short channel effects (SCEs). Super Lattices Microstruct. **85**, 357–369 (2015)
21. T. Tripathi, D.S. Chauhan, S.K. Singh, A novel approach to design SRAM cells for low leakage and improved stability. J. Low Power Electron. Appl. **8**(4), 41 (2018)
22. Y.B. Kim, Y.B. Kim, F. Lombardi, Low power 8T SRAM using 32 nm independent gate FinFET technology, in *2008 IEEE International SOC Conference* (IEEE, 2008), pp. 247–250
23. R.K. Maurya, B. Bhowmick, Review of FinFET devices and perspective on circuit design challenges. SILICON **14**(11), 5783–5791 (2022)
24. A. Samal, S.L. Tripathi, S.K. Mohapatra, A journey from bulk MOSFET to 3 nm and beyond. Trans. Electr. Electron. Mater. **21**(5), 443–455 (2020)
25. A. Guler, N.K. Jha, Three-dimensional monolithic FinFET-based 8T SRAM cell design for enhanced read time and low leakage. IEEE Trans. Very Large Scale Integr. (VLSI) Syst. **27**(4), 899–912 (2019)
26. M.N. Reddy, D.K. Panda, A comprehensive review on FinFET in terms of its device structure and performance matrices. SILICON **14**(18), 12015–12030 (2022)
27. V. Mahor, M. Pattanaik, Low leakage and highly noise immune FinFET-based wide fan-in dynamic logic design. J. Circuits Syst. Comput. **24**(05), 1550073 (2015)
28. V. Chauhan, D.P. Samajdar, Recent advances in negative capacitance FinFETs for low-power applications: a review. IEEE Trans. Ultrason. Ferroelectr. Freq. Control **68**(10), 3056–3068 (2021)
29. T. Cakici, K. Kim, K. Roy, FinFET based SRAM design for low standby power applications, in *8th International Symposium on Quality Electronic Design (ISQED'07)* (IEEE, 2007), pp. 127–132
30. A. R. Rajprabu, R. Rajnarayanan, Performance analysis of CMOS and FinFET logic. IOSR J. VLSI Signal Process. **2**(1), 1–6 (2013)
31. V.K. Sharma, Design and simulation of FinFET circuits at different technologies, in *2021 6th International Conference on Inventive Computation Technologies (ICICT)* (IEEE, 2021), pp. 1–6
32. A. Razavieh, P. Zeitzoff, E.J. Nowak, Challenges and limitations of CMOS scaling for FinFET and beyond architectures. IEEE Trans. Nanotechnol. **18**, 999–1004 (2019)
33. M.D. Rao, Y.V. Narayana, V.V.K.D.V. Prasad, Original research article FinFET based SRAM cells design in various topologies using different power reduction techniques. J. Auton. Intell. **7**(1) (2024)
34. N. Tiwari, V. Neema, K.J. Rangra, Y.C. Sharma, performance parameters of low power sram cells: a review. I-Manager's J. Circuits Syst. **6**(1), 25 (2017)
35. E.C. Apollos, Design principles of SRAM memory in nano-CMOS technologies. Int. J. Comput. Appl. **178**(11), 5–11 (2019)
36. M. Limachia, N. Kothari, Characterization of various FinFET based 6T SRAM cell configurations in light of radiation effect. Sadhana **45**(1), 31 (2020)

37. M.K. Gupta, P. Weckx, S. Cosemans, P. Schuddinck, R. Baert, D. Yakimets, D. Jang, et al., Device circuit and technology co-optimisation for FinFET based 6T SRAM cells beyond N7, in *2017 47th European Solid-State Device Research Conference (ESSDERC)* (IEEE, 2017), pp. 256–259
38. R. Banu, P. Shubham, Design and performance analysis of 6T SRAM cell in 22 nm CMOS and FinFET technology nodes, in *2017 International Conference on Recent Advances in Electronics and Communication Technology (ICRAECT)* (IEEE, 2017), pp. 38–42
39. R.N. Asli, S. Taghipour, A near-threshold soft error resilient 7T SRAM cell with low read time for 20 nm FinFET technology. J. Electron. Test. **33**(4), 449–462 (2017)
40. G. Sneha, B.H. Krishna, C.A. Kumar, Design of 7T FinFET based SRAM cell design for nanometer regime, in *2017 International Conference on Inventive Systems and Control (ICISC)* (IEEE, 2017), pp. 1–4
41. Y. Yang, H. Jeong, S.C. Song, J. Wang, G. Yeap, S.O. Jung, Single bit-line 7T SRAM cell for near threshold voltage operation with enhanced performance and energy in 14 nm FinFET technology. IEEE Trans. Circuits Syst. I Regul. Pap. **63**(7), 1023–1032 (2016)
42. M. Ansari, H. Afzali-Kusha, B. Ebrahimi, Z. Navabi, A. Afzali-Kusha, M. Pedram, A near-threshold 7T SRAM cell with high write and read margins and low write time for sub-20 nm FinFET technologies. Integration **50**, 91–106 (2015)
43. M. Garg, B. Singh, A comparative performance analysis of CMOS and FinFET based voltage mode sense amplifier, in *2016 8th International Conference on Computational Intelligence and Communication Networks (CICN)* (IEEE, 2017), pp. 544–547
44. C.H. Neelima, T. Ravinder, D. Sudha, Design of 8TSram Using FinFET Technology. TEST Eng. Manage. **82**, 3168–3171 (2020)
45. A. Guler, N.K. Jha, Three-dimensional monolithic FinFET-based 8T SRAM cell design for enhanced read time and low leakage. IEEE Trans. Very Large Scale Integr. (VLSI) Syst. **27**(4), 899–912 (2019)
46. R.S. Kushwah, S. Akashe, Analysis of leakage reduction technique on FinFET based 7T and 8T SRAM cells. Radioelectron. Commun. Syst. **57**(9), 383–393 (2014)
47. N.E. Alias, A. Hamzah, M.L.P. Tan, U.U. Sheikh, and., & Riyadi, M. A., Low-power and high performance of an optimized FinFET Based 8T SRAM cell design, in *2019 6th International Conference on Electrical Engineering.* (IEEE, Computer Science and Informatics (EECSI), 2019), pp.66–70
48. M. Monica, P. Chandra Mohan, A Novel 8T SRAM Cell using 16 nm FinFET Technology. SASTech-Tech. J. RUAS **16**(1), 5–8 (2017)
49. N. Sharma, Ultra-Low power dissipation in 9T SRAM design by using FinFET Technology, in *2016 International Conference on ICT in Business Industry & Government (ICTBIG)* (IEEE, 2016), pp. 1–5
50. Y. Yang, J. Park, S.C. Song, J. Wang, G. Yeap, S.O. Jung, Single-ended 9T SRAM cell for near-threshold voltage operation with enhanced read performance in 22-nm FinFET technology. IEEE Trans. Very Large Scale Integr. (VLSI) Syst. **23**(11), 2748–2752 (2014)
51. F. Moradi, M. Tohidi, Low-voltage 9T FinFET SRAM cell for low-power applications, in *2015 28th IEEE International System-on-Chip Conference (SOCC)* (IEEE, 2015), pp. 149–153
52. T.W. Oh, H. Jeong, K. Kang, J. Park, Y. Yank, S.O. Jung, Power-gated 9T SRAM cell for low-energy operation. IEEE Trans. Very Large Scale Integr. (VLSI) Syst. **25**(3), 1183–1187 (2016)
53. P.R. Vijapur, B.V. Uma, Comparative analysis of novel 9T static random access memory at different technologies of FinFET. Int. J. Pure Appl. Math. **118**(24) (2018)
54. S. Pal, A. Bhattacharya, A. Islam, Comparative study of CMOS-and FinFET-based 10T SRAM cell in subthreshold regime, in *2014 IEEE International Conference on Advanced Communications.* (IEEE, Control and Computing Technologies, 2014), pp.507–511
55. U. Mushtaq, V.K. Sharma, Design and analysis of INDEP FinFET SRAM cell at 7-nm technology. Int. J. Numer. Model.: Electron. Netw. Devices Fields, e2730 (2020)
56. P. Praveen, R.K. Singh, Design of enhanced reversible 9T SRAM design for the reduction in sub-threshold leakage current with14 nm FinFET technology. ACM Trans. Des. Autom. Electronic Syst. **28**(6), 1–29 (2023)

57. S. Maabi, E.S. Sayyah, M.H. Moaiyeri, S. Hessabi, *A Low-Power Hierarchical FinFET-Based SRAM*, pp. 54–60 (2016).
58. S.S. Ensan, M.H. Moaiyeri, S. Hessabi, A robust and low-power near-threshold SRAM in 10-nm FinFET technology. Analog Integr. Circ. Sig. Process **94**(3), 497–506 (2018)
59. N. Yadav, A.P. Shah, S.K. Vishvakarma, Stable, reliable, and bit-interleaving 12T SRAM for space applications: a device circuit co-design. IEEE Trans. Semicond. Manuf. **30**(3), 276–284 (2017)
60. K.K. Kishore, B.L. Radha, Design and implementation of 12-T SRAM Cell in 32nm FinFET technology. Int. J. Eng. Res. Electron. Commun. Eng. (IJERECE) **4**(8) (2017)
61. M.U. Mohammed, A. Nizam, L. Ali, M.H. Chowdhury, FinFET based SRAMs in Sub-10 nm domain. Microelectron. J. **114**, 105116 (2021)
62. G. Torrens, B. Alorda, C. Carmona, D. Malagon-Perianez, J. Segura, S. Bota, A 65-nm reliable 6T CMOS SRAM cell with minimum size transistors. IEEE Trans. Emerg. Top. Comput. **7**(3), 447–455 (2017)
63. K.S. Sreekala, S. Krishnakumar, RNM calculation of 6T SRAM cell in 32 nm process node based on current and voltage information. Indian J. Sci. Technol. **10**(29), 1–7 (2017)
64. T.V. Lakshmi, M. Kamaraju, A review on SRAM memory design using FinFET technology. Int. J. Syst. Dyn. Appl. (IJSDA) **11**(6), 1–21 (2021)
65. C.C. Wang, R.G.B. Sangalang, I.T. Tseng, A single-ended low power 16-nm FinFET 6T SRAM design with PDP reduction circuit. IEEE Trans. Circuits Syst. II Express Briefs **68**(12), 3478–3482 (2021). https://doi.org/10.1109/TCSII.2021.3123676
66. T.W. Oh, H. Jeong, J. Park, S.O. Jung, Pre-charged local bit-line sharing SRAM architecture for near-threshold operation. IEEE Trans. Circuits Syst. I Regul. Pap. **64**(10), 2737–2747 (2017). https://doi.org/10.1109/TCSI.2017.2702587
67. F. Frustaci, D. Blaauw, D. Sylvester, M. Alioto, Approximate SRAMs with dynamic energy-quality management. IEEE Trans. Very Large Scale Integr. (VLSI) Syst. **24**(6), 2128–2141 (2016). https://doi.org/10.1109/TVLSI.2015.2503733
68. N. Kaur, N. Gupta, H. Pahuja, B. Singh, S. Panday, Low power FinFET based 10T SRAM cell, in *2016 Second International Innovative Applications of Computational Intelligence on Power, Energy and Controls with their Impact on Humanity (CIPECH)* (IEEE, 2016), pp. 227–233
69. H. Farkhani, A. Peiravi, J.M. Kargaard, F. Moradi, Comparative study of FinFETs versus 22nm bulk CMOS technologies: SRAM design perspective, in *2014 27th IEEE International System-on-Chip Conference (SOCC)* (IEEE, 2014), pp. 449–454
70. G. Pasandi, S.M. Fakhraie, An 8T low-voltage and low-leakage half-selection disturb-free SRAM using Bulk-CMOS and FinFETs. IEEE Trans. Electron Devices **61**(7), 2357–2363 (2014). https://doi.org/10.1109/TED.2014.2321295
71. N. Yadav, S. Dutt, M. Pattnaik, G.K. Sharma, Double-gate FinFET process variation aware 10T SRAM cell topology design and analysis, in *2013 European Conference on Circuit Theory and Design (ECCTD)* (IEEE, 2013), pp. 1–4
72. R.K. Raushan, M.R. Ansari, U. Chauhan, M. Khalid, B. Mohapatra, Implementation of 12T and 14T SRAM bitcell using FinFET with optimized parameters. Trans. Electr. Electron. Mater. **22**(3), 328–334 (2021)
73. R.N. Asli, S. Taghipour, Reliable and high-performance asymmetric FinFET SRAM cell using back-gate control. Microelectron. Reliab. **104**, 113545 (2020)
74. S. Birla, N.K. Shukla, N. Singh, R.K. Raja, Performance analysis of 8T FinFET SRAM bit-cell for low-power applications, in *2020 5th International Conference on Computing, Communication and Security (ICCCS)* (IEEE, 2020), pp. 1–4
75. C. Duari, S. Birla, A.K. Singh, A dual port 8T SRAM cell using FinFET & CMOS logic for leakage reduction and enhanced read & write stability. J Integr Circuits Syst **15**(2), 1–7 (2020)
76. R.R. Vallabhuni, K.C. Koteswaramma, B. Sadgurbabu, Comparative validation of SRAM cells designed using 18nm FinFET for memory storing applications, in *Proceedings of the 2nd International Conference on IoT, Social, Mobile, Analytics & Cloud in Computational Vision & Bio-Engineering (ISMAC-CVB 2020)* (2020)

77. J. Verma, A. Passi, S. Sindhu, S. Gayathiri, Design 10-Transistor (10t) SRAM using FinFET technology. Int. J. Eng. Adv. Technol **9**(1), 566–572 (2019)
78. S. Saxena, R. Mehra, Low-power and high-speed 13T SRAM cell using FinFETs. IET Circuits Devices Syst. **11**(3), 250–255 (2016)
79. L.M. Dani, G. Singh, M. Kaur, FinFET based 6T SRAM cell for nano-scaled technologies. Int. J. Comput. Appl. **127**(13), 5–10 (2015)
80. D. Verma, S. Babu, S. Akashe, Comparison of conventional 6T SRAM cell and FinFET based 6T SRAM cell parameters at 45 nm technology. Int. J. Adv. Comput. Res. **5**(21), 367 (2015)

Robotics and IoT in Autonomous Vehicles: Background, Integration Trends and Recent Advances

Saumya Suman, Bharat Bhushan, Hari Mohan Rai, and Omar S. Saleh

Abstract In recent days, accidents are rapidly increasing. Internet of Things (IoT) enabled automobiles have shown significant measures in decreasing road accidents by the help of roadside IoT devices like cameras, sensors, laser radars and many more. IoT is a new paradigm which gives a set of services for the innovation of technology which is based on merging of many processes like sensing, computation, communication, and identification. These processes are already core essentials of a humanoid robot. By using technologies in the IoT enabled autonomous vehicles, a model can be trained such a way that it can incorporate for solving randomness in day-to-day life. These humanoid robots can learn from the dataset of the sensors, cameras, and lasers. A well-built robot can fully function as a ride sharing driver. Such technology is currently in the AI system of Tesla, but only in the form of AI in the form of IoT automobile. This paper highlights the basic classifications of IoT and discusses its various devices and how they are helping at present and will help in future. Further these devices are categorized in three parts- low-end, middle-end and high-end devices. Finally, the paper discusses robotics and autonomous vehicles, and the integration ofRobotics using IoT in self- driving autonomous vehicles.

Keywords Intelligent transportation · Robotics · Internet of Things · Wireless systems · Communication · Autonomous

---

S. Suman (✉) · B. Bhushan
Department of Computer Science and Engineering and Technology, Sharda University, Greater Noida, India
e-mail: Saumyaaa90@gmail.com

H. Mohan Rai
School of Computing, Gachon University, Seongnam, South Korea
e-mail: drhmrai@gachon.ac.kr

O. S. Saleh
Directorate of Studies, Planning and Follow Up, Ministry of Higher Education and Scientific Research, Baghdad, Iraq

R. Sharma and G. Jeon (eds.), *Data Analytics for Smart Robotics and Its Applications*, Intelligent Systems Reference Library 272,
https://doi.org/10.1007/978-3-031-87697-4_5

1 Introduction

A system of affiliated sensors and devices which are used for storing the data with each other, is known as IoT. Kevin Ashton, an explorer of British Technology as well as an Executive Director of Auto-ID Labs of MIT, who proposed IoT in 1999 [1]. From the old days of computing, IoT has a very long, and full of landscape history. Birth and Evolve of the IoT all started from 1970s and goes till 2020s, starting with the developing of first wireless networks to beginning of 5G networks and coming of new applications of technologies of IoT, it is predicted that there will be around 25 billion devices of IoT by 2025 comes in the under of GSMA [2]. IoT devices are hardware of open source which have active communities and majority of operating system used to support this. IoT examine its impacts on day-to-day life, upcoming business models and industry.

There are many challenges of IoT which are scalability, data management, security, encrypt communication and many. As this paper is about the self-driving or the autonomous vehicles, so there are challenges regarding this. But now after reaching the top factor of IoT, there are many challenges regarding loads of traffic and many models of traffic [3]. Requirements of traffic are growing and coming and by this we want the capabilities of network, security mechanisms, protocols, traffic models, etc. because of Internet and devices of day-to-day life becoming the traffic consumers and producers [4]. Some issues are related to technology combination, data storage incrementation, risks of security, privacy and processing demands, etc. There are many IoT related problems which must be and needs to be solved from many fields like environmental impacts, social impacts, enabling technologies, models of business and many [5]. There are many companies which are investing more in autonomous vehicles and trying to put many changes and doing advancements in the industry of automotive [6].

IoT is basically a combination of sensing, communicating, many standards, and other capabilities. IoT is solving many issues very fast, day-by-day its advancements are increasing with the full speed. It has a very great impact on automotive industry [7]. As the self-driving vehicle technology is improving very rapidly over the past decade. For solving the traffic requirements, we want models, protocols, and many. Numbers of applications of IoT are combined in many domains like health, traffic, smart metering, remote monitoring, agriculture, retailing, and many. It has extraordinary inherent which effects our livelihood [8]. CAEV which stands for Connected and Automated Electric Vehicle which is being researched very widely and own number of problems applications, it also interacts with RSU which stands for roadside unit [9]. CAEV automated charging facilities and reservation which will help in charging with human intervention with the help of sensors, actuators, some software, etc. CAEV charging managements system and CAEV taxi service is also helps to solve our day-to-day life problems [10].

The remainder of this paper is given below. Section 2 presents the introduction of IoT, its characteristics, its devices which also breaks down into three parts-low end, middle end and high end, and its challenges. Section 3 explores description of Robotics, Autonomous vehicles (including its classifications and fundamental contexts) and its relationship. This paper finally concludes with the recent advancements of Robotics and IoT in autonomous vehicles.

2 IoT

Computing, Sensing, Embedding, Networking, Identifying and Communicating technologies forms IoT. Fourth revolution of disruptive technologies is brought by IoT which gives faultless and set of services to everywhere, at any time to anything [11].

2.1 Characteristics

The life's quality day by day increasing when uses are using IoT applications services. The main foundations of IoT systems are Identifications, Communications, Sensing, and many. IoT is examined as global infrastructure according to Internet-oriented vision, which gives connections between physical and virtual object [12]. The IoT networks are said as interoperable and dissimilar means heterogenous and it has faultless connections which can be connected anytime, anywhere, with anyone. IoT has a special feature which can see, think, hear, can coordinate decisions, can share information, and can perform jobs these features. IoT need "technology stack," which is a combination of many technologies which enable its processes to connect faultlessly, and this technology stack contains hardware and software, immerse in it.

2.1.1 Sensors

IoT is a group of many sensors and its types which have their own requirements. These all sensors can be wearable, biological, chemical, RFID Tags, and many. Sensing is basically for grouping many data using many different sensors from the environment, and the data which is grouped are then analysed for actuating and improving decision making in IoT [13]. Temperature, light, pressure, air, motion, all get observed through sensors. It also used for collecting information by checking or sensing the its environment.

2.1.2 Data Storage

The important sources of all data are called as devices, objects, or things which is the centre of the pillars of IoT. Processing units act as a brain of IoT, and the processing units like microcontrollers, microprocessors, software packages, SoCs and FPGAs [14]. The data storage, ubiquity and analysis facilitated by edge platforms, cloud, cloudlets, and fog in IoT. IoT offers many services like data publishing, device monitoring, device controlling and device discovery. It also provides services like Information aggregation services, collaborative aware and ubiquitous services.

2.1.3 Connectivity

Communication and the interaction among the IoT devices and with internet show the connectivity in IoT. Wi-Fi, Bluetooth these all are the technologies which helps in this connectivity for connecting devices and internet. Connectivity is one of the key features in IoT devices for receiving and sharing data and information [15]. Traffic light, vehicles, sensors all get connected to control the traffic flow very easily.

2.1.4 Real-Time Access

Sharing or receiving the data or the resources easily and quickly is term as Real-Time Access. It helps in many areas like manufacturing, transportation, logistics, healthcare, etc. It is an instant process and an immediate action which is done to make the life easy and makes devices works accordingly [16]. Real-Time GPS tracking system is used for avoiding accidents. It also helps in decision making processes.

2.1.5 Security

IoT also Provide securities like security in devices, security in services and security in communication. It also helps in sharing the captured data across the whole globe with the help different communication standards [17]. Security and privacy should be considered from all the condition containing social and legal perspective. In every level of IoT infrastructure and management of trust should include security functionalities for winning the trust and of all the human and making them realize of privacy with proving the security feature of IoT [18]. Standardize perspective of interchanging information leads to more accentuate of security issues in a heterogenous surrounding.

2.2 Devices

The devices of IoT consists from smartphones to RFID readers, these wearable devices to gadgets, tablets, only to mentioning some. The devices play a very vital role in whole IoT development as their size are small, integrating to microcontrollers and microprocessors with other components, like memory, I/O, ready to program package

Table 1 Types of devices

Level	Devices	Description
Low-end Devices [20, 21]	TelosB	2.4 GHz low energy mote compliant
	OpenMote	Comes with a TI SoC combining an ARM Cortex-M3 microcontroller with an IEEE802
	LSN50	Open-source hardware with long-range unique attributes
Middle-end Devices [22, 23]	Netduino	Open-source electronics prototyping platform based on the NET Micro Framework
	Tessel 2	Hardware development board with Wi-Fi capacity for building scripts in Node. js
	ESP8266	A system on chip (SOC) which contains 32-bit microcontroller, Wi-Fi transceiver and some other components
High-end Devices [24]	Banana Pi	Single board computer, targets to be a cheap, small, and flexible enough computer for daily life
	Odroid	Brand of single board computers and tablet computers made by Hardkernel Co., Ltd., a South Korea company
	Wandboard	An ultra-low power complete computer with high performance multimedia capabilities based around the new upcoming Freescale i

[19]. Here these devices are being classified in three terms: Low-end, Middle-end and High-end IoT devices and some of its examples are described in Table. 1.

2.2.1 Low-End Devices

Low-end devices are a whole set of connected devices, which termed as constrained devices which is a resource challenged to run traditional OS whose examples are Windows 10 IoT Core or Linux. RAM which stands for Random Access Memory of low end have KB in tens [20]. Low End devices are made for those applications which contain processes like sensing and actuating. These applications are programmed by WSN (Wireless Sensor Networks) which functions poorly or can be programmed by using low-level firmware [21].

2.2.2 Middle-End Devices

Middle End IoT devices have many features and provide better processing capabilities like image recognition where multiple communication technologies and constrained resources are less [22]. It is mostly compared with low end devices and with high-end devices. Clock speed and RAM of middle end have MHz and KB in hundreds. It has higher KBs and clock speed of low end IoT devices [23].

2.2.3 High End Devices

High End Devices are those devices which are mainly Single Board Computers (SBS) with enough resources with powerful processing unit. Tentative execution like execution of heavy Machine Learning algorithms is also performed by these applications. One of the examples is Raspberry Pi which is mostly known for their best connectivity its interface [24]. Raspberry Pi also have camera inserted whose interface like CSI and DSI stands for Camera Serial Interface and Display Serial Interface. Embedded system and boards are mainly focused.

2.3 Challenges

Due to its significant impact on various aspects of people's lives and their technologies for efficient and beneficial data exchange among encapsulated devices [25]. When tackling IoT difficulties, the following important factors need to be considered: business models, services and applications, and the effects on the environment and society. Numerous problems are getting worse, some of which are brought on by connected devices, which raises traffic demand and the models that go along with it. Data capacity constraints, the requirement for data cleaning, and the possibility of security and privacy breaches are additional problems that result from the blending of several technologies [26]. All the mentioned issues and challenges that all needs face must be analyzed first and then resolved, as they are all based on the perspective of technology. Therefore, here are some challenges, such as standardization, availability and reliability, its architecture, its data storage, its processing, and its visualization. These challenges are also described in Table 2.

2.3.1 Standardization

Standardization is a heart of IoT development which should be considered in the future, where open standards are the essential component for IoT development. In the development and improvement of IoT applications, the diversity of standards is being recognized as a major challenge, as their architecture technologies and communication technologies are pillars for IoT development in the upcoming years or days [27]. Several standardisation organisations are active in creating the framework for Internet of Things standards, including the World Wide Web Consortium (W3C), the European Telecommunications Standards Institute (ETSI), and the International Telecommunication Union (ITU). These standardisation efforts are broad in scope and seek to provide perfect designs, connections, and open standards. However, integrating and combining different standards and contributions is difficult due to their inconsistency and lack of coherence, which significantly hinders the IoT ecosystem [28].

Table 2 Challenges of IoT

Challenge	Description
Standardization [27, 28]	Loss of universal standards across IoT platform and its devices which hinders scalability and interoperability
Architecture [29, 30]	Complex IoT architectures needs coordination between devices, networks, and cloud systems for seamless operation
Availability and Reliability [31, 32]	Ensuring continuous, reliable connectivity and system uptime across networks, often in diverse environments
Data storage, data processing and visualisation [33]	Massive data needs scalable storage solutions with strong security and it should be handled, analysed, and presented in actionable formats for users which requires advanced tools and skills
Scalability [34, 35]	IoT systems must scale efficiently for supporting many devices and it requires flexible infrastructure and software
Management and Self-Configuration [36]	Managing numerous devices and enabling self-configuration to decrease the intervention which are critical for efficiency
Network Performance and QoS [37, 38]	Ensure optimal performance of network and quality of service time-to-time, reliable data transmission

2.3.2 Architecture

IoT architecture must provide continuous mobility to ensure service continuity without interruptions and interoperability. To enable smooth communication across the system, it must also support diverse networks, data, and the characteristics of objects and applications. Implementing an open, cooperative, and standardised architecture that eliminates hardware infrastructure and application-specific logic is one of the main problems in IoT systems [29]. General, hardware/network, and software/process are the three main categories of architecture. To meet the needs of the Internet of Things, the generic architecture acts as a theoretical model. Hardware or network design must be compatible with a variety of communication protocols and network types to guarantee smooth connectivity between IoT devices. Additionally, activities like computing, aggregation, and managing massive amounts of data for service setup are made possible by software architecture [30].

2.3.3 Availability and Reliability

IoT applications should always be accessible for authorized purposes. This concept is known as availability, and it is a critical challenge that must be addressed for effective IoT management. The dynamic nature of IoT devices and the frequent changes in topology pose significant challenges, particularly in terms of mobility [31]. In mesh topologies, scheduling and routing present several difficulties, hence routing is essential to system dependability. Multi-hop routing, scalability, context awareness, dynamic topology changes, and other crucial features should all be supported by routing procedures. Appropriate protocols, self-healing mechanisms, and a monitoring system are required to guarantee system robustness. Communication technology that are not always available can cause service outages. Another difficulty is the users' mobility and the services they utilise, which necessitates a trade-off between dependability and energy usage. This problem has not been resolved [32].

2.3.4 Data Storage, Data Processing and Visualisation

To support the transfer of large data as a service to IoT applications, a comprehensive logical plan is very necessary. Techniques such as various data mining methods, including machine learning, AI, and intelligent decision-making algorithms, can be employed to extract valuable data and knowledge from raw data. However, there is a challenge in terms of cost constraints [33]. The vast volume of data created may be efficiently managed using cloud technology. Delivering massive datasets from sensors, smartphones, and many other devices to cloud infrastructure, however, comes with a number of difficulties, such as the requirement for internet connectivity, which can be expensive, time-consuming, congested, and limited in bandwidth, as well as security issues with data transmission.

2.3.5 Scalability

The capability to seamlessly integrate novel devices and services without compromising the performance of existing services and augmenting the IoT system [34]. Using a wide range of devices with processing power, bandwidth, and other resources is one of its main challenges. The creation of an IoT architecture that supports scalability is a major problem for future IoT development. Scalable devices should be positioned correctly for detection, and they should be allowed to interact and share resources among themselves with little assistance from humans. The implementation of a layered infrastructure is necessary to attain scalability and facilitate communication and resource sharing. A distributed network of devices should be used to address scalability difficulties in the architecture [35].

2.3.6 Management and Self-configuration

A key aspect of IoT utilization is the effective management of IoT devices and their applications. This includes configuration, monitoring, and control, all of which are made extremely difficult by the complexity, heterogeneity, traffic demands, and widespread use of IoT devices. The ability of an Internet of Things system to dynamically adjust to changing environmental conditions is known as self-configuration. Device management mechanisms, network management functionalities, and data management mechanisms are the three main management methods used [36]. Data management mechanisms are indispensable for providing numerous advantages, such as data analytics, security, data recovery, and raw data aggregation. Similarly, network management functionality is crucial for ensuring the effectiveness of network topology management, congestion control management, and device synchronization.

2.3.7 Network Performances and QoS

Issues related to traffic models and their corresponding loads directly impact the performance of the network and the Quality of Service (QoS). The ability to handle the diverse devices connected to processes and services is directly dependent on the performance of IoT systems [37]. Several factors, including bandwidth, throughput, latency, and others, are associated with Quality of Service (QoS) in the context of Internet of Things (IoT) systems. QoS in IoT systems is influenced by embedded technologies, traffic demands, protocols, and other considerations. Numerous studies have been conducted to address and evaluate the performance of IoT services, yet challenges persist. All components and communication protocols integrated into an IoT system play a crucial role in determining its overall performance [38].

3 Robotics and Autonomous Vehicles

Robotics and autonomous vehicles are framing the future by merging advanced technology with day-to-day life. Nowadays, industries including manufacturing, healthcare, and even home support rely heavily on robots that are made to mimic human behaviour. In reverse, autonomous cars are transforming transportation by navigating roadways without taking help from human assistance by utilising sensors, artificial intelligence, and sophisticated algorithms. The main goal of these advances is to make systems safer and more effective for everyone, not simply for convenience. Beyond automobiles, robotics also includes drones that transport packages and robots that explore dangerous areas that are off-limits to humans. Though they are still developing, these technologies have the potential to revolutionise the way we travel, work, and live.

3.1 Robotics

Engineering, computer science, and artificial intelligence are used in robotics to build machines that can contain activities either on their own or under human supervision. Robots can be anything from basic mechanical arms used in manufacturing to sophisticated AI-powered devices used in personal help, healthcare, and space exploration. The sense, process, and action capabilities are fundamental to robotics. It is like how humans think and act, robots utilise sensors to collect information about their surroundings, which is then processed by algorithms to make judgements [26]. Robots can learn, adapt, and even solve complicated issues on their own thanks to artificial intelligence, which increases their usefulness in a variety of contexts.

Robots are very beneficial to industries like manufacturing, where they do dangerous and repetitive work. Robots are used in healthcare to help with surgery, inventory control, and rehabilitation. Robots also play a major role in exploration; space rovers, for instance, gather information on planets where it is difficult for humans to live. These are becoming more and more ingrained in daily life thanks to smart assistants and hoover robots. In research, collaborative robots, or "cobots," and bio-inspired robots are made to work alongside people to increase productivity and convenience [39].

However, there are ethical issues with robotics, particularly in relation to safety, privacy, and employment. These concerns need to be carefully considered as robots get more sophisticated to strike a balance between the advancement of technology and the effects on society [40]. All things considered, robotics is still developing, offering revolutionary potential across a range of fields, and changing the way people engage with technology.

3.2 Autonomous Vehicle

Complex automotive systems, electric vehicle, combining connected vehicles and autonomous vehicles these all comes under autonomous and connected vehicles [41]. An autonomous vehicle is a type of vehicle in which it can drive itself without any help of human beings. This type of vehicle uses sensing techniques without any intervention of human. As it is a type of CAEV, and it has five vital element which includes—System of Localisation and mapping, System of Perception, System of Communication, System of Battery storage and Policy of Driving [42]. It also used to communicate with RSU which stands for roadside unit, structure of cloud and different CAEVs which is use to do computing work collectively.

3.2.1 Classifications

Vehicle Automation is classified into different levels which ranges from Level 0 to Level 5 (from no automation to full automation) as depicted in Fig. 1. And this classification is updated by SAE which stands for Society of Automotive Engineers in 2016 [43]. And these levels are mentioned below.

- Level 0—No Automation: Driving works which contains road and environment monitoring across the car and its controlling, these all are the responsibility of driver which is focused on. All driving functions, including steering, braking, and acceleration, are fully under in the control of driver.
- Level 1—Driver Assistance: Driver- assistance system can either assist steering or acceleration or deceleration, but cannot do both by the human driver. The driver has full control of all driving tasks, such as steering, braking, and acceleration. There are no features for automation as here the driver should control and must keep an eye on the surroundings even when the car can help with a single job, such as lane-keeping or cruise control [44].
- Level 2—Partial Automation: It involves advanced levels of assistance of driver it needs continuous surveillance of human driver, as they monitor the car on the road and environment as well as acceleration/deceleration and steering control can be taken care together by driver assistance system. Adaptive cruise control with lane centring allows the car to regulate steering and acceleration/deceleration simultaneously, but the driver needs to be prepared to take over [45].

Fig. 1 Levels of vehicle

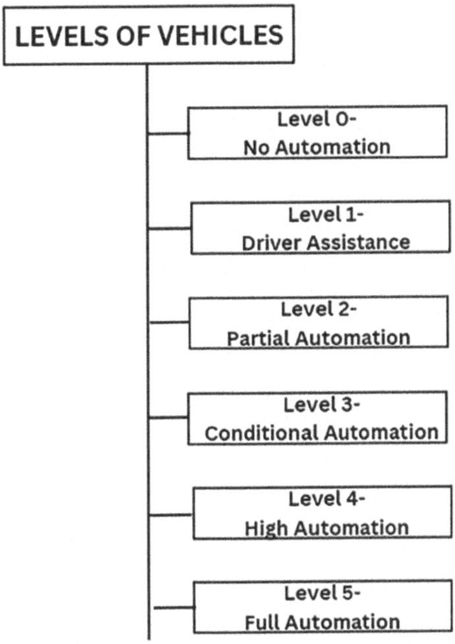

- Level 3—Conditional Automation: It needs limited management of the driver. Human driver will acknowledge properly as they have expectations of dynamic driving work by autonomous system of driving. Under some circumstances, the car can handle all driving duties, but the driver must take over when signalled. On-demand human involvement is necessary.
- Level 4—High Automation: This level is liberated because human driver will not acknowledge to request to autonomous system of driving. In some circumstances or locations, the car can operate entirely on its own without assistance from a driver, but in other circumstances, it can still have a steering wheel [46].
- Level 5—Full Automation: Presence of driver is not necessary. Autonomous system of driving handles all driving assignment by their own in all conditions of environment and roadways. Without a driver or steering wheel, the car is fully autonomous in all driving conditions. It is never necessary for passengers to take over.

3.2.2 Fundamental Contexts

All fundamental contexts affect the action of assistance either in explicit or implicit form. Intelligent transportation system contains one of the major problems that is termed as Context- awareness [47]. Some drivers fail to see a traffic sign, some become careless in long driving trips, and many problems occurs. So, for solving this failures ADAS (stands for Advanced Driver Assistance System) application is introduced. Some fundamental contexts are as follows:

3.2.3 Driver Based Contexts

The data processing is performed by driver's side. Vehicle Intelligence is utilized to monitor the human project and its conditions, as well as the status of the driver and/or passengers during a journey [48]. ADAS have a widely recognized application in the form of Drowsiness Detection. This technology processes visual data to assess the driver's mental and physical state.

3.2.4 Environment and Road Infrastructure Contexts

Vehicles must aware of their surroundings, including the environment and road infrastructure, regardless of their type, shape, and condition. Road safety is also depended on the weather condition road condition which affect the quality of driving [49]. Number of accidents can be reduced if the driver fully follows the traffic rules, paying attention to traffic signs and other relevant information. ADAS applications aim to understand the environment context which is more challenging. The vehicular environment contains external objects like traffic signal, pedestrians, and many other vehicles should analyse and sensed for taking immediate action against environmental, road conditions [50].

3.2.5 Driving Based Contexts

We should acquire knowledge regarding vehicle motions and driver habits for the purpose of evaluating the quality of human driver architecture in vehicle intelligence [51]. These sensors, which function as an accelerometer and gyroscope, enable easy detection of various driving contexts, including braking, turning, starting, and stopping. Driver-environment interaction plays a crucial role in driving-based contexts. Turning is also of paramount importance in driving-based contexts, as it involves numerous events related to direction, such as 'left or right turning,' which must be accurately acquired [52].

3.2.6 Vehicle Based Contexts

Fault and errors are detected by interface of Current vehicular diagnostic. Driving vehicles can do actions by their own if any type of fault or failure or crash of system occurs. Hence, some preceding actions should be taken care for this type of unexpected crash of mechanical parts of vehicle, so for this monitoring should be done previously and all the time [53]. OBD which stands for On-Board diagnostics which uses sensors to create problem codes which are presently solving this situation.

3.3 Relationship Between Robotics and Autonomous Vehicles

Autonomous cars and robotics have an essential relationship because robotics presents the essential framework for decision-making, sensing, and processing which gives permission to vehicles to work without human drivers. Self-driving cars, also known as autonomous vehicles which depend on robotic systems to sense their surroundings, make judgements in real time, and travel safely [54]. Combination of sensors, algorithms, and artificial intelligence—all components of robotics which lays the foundation for this interaction. Cameras, radar, lidar, and ultrasonic sensors are used by autonomous cars to monitor road markings, read traffic signs, and identify obstructions [55]. The continuous feedback from these sensors is processed by robotic algorithms to determine when to stop, accelerate, or turn. Robotics provides the essential systems required for autonomy, and autonomous vehicles are advanced robots on wheels. Autonomous vehicles are becoming more dependable, responsive, and able to safely traverse challenging, real-world conditions as robotics technology develops [56].

4 Robotics and IoT in Autonomous Vehicles

Robotics and IoT has a sympathetic sector which contain the industry for automative vehicles. There are some challenges like efficiency of fuel consumption, challenges related to mobility, self-propelled safety. There is an exchange of data and resources among the smart city, CAEV and its neighbouring which is empowered by IoT [57]. IoT devices task rapidly replacing human work and making the work of human very easy and enhancing it by providing many features to it and there are some applications and services from where we will be able to learn about such replacement and growth of IoT technology. IoT features various applications and services which integrates IoT technology and autonomous driving like automated parking facility, platooning, driver monitoring, managing fleet [58]. The recent advancements are discussed below and summarized in Table 3.

4.1 Traffic Management

An important technology is introduced for managing and analysing the intelligence of traffic management, and that technology is IoT. ITS which stands for Intelligent Transport System, it monitors the real-time of whole transport system. It is used by traffic managers by getting the route planning and information of management and drivers. Traffic congestion, travel time, and carbon dioxide emissions all can be reduced, and this power has with ITS.

Cristoforis et al. [59] contributed a behavior-based platform for designing robotics-focused classes, programming robots, and other outreach initiatives. More than 90% of pupils found it straightforward to use, according to evaluation statistics. Akhtar et al. [60] discussed about the enhancement of congestion level which is based on how the traffic management system works using IoT and how it suppresses the traffic light timings which depends upon the real-time congestion level of measurement of traffic using sensors like ultrasonic at the road crossings. Evripidou et al. [61] proposed a set of six learning objectives that can be used as a survey for the newest ER platforms and as a foundation for a workable paradigm for designing robotic activities. Jarwan et al. [62] contribution is to develop the schemes of data management to handle with unavoidable loss of data by distinguishing some part of data which have larger value to overlaying applications, also develop a novel metric which is referred to as information content (IC) and calibrating the amount of data information and resources. Dui et al. [63] contributed for solving the issue of processing traffic flow using IoT, and create ITS automation for global control challenge and this issue is solved by proposing traffic model based on IoT for enhancing the recognition of responsibility of transport and road system. Ibnkahla et al. [64] discovered Backhaul (BH) selection is not addressed in literature and there is vital requirement to discover the brilliant way to enable the edge devices to get knowledge of how to deal with unknown-state which is partially observable components of BH network, which is

Table 3 Recent advancements in regards of robotics and IoT in autonomous vehicle

IoT in self-driving vehicles	References	Year	Contribution
Traffic Management	Cristoforis et al. [59]	2018	Behavior-based platform for creating lessons on a robotics-focused classes and programming robots
	Akhtar et al. [60]	2020	Congestion level improvement that is dependent on the Internet of Things-based traffic management system
	Evripidou et al. [61]	2020	An evaluation of the most recent ER platforms
	Jarwan et al. [62]	2022	Creation of data management strategies to deal with inevitable data loss
	Dui et al. [63]	2023	Resolving the traffic flow processing problem and developing ITS automation
	Ibnkahla et al. [64]	2023	A clever method for allowing edge devices to learn about certain BH network components
	Hazarika et al. [65]	2024	Object detection method that finds and counts all the cars
Fleet Management	Minovski et al. [66]	2020	Quality of Experience (QoE) field is used to hide the relationship between intelligent technology and people
	Zhang et al. [67]	2021	Analysis of the lifecycle costs of battery charge levels within a specified range
	Ding et al. [68]	2022	Important characteristics of the actuator, such as air pressure, heating current, and SMA pre displacement
	Feng et al. [69]	2023	Combining sensor-based schemes to enhance real-time sensing performance
	Akhunzada et al. [70]	2024	Evaluation criterion to use to evaluate the mechanism's presentation in a methodical manner
	Ghoushchi et al. [71]	2024	Techniques for assessing the potential for IoT manufacturing in driverless cars

(continued)

Table 3 (continued)

IoT in self-driving vehicles	References	Year	Contribution
	Harlow et al. [72]	2024	Appealing alternative to cameras, lidars, and other near-visual sensors
Driver Monitoring	Hussain et al. [73]	2021	System for fleet maintenance that tracks the condition of the vehicles and reports any anomalies
	Kong et al. [74]	2021	Achieve the immutable distribution and verifiable aggregation of performance data
	Chen et al. [75]	2022	Identifying the emotional sensitivity of drivers of autonomous cars and researching the architecture of the Internet of Things
	Zhang et al. [76]	2023	Various aspects of driver monitoring from the standpoint of application requirements
	Shariff et al. [77]	2023	When training a submanifold sparse neural network (SSNN) model, driver distraction is a crucial use case
	Yuan et al. [78]	2023	Examined two datasets from the actual world, one with and one without system heterogeneity
	Oğuz et al. [79]	2024	System that lets academics studying swarm robotics test their algorithms on real unmanned aerial vehicles
	Holmes et al. [80]	2024	In matrix-weighted (anisotropic) state-estimation problems and tightness
Safe Driving	Schmaus et al. [81]	2018	Interfaces between humans and robots to endure the harsh conditions of space deployment
	Bakibillah et al. [82]	2019	Data from the vehicle's self-driving system is shown to determine the optimal speed
	Zhu et al. [83]	2020	Transportation system artificial intelligence to improve the "intelligence" of IoT-enabled ITS

(continued)

Table 3 (continued)

IoT in self-driving vehicles	References	Year	Contribution
	Gao et al. [84]	2020	Inverse Distance Weighted (IDW) integration technique with grounded equivalent
	Choi et al. [85]	2020	SAMs are intended for use as lubricants in vacuum vapour deposition, which creates a strongly bonded hydrocarbon coating, and magnetic disc drives
	Prakash et al. [86]	2022	System may injure drivers and passengers because it starts with a 15–30 s delay and unlocks swiftly
	Du et al. [87]	2024	Neural network-based road obstacle avoidance and detection
Environmental Sustainability	Doluca et al. [88]	2019	Notable increase in sustainability and environmental innovation projects
	Navarro et al. [89]	2020	Locating a touch using soft capacitive sensing and pneumatic sensing
	Das et al. [90]	2023	Enables direct communication and network connections for automobiles
	Liu [91]	2023	Cooperative perception performance while considering the constraints of the communications channel
	Dawson et al. [92]	2023	Overview of credential learning theory and practice will be given
	Shukla et al. [93]	2024	Comprehensive framework that categorises and identifies the key technologies for evaluating Construction 4.0's sustainability
	Bhatta et al. [13]	2024	SP-TVS-based vehicle energy harvesting module for self-sufficient transportation

the primary goal which is discovered and mentioned by them. Hazarika et al. [65] contributed for a vision based Dynamic Traffic Light System (DTLS) is proposed by using the You Only Look Once (YOLO) which used for object detection algorithm that detects and calculate the total number of vehicles on the roads of a traffic signal junction, as traffic junctions are turned based on the computed traffic for decreasing them total delay occurred by vehicles at the junction.

4.2 Fleet Management

Maintenance of Fleet management needs appropriate knowledge as well as skills for technical because it is one of the biggest challenges coming from the past. In first and third world countries they are variety of vehicles available which contains built-in sensors acquired through an OBD which stands for onboard diagnostics system which aware drivers of any critical condition. To overcome these issues IoT and cloud computing monitors the health condition of vehicles and report any anomaly with data of the nearest centre for its maintenance. Using dedicated sensors system monitors all the subsystems of the vehicle.

Minovski et al. [66] discussed the transformation in Quality of Experience (QoE) field to conceal the relation of brilliant machines and human beings and further also clarify the case study of describing the QoIoT (Quality of IoT-experience). Zhang et al. [67] discovered lifecycle cost analysis of state of battery charge among the predefined range and calibrate its cost-effective feature as battery aging is almost fully depend on the state of battery charge. Ding et al. [68] suggested that to make it easier to choose the actuator's crucial parameters—like SMA pre displacement, heating current, and air pressure—based on performance requirements, an understandable constitutive and heat transfer model was created. Feng et al. [69] contributed by proposing a scheme for improving the sensing performance in real time, by grouping sensor-based scheme to gain vigorous inference and time- restriction inference issue with heterogenous IoT machines in brilliant system of transportation. Akhunzada et al. [70] contributed by used evaluation criterion to access in a systematic way for assessing the presentation of mechanism which also contains a comparison with ensemble which which was recently presented and hybrid Deep Learning and standard DL-algorithms. Ghoushchi et al. [71] proposed by using two processes: the FF-WASPAS which stands for Fermatean Fuzzy Weighted Aggregated Sum Product Assessment and F-LMAW which stands for Fuzzy Logarithm of Incremental Weights, these are the methods for evaluating the scenario of production of IoT in autonomous vehicles and Augmented Intelligence. Harlow et al. [72] discovered continuous wave modulated by frequency mmAn attractive substitute for lidars, cameras, and other near-visual sensors are wave radars that operate in the 76–81 GHz range. Radar is now more widely accessible in new packaging classes, is more practical for robotics, and can avoid visual clutter like smoke, dust, and fog thanks to its longer wavelengths.

4.3 Driver Monitoring

According to World Health Organisation there are more than 1.3 million people who die every year in road accidents and the people aged between 15 and 29 years. The main reason for this includes drowsiness, careless driving, improper law enforcement, sleepiness, lack of road safety awareness and there are many. There are existing driver monitoring methods which are classified into behaviour analysis-based approaches and vehicle sensor-based approaches. Unlike these existing approaches, there is an integral IoT-based remote monitoring solution.

Hussain et al. [73] introduced Car e-Talk, a fleet maintenance system that monitors vehicle health and reports any irregularities along with information on the closest maintenance facility using cloud computing and Internet-of-things technologies. Kong et al. [74] suggested a method initially uses a blockchain with proof-of-stake (PoS) consensus to accomplish the verifiable aggregation and immutable distribution of performance information. Furthermore, the suggested method uses an oblivious transfer (OT) protocol to come closer to the target performance record after achieving rapid identification with a Bloom filter to obtain a driver's aggregated performance record from the blockchain. Chen et al. [75] contributed in collecting details about the Face identity that how it is important for detecting emotional sensitivity of autonomous vehicle drivers and studied the IoT architecture which directs the physiological signal and track the driver's emotions. Zhang et al. [76] discovered that there is some typically different stated of driver monitoring from the perspective of demands of applications for driver state monitoring in conditionally automated driving, so the research is done on according to the types of sensing data. Shariff et al. [77] contributed to validate the effectiveness of proposed machine learning pipeline meant on data that they have founded the Driver Distraction as a critical use case, as submanifold sparse neural network (SSNN) model is trained on synthetic event data generated from publicly available Drive and Act and Driver Monitoring Dataset (DMD) using a video-to-event conversion algorithm (V2E). Yuan et al. [78] suggested a federated transfer-ordered-personalized learning (FedTOP) architecture, which is tested on two real-world data sets with and without system heterogeneity. Oğuz et al. [79] discovered that for offering a platform that enables swarm robotics researchers to test their algorithms on actual unmanned aerial vehicles (UAVs) without the need to build their own specially designed UAVs or wait for more sophisticated sensing technologies to become available commercially. Holmes et al. [80] discovered the testing of semidefinite relaxations' tightness in matrix-weighted (anisotropic) state-estimation issues and expose any potential drawbacks.

4.4 Safe Driving

In traffic scenarios, autonomous self-driving system should decide and control the real driving situations, making scenarios a difficult aspect of virtual simulation test.

The rapid development of artificial intelligence started the deployment and development of autonomous vehicles which has potential in taking the edge off traffic congestion and excluding and ignoring the collisions.

Schmaus et al. [81] discovered the robots must be able to communicate, operate autonomously, and build human–robot interfaces to withstand the demanding circumstances of space deployment. Bakibillah et al. [82] discovered a learning-based event-driven ecological (eco) driving system (EDS) that uses a vehicle's self-driving data to determine the ideal velocity is presented in this study. Zhu et al. [83] proposed works on real and artificial intelligence of transportation system for enhancing the "intelligence" of IoT-enabled ITS. Gao et al. [84] contributed a GPU which stands for Graphics Processing Unit – grounded equivalent Inverse Distance Weighted (IDW) integration algorithm which helps to boost the security and safety of self-dependent driving. Choi et al. [85] discovered the potential for SAMs to be used as lubricants for magnetic disc drives, or as a part of a lubrication system. Prakash et al. [85] produced a begins with delay of 15–30 s, if system unlocks quickly, then it may cause damage to passengers and drivers. Prakash et al. [86] produced a begins with delay of 15–30 s, if system unlocks quickly, then it may cause damage to passengers and drivers. Du et al. [87] proposed system of avoidance and detection of road obstacle based on neural network, also highlighted the trends of this network and problems which comes in innovation of neural networks.

4.5 Environmental Sustainability

According to U.S. EIA which stands for Energy Information Administration, the total global oil consumption in 2022 was 87.1 million barrels per day. There will only 51 years of oil left in the future, if current consumption rate continues, causing rapid depletion of non-renewable energy sources. Global vehicle carbon emissions from fossil fuels totalled 10.654 Gt in 2017, according to the U.S. Doluca et al. [88] discovered that there has been a significant rise in environmental innovation and sustainability initiatives. Navarro et al. [89] suggested that using soft capacitive sensing to locate a touch and pneumatic sensing to evaluate the deformation and force intensity in the context of intrinsic sensing. Das et al. [90] discovered the most recent, reliable, and secure Cellular Vehicle-to-Everything (C-V2X) technology allows vehicles to connect via network and direct communication modes. Liu et al. [91] suggested for cooperative perception performance while taking communications channel limitations into consideration and the channel impairments and various fusion techniques are assessed. Dawson et al. [92] discovered that clear introduction will be provided for the theory and practice of certificate learning for individuals. Shukla et al. [93] proposed that main goal is to close these gaps by putting forth a thorough multifaceted framework for assessing the sustainability of Construction 4.0 that classifies and identifies the main technologies. Bhatta et al. [13] highlighted the affirmation of VEHM which stands for vehicle energy harvesting module made from SP-TVS.

5 Conclusion

The past record of road accidents without IoT enabled automobiles has not been great so far, which shows little to almost no promise in improvement to robotics field. A solution to such problem requires the integration of IoT enabled automobiles with Robots are in need to provide proper assistance to human resources. These autonomous vehicles and robots are already equipped with high quality sensors, laser radars and many more. IoT is becoming the new paradigm by providing a set of services through innovation, by merging multiple processes like sensing, computation, communication, and identification. With the already existing databases of picture recognition, a robotic vision can also be created. This highlights the computer vision of the robotics. Emerging technologies from the IoT proves to be beneficial for humanoid robots and to make an integration proves only to be useful. The road navigation system in today's IoT enabled automobiles are so powerful that they use GPS system to pinpoint their location and to make real-time actions when the driver is not active. Through the mentioned feature one can train to make an algorithm for the humanoid robot and ensures highly effective robots. IoT enabled automobiles ensures the high-quality sensors and fast responsive lasers which are highly accurate and instant to respond. These features of IoT enabled automobiles ensures a highly functioning robot that can predict and react to situations like this to have a safer and better road.

References

1. A. Abusitta, G.H.S. de Carvalho, O.A. Wahab, T. Halabi, B.C.M. Fung, S.A. Mamoori, Deep learning-enabled anomaly detection for IOT Systems. Internet Things **21**, 100656 (2023). https://doi.org/10.1016/j.iot.2022.100656
2. S. Dhar, A. Khare, A.D. Dwivedi, R. Singh, Securing IOT devices: a novel approach using blockchain and quantum cryptography. Internet Things **25**, 101019 (2024). https://doi.org/10.1016/j.iot.2023.101019
3. M. Beyrouti, A. Lounis, B. Lussier, A. Bouabdallah, A.E. Samhat, Vulnerability-oriented risk identification framework for IOT risk assessment. Internet Things **27**, 101333 (2024). https://doi.org/10.1016/j.iot.2024.101333
4. J. Li, J. Jin, L. Lyu, D. Yuan, Y. Yang, L. Gao, C. Shen, A fast and scalable authentication scheme in IOT for smart living. Futur. Gener. Comput. Syst. **117**, 125–137 (2021). https://doi.org/10.1016/j.future.2020.11.006
5. P. Gkikopoulos, P. Kropf, V. Schiavoni, J. Spillner, Reliable IOT analytics at scale. J. Parallel Distrib. Comput. **187**, 104840 (2024). https://doi.org/10.1016/j.jpdc.2024.104840
6. Q. Zhang, H. Zhong, W. Shi, L. Liu, A trusted and collaborative framework for Deep Learning in IOT. Comput. Netw. **193**, 108055 (2021). https://doi.org/10.1016/j.comnet.2021.108055
7. L. Liu, S. Lu, R. Zhong, B. Wu, Y. Yao, Q. Zhang, W. Shi, Computing systems for autonomous driving: state of the art and challenges. IEEE Internet Things J. **8**(8), 6469–6486 (2021). https://doi.org/10.1109/jiot.2020.3043716
8. S. Grigorescu, B. Trasnea, T. Cocias, G. Macesanu, A survey of Deep Learning techniques for autonomous driving. J. Field Robot. **37**(3), 362–386 (2019). https://doi.org/10.1002/rob.21918

9. S.S. Haghshenas, G. Guido, A. Vitale, S.J. Ghoushchi, Quantitative and qualitative analysis of internet of things (IOT) in Smart Cities and its applications, in *2022 IEEE International Conference on Dependable, Autonomic and Secure Computing, International Conference on Pervasive Intelligence and Computing, International Conference on Cloud and Big Data Computing, International Conference on Cyber Science and Technology Congress (DASC/PiCom/CBDCom/CyberSciTech)* (2022). https://doi.org/10.1109/dasc/picom/cbdcom/cy55231.2022.9927793
10. D. Parekh, N. Poddar, A. Rajpurkar, M. Chahal, N. Kumar, G.P. Joshi, W. Cho, A review on autonomous vehicles: progress, methods and challenges. Electronics **11**(14), 2162 (2022). https://doi.org/10.3390/electronics11142162
11. M.S. Mohammed, A.M. Abduljabar, M.M. Faisal, B.M. Mahmmod, S.H. Abdulhussain, W. Khan, P. Liatsis, A. Hussain, Low-cost autonomous car level 2: design and implementation for conventional vehicles. Results Eng. **17**, 100969 (2023). https://doi.org/10.1016/j.rineng.2023.100969
12. M. Masud, G. Muhammad, H. Alhumyani, S.S. Alshamrani, O. Cheikhrouhou, S. Ibrahim, M.S. Hossain, Deep learning-based intelligent face recognition in IOT-cloud environment. Comput. Commun. **152**, 215–222 (2020). https://doi.org/10.1016/j.comcom.2020.01.050
13. T. Bhatta, O. Faruk, M.R. Islam, H.S. Kim, S.S. Rana, G.B. Pradhan, A. Deo, D.-S. Kwon, I. Yoo, J.Y. Park, Polymeric multilayered planar spring-based hybrid nanogenerator integrated with a self-powered vibration sensor for automotive vehicles IOT applications. Nano Energy **127**, 109793 (2024). https://doi.org/10.1016/j.nanoen.2024.109793
14. J. Heidary Dahooie, A. Mohammadian, A.R. Qorbani, T. Daim, A portfolio selection of internet of things (iots) applications for the sustainable urban transportation: a novel hybrid multi criteria decision making approach. Technol. Soc. **75**, 102366 (2023). https://doi.org/10.1016/j.techsoc.2023.102366
15. S. Hemesath, M. Tepe, Framing the approval to test self-driving cars on public roads. The effect of safety and competitiveness on citizens' agreement. Technol. Soc. **72**, 102177 (2023). https://doi.org/10.1016/j.techsoc.2022.102177
16. T. Kavitha, N. Pandeeswari, R. Shobana, V.R. Vinothini, K. Sakthisudhan, A. Jeyam, A.J. Malar, Data congestion control framework in wireless sensor network in IOT enabled intelligent transportation system. Measurement Sens. **24**, 100563 (2022). https://doi.org/10.1016/j.measen.2022.100563
17. S. Mehar, S. Zeadally, G. Rémy, S.M. Senouci, Sustainable transportation management system for a fleet of electric vehicles. IEEE Trans. Intell. Transp. Syst. **16**(3), 1401–1414 (2015). https://doi.org/10.1109/TITS.2014.2367099
18. Y.H. Kim, W.A. Cahyadi, Y.H. Chung, Experimental demonstration of VLC-based vehicle-to-vehicle communications under fog conditions. IEEE Photonics J. **7**(6), 1–9 (2015). Art no. 7905309. https://doi.org/10.1109/JPHOT.2015.2499542
19. D. Han, B. Bai, W. Chen, Secure V2V communications via relays: resource allocation and performance analysis. IEEE Wireless Commun. Lett. **6**(3), 342–345 (2017). https://doi.org/10.1109/LWC.2017.2690292
20. R. Vosooghi, J. Kamel, J. Puchinger, V. Leblond, M. Jankovic, Robo-taxi service fleet sizing: assessing the impact of user trust and willingness-to-use. Transportation **46**(6), 1997–2015 (2019).https://doi.org/10.1007/s11116-019-10013-x
21. O. Kaiwartya, A.H. Abdullah, Y. Cao, A. Altameem, M. Prasad, C.-T. Lin, X. Liu, Internet of vehicles: motivation, layered architecture, network model, challenges, and future aspects. IEEE Access **4**, 5356–5373 (2016). https://doi.org/10.1109/access.2016.2603219
22. I. Alam, S. Kumar, P.K. Kashyap, A seven-layered model architecture, network model, protocol stack, security, application, issues, and challenges in internet of vehicle. Recent Patents Eng. **15**(4) (2021). https://doi.org/10.2174/1872212114999200719144002
23. A. Singh, L. Gaba, A. Sharma, Internet of vehicles: proposed architecture, network models, open issues and challenges. Amity Int. Conf. Artif. Intell. (AICAI) **2019**, 632–636 (2019). https://doi.org/10.1109/aicai.2019.8701312

24. P.A. Singleton, J. De Vos, E. Heinen, B. Pudāne, Potential health and well-being implications of autonomous vehicles. Adv. Transp. Policy Plann. 163–190 (2020). https://doi.org/10.1016/bs.atpp.2020.02.002
25. E. Dogan, F. Costantini, R. Le Boennec, Ethical issues concerning automated vehicles and their implications for transport. Adv. Transp. Policy Plann. 215–233 (2020). https://doi.org/10.1016/bs.atpp.2020.05.003
26. M. Ajana El Khaddar, Middleware solutions for the internet of things: a survey. Middleware Archit. (2021). https://doi.org/10.5772/intechopen.100348
27. M.A. Razzaque, M. Milojevic-Jevric, A. Palade, S. Clarke, Middleware for internet of things: a survey. IEEE Internet Things J. **3**(1), 70–95 (2016). https://doi.org/10.1109/jiot.2015.2498900
28. S. Dias Vasconcelos, J. Filho da Costa Castro, F. Gouveia, A. Venancio de Moura Lacerda Filho, R. Fonseca Buzo, L. Henrique Alves de Medeiros, L. Rodrigues Limongi, D. da Costa Marques, A. Lopes Fernandes, J. Chai, N. Kellyano Leite Dantas, C. Zhang, P. Rosas, N. Medina, Assessment of electric vehicles charging grid impact via predictive indicator. IEEE Access **12**, 163307–163323 (2024). https://doi.org/10.1109/access.2024.3482095
29. J.F. Castro, D.C. Marques, L. Tavares, N.K. Dantas, A.L. Fernandes, J. Tuo, L.H. de Medeiros, P. Rosas, Energy and demand forecasting based on logistic growth method for electric vehicle fast charging station planning with PV solar system. Energies **15**(17), 6106 (2022). https://doi.org/10.3390/en15176106
30. S.M. Alshareef, W.G. Morsi, Impact of fast charging stations on the voltage flicker in the electric power distribution systems. In: *2017 IEEE Electrical Power and Energy Conference (EPEC)* (2017). https://doi.org/10.1109/epec.2017.8286226
31. M. Farhoodnea, A. Mohamed, H. Shareef, H. Zayandehroodi, Power quality impact of renewable energy based generators and electric vehicles on distribution systems. Proc. Technol. **11**, 11–17 (2013). https://doi.org/10.1016/j.protcy.2013.12.156
32. G. Wager, J. Whale, T. Braunl, Driving electric vehicles at highway speeds: the effect of higher driving speeds on energy consumption and driving range for electric vehicles in Australia. Renew. Sustain. Energy Rev. **63**, 158–165 (2016). https://doi.org/10.1016/j.rser.2016.05.060
33. L. Timilsina, P.R. Badr, P.H. Hoang, G. Ozkan, B. Papari, C.S. Edrington, Battery degradation in electric and hybrid electric vehicles: a survey study. IEEE Access **11**, 42431–42462 (2023). https://doi.org/10.1109/access.2023.3271287
34. J. Wang, H. Wang, Y. Fan, Techno-economic challenges of fuel cell commercialization. Engineering **4**(3), 352–360 (2018). https://doi.org/10.1016/j.eng.2018.05.007
35. K.M. Alam, M. Saini, A. El Saddik, Toward social internet of vehicles: concept, architecture, and applications. IEEE Access **3**, 343–357 (2015). https://doi.org/10.1109/access.2015.2416657
36. J.A. Solorio, J.M. Garcia-Bravo, B.A. Newell, Voice activated semi-autonomous vehicle using off the shelf home automation hardware. IEEE Internet Things J. **5**(6), 5046–5054 (2018). https://doi.org/10.1109/jiot.2018.2854591
37. C. Chen, L. Liu, T. Qiu, Z. Ren, J. Hu, F. Ti, Driver's intention identification and risk evaluation at intersections in the internet of vehicles. IEEE Internet Things J. **5**(3), 1575–1587 (2018). https://doi.org/10.1109/jiot.2017.2788848
38. A. Djenna, D. Eddine Saidouni, Cyber attacks classification in IOT-based-healthcare infrastructure. In: *2018 2nd Cyber Security in Networking Conference (CSNet)*, pp. 1–4 (2018). https://doi.org/10.1109/csnet.2018.8602974
39. M.M. Rana, Attack resilient wireless sensor networks for smart electric vehicles. IEEE Sens. Lett. **1**(2), 1–4 (2017). https://doi.org/10.1109/lsens.2017.2690907
40. H. Zhu, S. Feng, F. Yu, Parking detection method based on finite-state machine and collaborative decision-making. IEEE Sens. J. **18**(23), 9829–9839 (2018). https://doi.org/10.1109/jsen.2018.2871869
41. J. Wu, M. Dong, K. Ota, J. Li, W. Yang, M. Wang, Fog-computing-enabled cognitive network function virtualization for an information-centric future internet. IEEE Commun. Mag. **57**(7), 48–54 (2019). https://doi.org/10.1109/mcom.2019.1800778

42. C. Zhan, H. Lai, Energy minimization in internet-of-things system based on rotary-wing UAV. IEEE Wireless Commun. Lett. **8**(5), 1341–1344 (2019). https://doi.org/10.1109/lwc.2019.2916549
43. G. Mohindru, K. Mondal, H. Banka, Internet of things and data analytics: a current review. WIREs Data Mining Knowl. Discovery **10**(3) (2019). https://doi.org/10.1002/widm.1341
44. A.N. Abougreen, Big medical data analytics under internet of things. Internet Things, 25–44 (2021). https://doi.org/10.1007/978-3-030-66633-0_2
45. D. Singh, A.K. Maurya, R.K. Dewang, N. Keshari, A review on internet of multimedia things (IOMT) routing protocols and quality of service. Internet Multimedia Things (IoMT) 1–29 (2022). https://doi.org/10.1016/b978-0-32-385845-8.00006-x
46. D. Sehrawat, N.S. Gill, Smart sensors: analysis of different types of IOT sensors. In: *2019 3rd International Conference on Trends in Electronics and Informatics (ICOEI)*, pp. 523–528 (2019). https://doi.org/10.1109/icoei.2019.8862778
47. K. Fizza, A. Banerjee, K. Mitra, P.P. Jayaraman, R. Ranjan, P. Patel, D. Georgakopoulos, QoE in IOT: a vision, survey and future directions. Discover Internet Things **1**(1) (2021). https://doi.org/10.1007/s43926-021-00006-7
48. J. Guth, U. Breitenbucher, M. Falkenthal, F. Leymann, L. Reinfurt, Comparison of IOT platform architectures: a field study based on a reference architecture. In: 2016 Cloudification of the Internet of Things (CIoT) (2016). https://doi.org/10.1109/ciot.2016.7872918
49. J. Yang, Q. Ni, G. Luo, Q. Cheng, L. Oukhellou, S. Han, A trustworthy internet of vehicles: the Dao to safe, secure, and collaborative autonomous driving. IEEE Trans. Intell. Veh. **8**(12), 4678–4681 (2023). https://doi.org/10.1109/tiv.2023.3337345
50. F. Tian, Z. Li, F.-Y. Wang, L. Li, Parallel learning-based steering control for autonomous driving. IEEE Trans. Intell. Veh. **8**(1), 379–389 (2023). https://doi.org/10.1109/tiv.2022.3173448
51. L. Chen, Y. Li, C. Huang, B. Li, Y. Xing, D. Tian, L. Li, Z. Hu, X. Na, Z. Li, S. Teng, C. Lv, J. Wang, D. Cao, N. Zheng, F.-Y. Wang, Milestones in autonomous driving and intelligent vehicles: survey of surveys. IEEE Trans. Intell. Veh. **8**(2), 1046–1056 (2023). https://doi.org/10.1109/tiv.2022.3223131
52. Y. Liu, X. Zhang, L. Huo, J. Wu, M. Guizani, Swarm learning and knowledge distillation empowered self-driving detection against threat behavior for intelligent IoT. IEEE Trans. Mob. Comput. **23**(6), 7117–7134 (2024). https://doi.org/10.1109/TMC.2023.3330514
53. A.I. Aygun, M.S. Hasan, A. Joshi, S. Kamalasadan, An optimal hybrid management of electric vehicle fleet charging and load scheduling in active electric distribution system. IEEE Trans. Ind. Appl. **60**(4), 5304–5316 (2024). https://doi.org/10.1109/TIA.2024.3384343
54. S. Ge, F.-Y. Wang, J. Yang, Z. Ding, X. Wang, Y. Li, S. Teng, Z. Liu, Y. Ai, L. Chen, Making standards for smart mining operations: intelligent vehicles for autonomous mining transportation. IEEE Trans. Intell. Veh. **7**(3), 413–416 (2022). https://doi.org/10.1109/tiv.2022.3197820
55. D. Isele, Interactive decision making for autonomous vehicles in dense traffic. In: *2019 IEEE Intelligent Transportation Systems Conference (ITSC)* (2019). https://doi.org/10.1109/itsc.2019.8916982
56. K.S. Song, Design of large diameter mine countermeasure hybrid power unmanned underwater vehicle. Autonomous Vehicle (2016). https://doi.org/10.5772/63795
57. B. Xia, P. Xie, J. Wang, Smart mobility with agent-based foundation models: towards interactive and collaborative intelligent vehicles. IEEE Trans. Intell. Veh. **9**(7), 5130–5133 (2024). https://doi.org/10.1109/tiv.2024.3457759
58. F.-Y. Wang, C. Lv, Foundation vehicles: from foundation intelligence to foundation transportation for future mobility. IEEE Trans. Intell. Veh. 1–5 (2024). https://doi.org/10.1109/tiv.2023.3330194
59. P. De Cristoforis, S. Pedre, M. Nitsche, T. Fischer, F. Pessacg, C. Di Pietro, A behavior-based approach for educational robotics activities. IEEE Trans. Educ. **56**(1), 61–66 (2013). https://doi.org/10.1109/TE.2012.2220359

60. M. Akhtar, M. Raffeh, F. ul Zaman, A. Ramzan, S. Aslam, F. Usman, Development of congestion level based dynamic traffic management system using IoT. In: *2020 International Conference on Electrical, Communication, and Computer Engineering (ICECCE)*, Istanbul, Turkey, pp. 1–6 (2020). https://doi.org/10.1109/ICECCE49384.2020.9179375
61. S. Evripidou, K. Georgiou, L. Doitsidis, A.A. Amanatiadis, Z. Zinonos, S.A. Chatzichristofis, Educational robotics: platforms, competitions and expected learning outcomes. IEEE Access **8**, 219534–219562 (2020). https://doi.org/10.1109/access.2020.3042555
62. A. Jarwan, A. Sabbah, M. Ibnkahla, Information-oriented traffic management for energy-efficient and loss-resilient IoT systems. IEEE Internet Things J. **9**(10), 7388–7403 (2022). https://doi.org/10.1109/JIOT.2021.3132925
63. H. Dui, S. Zhang, M. Liu, X. Dong, G. Bai, IoT-enabled real-time traffic monitoring and control management for intelligent transportation systems. IEEE Internet Things J. **11**(9), 15842–15854 (2024). https://doi.org/10.1109/JIOT.2024.3351908
64. M. Ibnkahla, A. Jarwan, Edge-based federated deep reinforcement learning for IoT traffic management. IEEE Internet Things J. **10**(5), 3799–3813 (2023). https://doi.org/10.1109/JIOT.2022.3174469
65. A. Hazarika, N. Choudhury, M.M. Nasralla, S.B.A. Khattak, I.U. Rehman, Edge ML technique for smart traffic management in intelligent transportation systems. IEEE Access **12**, 25443–25458 (2024). https://doi.org/10.1109/ACCESS.2024.3365930
66. D. Minovski, C. Åhlund, K. Mitra, Modeling quality of IoT experience in autonomous vehicles. IEEE Internet Things J. **7**(5), 3833–3849 (2020). https://doi.org/10.1109/JIOT.2020.2975418
67. L. Zhang, Z. Zeng, X. Qu, On the role of battery capacity fading mechanism in the lifecycle cost of electric bus fleet. IEEE Trans. Intell. Transp. Syst. **22**(4), 2371–2380 (2021). https://doi.org/10.1109/TITS.2020.3014097
68. Q. Ding, J. Chen, W. Yan, K. Yan, A. Kyme, S.S. Cheng, A high-performance modular SMA actuator with fast heating and active cooling for medical robotics. IEEE/ASME Trans. Mechatron. **27**(6), 5902–5913 (2022). https://doi.org/10.1109/TMECH.2022.3190930
69. X. Feng et al., Time-constrained ensemble sensing with heterogeneous IoT devices in intelligent transportation systems. IEEE Trans. Intell. Transp. Syst. **24**(11), 12949–12960 (2023). https://doi.org/10.1109/TITS.2022.3170028
70. A. Akhunzada, A.S. Al-Shamayleh, S. Zeadally, A. Almogren, A.A. Abu-Shareha, Design and performance of an AI-enabled threat intelligence framework for IOT-enabled autonomous vehicles. Comput. Electr. Eng. **119**, 109609 (2024). https://doi.org/10.1016/j.compeleceng.2024.109609
71. S. Jafarzadeh Ghoushchi, S. ShaffieeHaghshenas, S. Vahabzadeh, G. Guido, Z.W. Geem, An integrated MCDM approach for enhancing efficiency in connected autonomous vehicles through augmented intelligence and Iot integration. Results Eng. **23**, 102626 (2024). https://doi.org/10.1016/j.rineng.2024.102626
72. K. Harlow, H. Jang, T.D. Barfoot, A. Kim, C. Heckman, A new wave in robotics: survey on recent MmWave radar applications in robotics. IEEE Trans. Rob. **40**, 4544–4560 (2024). https://doi.org/10.1109/TRO.2024.3463504
73. S. Hussain, U. Mahmud, S. Yang, Car e-talk: an IoT-enabled cloud-assisted smart fleet maintenance system. IEEE Internet Things J. **8**(12), 9484–9494 (2021). https://doi.org/10.1109/JIOT.2020.2986342
74. Q. Kong, R. Lu, F. Yin, S. Cui, Blockchain-based privacy-preserving driver monitoring for MaaS in the vehicular IoT. IEEE Trans. Veh. Technol. **70**(4), 3788–3799 (2021). https://doi.org/10.1109/TVT.2021.3064834
75. Z. Chen, X. Feng, S. Zhang, Emotion detection and face recognition of drivers in autonomous vehicles in IOT platform. Image Vis. Comput. **128**, 104569 (2022). https://doi.org/10.1016/j.imavis.2022.104569
76. Y. Qu, H. Hu, J. Liu, Z. Zhang, Y. Li, X. Ge, Driver state monitoring technology for conditionally automated vehicles: review and future prospects. IEEE Trans. Instrumentation Measurement **72**, 1–20 (2023). Art no. 3000920. https://doi.org/10.1109/TIM.2023.3301060

77. W. Shariff et al., Neuromorphic driver monitoring systems: a computationally efficient proof-of-concept for driver distraction detection. IEEE Open J. Veh. Technol. **4**, 836–848 (2023). https://doi.org/10.1109/OJVT.2023.3325656
78. L. Yuan, L. Su, Z. Wang, Federated transfer–ordered–personalized learning for driver monitoring application. IEEE Internet Things J. **10**(20), 18292–18301 (2023). https://doi.org/10.1109/JIOT.2023.3279273
79. S. Oğuz et al., An open-source UAV platform for swarm robotics research: using cooperative sensor fusion for inter-robot tracking. IEEE Access **12**, 43378–43395 (2024). https://doi.org/10.1109/ACCESS.2024.3378607
80. C. Holmes, F. Dümbgen, T. Barfoot, On semidefinite relaxations for matrix-weighted state-estimation problems in robotics. IEEE Trans. Rob. **40**, 4805–4824 (2024). https://doi.org/10.1109/TRO.2024.3475220
81. P. Schmaus et al., Preliminary insights from the METERON SUPVIS justin space-robotics experiment. IEEE Robot. Autom. Lett. **3**(4), 3836–3843 (2018). https://doi.org/10.1109/LRA.2018.2856906
82. A.S.M. Bakibillah, M.A.S. Kamal, C.P. Tan, T. Hayakawa, J.-I. Imura, Event-driven stochastic eco-driving strategy at signalized intersections from self-driving data. IEEE Trans. Veh. Technol. **68**(9), 8557–8569 (2019). https://doi.org/10.1109/TVT.2019.2931519
83. F. Zhu, Y. Lv, Y. Chen, X. Wang, G. Xiong, F.-Y. Wang, Parallel transportation systems: toward IoT-enabled smart urban traffic control and management. IEEE Trans. Intell. Transp. Syst. **21**(10), 4063–4071 (2020). https://doi.org/10.1109/TITS.2019.2934991
84. K. Gao, B. Wang, L. Xiao, G. Mei, Incomplete road information imputation using parallel interpolation to enhance the safety of autonomous driving. IEEE Access **8**, 25420–25430 (2020). https://doi.org/10.1109/ACCESS.2020.2971059
85. J. Choi, T. Kato, Self-assembled monolayers as lubricants for magnetic disk drives. IEEE Trans. Magn. **41**(2), 599–603 (2005). https://doi.org/10.1109/TMAG.2004.838065
86. G. Prakash, A.G. GM, IOT based collision detection warning system, emergency door and seat belt unlock in E-Vehicle. In: *2022 International Conference on Computer, Power and Communications (ICCPC)*, Chennai, India, pp. 463–467 (2022). https://doi.org/10.1109/ICCPC55978.2022.10072182
87. S. Du, J. Zhang, Y. Wang, Z. Li, Integration of computer vision and IOT into an automatic driving assistance system for "electric vehicles." IEEE Trans. Ind. Inf. **20**(3), 4765–4772 (2024). https://doi.org/10.1109/TII.2023.3326546
88. H. Doluca, B. Holzner, M. Wagner, Corporate sustainability and environmental innovations: practical implications from a cross-country analysis over 15 years. IEEE Eng. Manage. Rev. **47**(2), 115–122 (2019). https://doi.org/10.1109/EMR.2019.2903007
89. S.E. Navarro et al., A model-based sensor fusion approach for force and shape estimation in soft robotics. IEEE Robot. Autom. Lett. **5**(4), 5621–5628 (2020). https://doi.org/10.1109/LRA.2020.3008120
90. D. Das, S. Banerjee, P. Chatterjee, U. Ghosh, U. Biswas, A secure blockchain enabled V2V communication system using smart contracts. IEEE Trans. Intell. Transp. Syst. **24**(4), 4651–4660 (2023). https://doi.org/10.1109/TITS.2022.3226626
91. C. Liu, Y. Chen, J. Chen, R. Payton, M. Riley, S.-H. Yang, Cooperative perception with learning-based V2V communications. IEEE Wireless Commun. Lett. **12**(11), 1831–1835 (2023). https://doi.org/10.1109/LWC.2023.3295612
92. C. Dawson, S. Gao, C. Fan, Safe control with learned certificates: a survey of neural Lyapunov, barrier, and contraction methods for robotics and control. IEEE Trans. Rob. **39**(3), 1749–1767 (2023). https://doi.org/10.1109/TRO.2022.3232542
93. S. Balasubramanian, V. Shukla, N. Islam, S. Manghat, Construction industry 4.0 and sustainability: an enabling framework. IEEE Trans. Eng. Manage. **71**, 1–19 (2024). https://doi.org/10.1109/TEM.2021.3110427

Advancing IoT-Aided Human–Robot Interaction

Rajeev Sharma and Satish Khatak

Abstract Stand-alone robots have been the norm for decades. Recently, however, the Internet of Things (IoT) and cloud technology have allowed them to connect and external knowledge resources. IoT becomes a reality as a result of these connected robots. As robots become more connected, humans will be able to interact with them in novel and unforeseen ways and gain value from them. In order to determine the current state of IoRT, Human–Robot Interaction (HRI) must be examined as a fresh emerging field. Additionally, for IoT-aided robotics to evolve, several fundamental issues must be resolved, design methods must be associated, and solid structural choices must be considered. IoRT applications, implications, and open problems are discussed in this chapter. A wide range of services are available in IoT-aided robotics today, such as communication networks, distributed and pervasive computing, consensus methods based on semantics, network security, etc.

Keywords Robot interaction · Internet of Things · Smart city · Intelligent system · Human interaction

1 Introduction

As sensors, devices, and algorithms continue to develop, robotics is attracting users from a wide range of backgrounds, including academia and industry [1–3]. Providing robots with the ability to communicate with user devices will make them more appealing to a wider audience and more accessible. Using modern communication technologies, IoRT enables this connection to be possible [4]. Therefore, robot software compatible with different types of users' resources (such as

R. Sharma (✉) · S. Khatak
Department of ECE, The Technological Institute of Textile and Sciences, Bhiwani, Haryana 127021, India
e-mail: rajeevsharma78@yahoo.com

S. Khatak
e-mail: khatak1977@yahoo.co.in

R. Sharma and G. Jeon (eds.), *Data Analytics for Smart Robotics and Its Applications*, Intelligent Systems Reference Library 272,
https://doi.org/10.1007/978-3-031-87697-4_6

devices and software) becomes important for supporting to their research as well as work. The majority of academic-oriented software and distributed frameworks for robotic programming and application development run best on desktop computers running Linux-based desktop operating systems (OSs). However, many academic-oriented applications and software are designed to run on desktop computers running Windows-based operating systems [5, 6]. Due to this limitation, many advanced and novice users cannot access the internet if they use another operating system. In addition, cross-disciplinary projects are often accompanied by software modules written in different programming languages (OS). By using new sensors and algorithms, robots are able to perform better and perform more tasks. There are, however, few official programming languages supported in robotic frameworks, with C++ being the main option [5]. YARP is a relevant framework for state-of-the-art robots [7]. In addition to ROS 2.0, ROS 2.0 officially supports C++, Python 2, and Lua [8]. Several robotics middleware/frameworks integrate bridge servers to communicate with devices and languages that are not supported by the middleware/framework [9]. Middleware-dependent socket messages are serialized so that devices and modules without middleware support can access them. Javascript Object Notation (JSON) is a popular format for serializing and streaming data. Messages are generally sent using POSIX sockets or Web sockets in the Requests/Process/Reply pattern (also called Client/Server).

Bridge servers are often used to perform this transformation between protocols. State-of-the-art packages such as rosbridge [10, 11], and yarp.js can be found in this area [12]. The protocol and implementation are described separately in the official ROS documentation. ROS supports sending JSON messages using the rosbridge serialization protocol. All three components of Rosbridge are included in these packages, including Rosbridge's library, rosapi, and server. Whenever "rosbridge" is mentioned in this article, it will refer to the implementation, not the protocol itself. A robotic network can communicate with external devices using modern Web communication protocols using yarp.js, as described in [12]. A bridge server implementation is also required for Yarp.js, just like Rosbridge [12]. Figure 1 illustrates how IoT-enabled robotics can be used.

A new term is being introduced in robotics, sensor, and cloud technologies: Internet of Robotic Things (IoRT) and Cloud Robotics (CR). An example of one of these applications would be to provide patients with dementia with a fully monitored environment [4]. As a result of analyzing data collected from sensors and wearables, a smart IoRT environment may infer an actuation strategy for achieving the desired environment state from the collected data [13]. In the near future, significant growth is expected for the IoRT concept, especially among companies that are keen to exploit it [14]. The IoRT value chain will be far more human-centric than traditional IoT environments. It has been well established that robot-human interaction (HRI) presents many benefits to humans. Robotics is concerned with the design, development, and evaluation of robots that are used or interacted with by humans [15]. Human–robot interaction (HRI) has traditionally focused on providing some form of empathy to robots through interactions with humans. An automated care home could then use this technology for assisted living. As a result of IoT and more

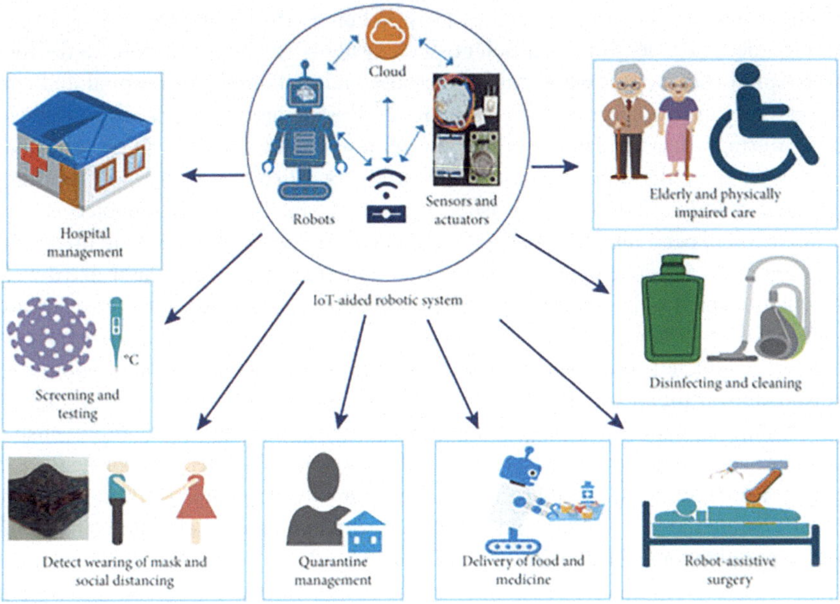

Fig. 1 The application of IoT-aided robotic system

specifically IoRT, interaction between humans and robots has become increasingly complex [16]. A survey paper on HRI-specific IoRT-enabled applications will be presented in the key sectors where robotics is gaining traction. We begin with a brief overview of IoRT's key topics in this paper. In addition to describing the research strategy, some trends in the topics discussed are discussed. Afterwards, an overview of human–robot interaction with IoT is presented. An overall conclusion and a few recommendations for future research conclude the paper.

Related Work

IoT and mechanical autonomy have never collaborated to their full potential in many ways. Our approach to solving problems is changing as a result of IoT arrangements. A number of IoT applications today affect how we live, such as smart homes, wearables, smart cities, keen networks, mechanical web, connected vehicles, connected wellness, smart retail, and clever flexible chains. We have been able to respond to basic problems faster with the Internet of Things because it provides ongoing, quantitative, and conclusive information. This has led to a significant reduction in reaction time and, in a couple of cases, we have been able to eliminate the need for mankind oversight. In addition, mechanical autonomy has been excluded from IoRT research, so now is a good time to innovate in this field.

It is expensive to send mechanical technology-based arrangements. It is generally understood that bots are used in health care and are widely in practice at present [17, 18]. Using these AI-based processes, patients with disabilities, senior citizens, and mobility challenges can be supported [19]. When medical equipment is observed and recorded, or when it is insufficient, such systems can be developed, so doctors and hospital staff do not waste time observing and maintaining a hospital and can take cares of even a large number of patients. Robots are not supposed to replace people during a disaster. However, they are supposed to be able to perform tasks that humans are not able to or cannot perform safely, or at least assist them in handling increased workloads during a disaster. Teleporting intelligent IoT-enabled robots allows medical personnel to assist sick or isolated patients without risking infection.

Researchers and developers are currently conducting extensive research and development on human–robot interaction (HRI). It is estimated that hundreds of papers are published each year and that numerous professional societies, as well as ad hoc meetings, are conducted in technical fields including mechanical, electrical, computer, control, and artificial intelligence. In 2015, the IEEE Robotics and Automation Society held its annual symposium on human–robot interaction. A comprehensive but slightly dated review of the literature can be found in [15]. HRI has been neglected by others and embraced by others, such as those involved in human–computer interaction, despite the fact that the interaction between humans and computers has long been a popular topic in human factors. However, the development and design of robots require research on human interaction. A human factors professional should work closely with engineers in these fields on research, design, and evaluation, as well as improving their knowledge of dynamics, control, and computer science (artificial intelligence). Humans need to trust that their robot teammates will protect each other's interests and welfare for the human–robot team to be successful. In order to perform combat missions, robots must be trusted [20]. People must trust robots in these environments, since trust determines whether they will accept robots' suggestions, obey robots' commands, and thus take advantage of them [21]. Human decision-making in uncertain and risky situations is heavily influenced by trust [22]. An individual may intervene earlier if he or she does not trust a robot when it is on the verge of completing a task the less trust he or she has in the robot [23, 24].

Warfighters in the field, however, have described how robotics and humans can actually develop trust in stressful situations. Sgt is the name given to the robot of one explosives disposal unit. In appreciation of its bomb disposal abilities, the Pentagon promoted Talon and even awarded it Purple Hearts[25]. Despite this, other soldiers' accounts demonstrate the difficulties in trusting robots during times of crisis. Developed and deployed in Iraq in 2007, SWORD supports combat operations through special weapon observation and reconnaissance [26]. Because SWORD's movements were unpredictable due to technological malfunctions, it never was used in the field even though it was fully operational [26]. The Internet of Robots [27], is an emerging research trend, as well as IoT-aided robotics applications. In this position paper, we attempt to shed some light on the implications of these technologies and open issues, as well as the target domains they entail.

The Internet of Things will be used for robotic applications that allow humans and robots to interact cooperatively. Participants have the autonomy to determine secure communication principles based on how they intend to exchange information and provide services based on this framework. Therefore, IoT-assisted robotics applications cover a wide range of research fields, such as short range communication technologies, semantic-oriented services, consensus theory, protocols, applications, information-centered networks, security, and everything else necessary for creating a smart, pervasive, and secure environment. The development of intelligent transportation systems (ITS) that provide road-related information requires an understanding of different road conditions. Shared traffic information with traffic management systems has been shown to save time and energy for drivers and commuters [28–30]. Automated traffic congestion and accident prediction reduce road-related problems through the proposed system. Traffic management systems incorporate advanced Internet-of-Thing technologies such as cameras, RFIDs (Radio-Frequency Identification), GPSs (Global Positioning Systems), ultrasonics, IR sensors, and more. A wireless or cellular network connects these sensors to processing centers. By using IoT clouds, vehicles, commuters, and drivers can receive, process, and receive the traffic data. There has been a significant amount of research on real-time monitoring systems in intelligent traffic management. In this article, we will discuss IoT sensor-based systems [31–34] and vision-based systems [35–38].

A variety of IoT sensors are proposed as a means of implementing this type of solution. Based on IR sensors [31], proposes a clustered approach to monitoring road conditions in real-time. An Android device performs algorithmic computations and stores all gathered data so it can be clustered. An application on a mobile device transmits the result of the processed data to drivers using Bluetooth technology. Mobility is improved through the provision of real-time information about traffic congestion and unexpected traffic events from roadside messaging devices. Based on the internet of things, the author's model [30], gathers, processes, and stores road information in real-time. This system enables roadside message units to communicate real-time information on traffic congestion and unusual traffic incidents, which will enhance mobility. With a magnetic sensor node, real-time road data can be collected and stored in a MongoDB database. Microcontrollers with Wi-Fi capabilities collect data and forward it to IoT platforms for further processing. A message unit that displays early-warning messages can help commuters and drivers avoid traffic jams during rush hour, especially. Table 1 shows the various existing models.

2 Modalities Used in Human–Robot Interaction

A variety of modalities are used to interact between humans and robots, including audio, visual, haptic, kinaesthetic, and proprioceptive interactions [39–43]. It is possible to use voice commands, visual gestures, and physical touches separately or in combination. Numerous researchers are investigating linguistic processing, cognitive architectures, and social signal processing to enhance the interaction between robots and their perception of their intelligence.

Table 1 Shows the existing model

Type	Model	Description
2D laser range finder	Hokuyo scanning range finder	Identifying invaders and obstacles, sensing the environment, detecting human body size and position
	SICK Laser LMS-2xx	Monitoring, identifying, categorizing, and controlling sizes, shapes, and locations during construction
3D sensors	Mesa imaging Swiss Ranger	Using time of flight distance measurement to generate high-quality 3D images in real time
	Microsoft Kinect Forecast 3D Laser	Identifying people's motions Navigating around obstacles and detecting them
Cameras	IEEE 1394 digital camera	A stereoscopic image is captured and processed
	Point Grey stereo cameras	A stereoscopic image is captured and processed
	Video stereo camera	A stereoscopic image is captured and processed
RFID	UHF RFID Reader	Persons and objects can be identified
Meteorological sensor	Gill Windsonic wind sensor	The direction and speed of the wind can be identified
Pose estimation	Applanix POS-LV imu/GPS interface	Even in the harshest GPS conditions, we are able to measure position and pose

2.1 *Audio Modality*

Communication between humans and robots is mediated by audio modality, which relies on audio to transmit messages. The ability to recognize and process natural language (NLP) is crucial for robots that are capable of understanding and responding to human speech.

Technology that recognizes and synthesizes human speech can be used by robots that use audio modalities [42, 44–47]. Speech recognition involves the robot understanding spoken commands, while speech synthesis involves it creating text-based responses and instructions. This robot has also been used to tutor languages in a number of applications, including voice assistants and robots controlled by voice [48–50].

2.2 Visual Modality

In addition to understanding body language and facial expressions, robots are also capable of understanding gestures and gaze direction with the visual modality. The visual perception of robots is enabled by camera sensors and computer vision algorithms [40, 51]. A wide range of objects, faces, gestures, and motions can be tracked and recognized by these algorithms. Human–robot interaction, surveillance, and robot navigation use this modality. Using computer vision and deep learning techniques, robots can now recognize and interpret visual cues, which allows humans to interact with them more effectively [52]. In visual modalities, facial expression recognition is one area of research, which can enable robots to respond appropriately to human emotions. Interactions with humans can be made more natural and intuitive by doing this. A robot can also recognize gestures such as nodding and pointing by using gesture recognition software. Navigating or manipulating objects can benefit from this. In addition to visual saliency detection, robots can also recognize objects in their environment using object recognition and visual saliency detection.

2.3 Haptic Modality

Humans and robots can communicate through touch, with robots capable of detecting touch, responding to it, and applying force or vibration as necessary. A number of advanced haptic interfaces have emerged over the past few years as haptic technology has advanced in recent years [39, 53]. As a result of this technology, robotic surgery, prosthetic devices, and tactile communication can be performed with robots using tactile cues. Research into force feedback in haptics is aimed at improving interaction between humans and robots by making it more natural and intuitive. Robots also can sense textures, shapes, and temperatures using tactile sensing.

2.4 Kinaesthetic Modality

As part of this, the robot is capable of determining posture, gait, and joint angle of the human body and responding appropriately. Motion can be sensed and controlled by robots that use kinaesthetic modalities. Sensors measure the joint positions and movements of robots, and actuators control the joints[43]. In addition to search and rescue robots, industrial robots and bipedal robots use this mode.

2.5 *Proprioceptive Modality*

A proprioceptive organism senses its body's position, orientation, and movement [41]. A robot's positioning and movement are sensed and responded to using proprioception in human–robot interaction. In addition to sensing their internal conditions, proprioceptive robots can also sense the environment [54]. They may be able to provide information about their joints and body forces, for example. This sensor can control the robot's movements, and problems can be diagnosed through it [55]. The researchers developed, based on proprioceptive-tactile data, a neural network model for learning proprioceptive-tactile representations on simulated humanoids. The model has limitations, however, that require further investigation.

The combination of these multiple modalities enables robots to interact with humans naturally, giving them a wide range of capabilities. Additionally, modalities that are similar to human interaction, such as visual and auditory, are preferable for better human–robot interaction because they make the process more intuitive and straightforward.

3 Human-Centered Projects Supported by NEP+ Tools

Traditionally, robots are designed primarily for machines, not for people [56]. Typically, this approach involves teams of engineers without a deep understanding of how their systems will be used in real life [57]. As a result, end-user needs and desires are disconnected from technology, and societal issues cannot be effectively addressed [58]. User-centered methodologies emphasize alternatives to traditional approaches that put the user at the center of their lives [58]. As part of these methods, domain experts are acknowledged as key contributors to the development of better design methodologies and the adoption of emerging technologies [59]. As a part of human-centered methods, domain experts and users are empowered to be part of the co-design process at different stages of the development process. Here are two real-life projects that demonstrate how NEP+ empowers developers to design robotics systems and platforms.

3.1 *The NEP Libraries Power Yokobo the Robot*

The Yokobo robot enhances couples' relationships by being non-anthropomorphic and non-vocal [60]. Kawaii Kansei Designs Excellence Award was presented to KawaiiKansei in 2021 for its innovative and user-centered design. A Raspberry Pi unit senses the environment, while another controls the robot. NEP libraries enable data exchange [61]. Junior Japanese students with mechanical and electronics backgrounds assembled Yokobo's components, but had little experience with robotics software architecture. The application also offers a number of other features (described in [62]), facilitates Yokobo's integration with smart homes.

3.2 Supporting Human-Centered Research Using RIZE

Human–robot interaction (HRI) has been a pivotal field for RIZE, which bridges disciplinary and knowledge gaps. The project has been collaboratively designed with service users, social scientists, designers, and roboticists. With RIZE, Design Thinking (DT) was used to support co-design activities [63]. RIZE was used to create robot behaviors for a study on the perception of robots among retired people [64]. A participatory design (PD) methodology has also been used in RIZE workshops, in which end-users are involved in the conceptualization and design stages of the project [65]. Workshops offer users an opportunity to learn more about robotic technology and participate in user-centered discussions. RIZE not only offers insight into its software architecture, but also participates in co-design activities from a human-centered perspective [64].

Discussion

Figures 2 and 3 demonstrate results from L-remote and R-remote scenarios, respectively. In order to obtain these values, 200 samples of each message size were obtained. An example of a connection between nodes executing in different computers can be seen in Fig. 3. It clearly shows that NEP over Zero MQ provides a better latency result than ROS using the official Rosbridge suite when used as a back-end option. Additionally, NEP's local host performance significantly outperforms ROS and Rosbridge's latency results. This problem can be greatly reduced with NEP, whether you are using it locally or remotely (in a LAN).As a result of the research, NEP framework can be used to support academic projects in a relevant way, thus answering the research question.

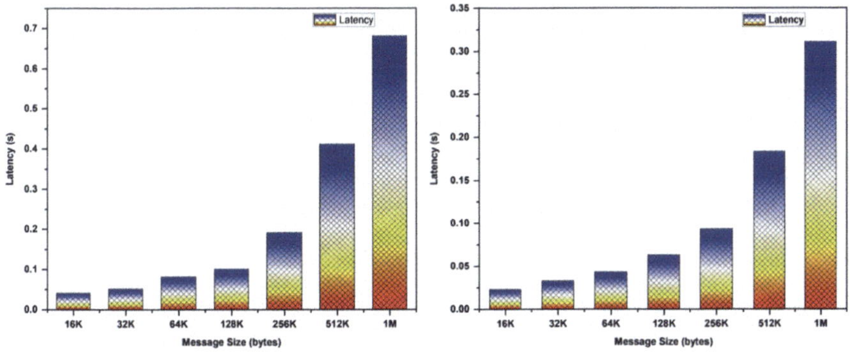

Fig. 2 An analysis of latency results for scenarios R-remote and ROS-rosbridge

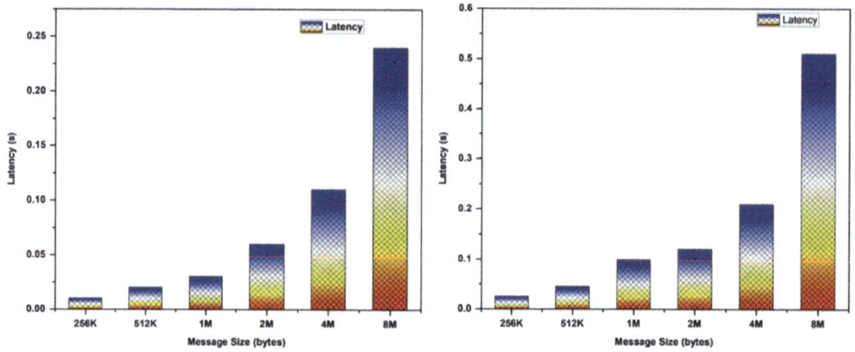

Fig. 3 Analysis of latency results for NEP and ROS-rosbridge scenarios with scenario L-local

Conclusion: Robotics has evolved from isolated, stand-alone devices to intelligent entities connected to the Internet of Robotic Things (IoRT), creating unprecedented opportunities for human–robot interaction (HRI).IoRT has enormous potential across various robotic domains, and its impact on human interactions is highlighted in this survey, which explores state-of-the-art in IoT-enabled HRI. IoRT must overcome a variety of challenges, including network security, data privacy, and the development of robust design methods, in order to achieve widespread adoption. For the IoRT to reach its full potential, we must address these issues so that connected robots and humans collaboratively improve productivity, safety, and quality of life across a variety of industries.

References

1. E. Coronado, X. Indurkhya, G. Venture, Robots meet children, development of semi-autonomous control systems for children-robot interaction in the wild, in *2019 IEEE 4th International Conference on Advanced Robotics and Mechatronics (ICARM)* (IEEE, Toyonaka, Japan, 2019), pp. 360–365. https://doi.org/10.1109/ICARM.2019.8833751.
2. E. Coronado, F. Mastrogiovanni, G. Venture, in *Design of a Human-Centered Robot Framework for End-User Programming and Applications*, ed. by V. Arakelian, P. Wenger. ROMANSY 22—Robot Design, Dynamics and Control. CISM International Centre for Mechanical Sciences, vol. 584 (Springer International Publishing, Cham, 2019), pp. 450–457. https://doi.org/10.1007/978-3-319-78963-7_56
3. A. Singh et al., Smart traffic monitoring through real-time moving vehicle detection using deep learning via aerial images for consumer application. IEEE Trans. Consum. Electron. 1–1 (2024). https://doi.org/10.1109/TCE.2024.3445728
4. P. Simoens, M. Dragone, A. Saffiotti, The Internet of Robotic things: a review of concept, added value and applications. Int. J. Adv. Robot. Syst. **15**(1), 1729881418759424 (2018). https://doi.org/10.1177/1729881418759424
5. E. Tsardoulias, P. Mitkas, Robotic frameworks, architectures and middleware comparison (2017). arXiv. https://doi.org/10.48550/ARXIV.1711.06842

6. N. Hussain, P. Rani, Comparative studied based on attack resilient and efficient protocol with intrusion detection system based on deep neural network for vehicular system security, in *Distributed Artificial Intelligence* (CRC Press, 2020), pp. 217–236
7. G. Metta, P. Fitzpatrick, L. Natale, YARP: yet another robot platform. Int. J. Adv. Robot. Syst. **3**(1), 8 (2006). https://doi.org/10.5772/5761
8. C.S.V. Gutiérrez, L.U.S. Juan, I.Z. Ugarte, V.M. Vilches, Towards a distributed and real-time framework for robots: Evaluation of ROS 2.0 communications for real-time robotic applications (2018). arXiv: arXiv:1809.02595. https://doi.org/10.48550/arXiv.1809.02595
9. N. Hussain, P. Rani, N. Kumar, M.G. Chaudhary, A deep comprehensive research architecture, characteristics, challenges, issues, and benefits of routing protocol for vehicular ad-hoc networks. Int. J. Distrib. Syst. Technol. IJDST **13**(8), 1–23 (2022)
10. C. Crick, G. Jay, S. Osentoski, B. Pitzer, O.C. Jenkins, "Rosbridge: ROS for non-ROS users, in *Robotics Research*, ed. by H.I. Christensen, O. Khatib. Springer Tracts in Advanced Robotics, vol. 100 (Springer International Publishing, Cham, 2017), pp. 493–504. https://doi.org/10.1007/978-3-319-29363-9_28
11. N. Hussain, P. Rani, H. Chouhan, U.S. Gaur, Cyber security and privacy of connected and automated vehicles (CAVs)-based federated learning: challenges, opportunities, and open issues. Fed. Learn. IoT Appl. (2022), pp. 169–183
12. C. Ciliberto, Connecting YARP to the web with Yarp.js. Front. Robot. AI **4**, 67 (2017). https://doi.org/10.3389/frobt.2017.00067
13. R.S. Batth, A. Nayyar, A. Nagpal, Internet of Robotic Things: driving intelligent robotics of future—concept, architecture, applications and technologies, in *2018 4th International Conference on Computing Sciences (ICCS)* (IEEE, Jalandhar, 2018), pp. 151–160. https://doi.org/10.1109/ICCS.2018.00033
14. B.-T. Akimana, M. Bonnaerens, J. Van Wilder, B. Vuylsteker, A survey of human-robot interaction in the internet of things, 2016, Accessed: 31 Oct 2024. [Online]. Available: https://www.researchgate.net/profile/Bjorn-Vuylsteker/publication/318722691_A_Survey_of_Human-Robot_Interaction_in_the_Internet_of_Things/links/5979adbdaca272177c1f4abc/A-Survey-of-Human-Robot-Interaction-in-the-Internet-of-Things.pdf
15. M.A. Goodrich, A.C. Schultz, Human-robot interaction: a survey. Found. Trends® Hum.-Comput. Interact. **1**(3), 203–275 (2007). https://doi.org/10.1561/1100000005
16. P. Rani, N. Hussain, R.A.H. Khan, Y. Sharma, P.K. Shukla, Vehicular intelligence system: time-based vehicle next location prediction in software-defined internet of vehicles (SDN-IOV) for the smart cities. Intell. Things AI-IoT Based Crit.-Appl. Innov. 35–54 (2021)
17. H.H. Lund, Modern artificial intelligence for human-robot interaction. Proc. IEEE **92**(11), 1821–1838 (2004). https://doi.org/10.1109/JPROC.2004.835362
18. R. Bamdale, S. Sahay, V. Khandekar, Natural human robot interaction using artificial intelligence: a survey, in *2019 9th Annual Information Technology, Electromechanical Engineering and Microelectronics Conference (IEMECON)* (IEEE, Jaipur, India, 2019), pp. 297–302. https://doi.org/10.1109/IEMECONX.2019.8877044
19. G. Wisskirchen et al., Artificial intelligence and robotics and their impact on the workplace. IBA Glob. Employ. Inst. **11**(5), 49–67 (2017)
20. V. Groom, C. Nass, Can robots be teammates?: Benchmarks in human–robot teams. Interact. Stud. Soc. Behav. Commun. Biol. Artif. Syst. **8**(3), 483–500 (2007). https://doi.org/10.1075/is.8.3.10gro
21. A. Freedy, E. DeVisser, G. Weltman, N. Coeyman, Measurement of trust in human-robot collaboration, in *2007 International Symposium on Collaborative Technologies and Systems* (IEEE, Orlando, FL, 2007), pp. 106–114. https://doi.org/10.1109/CTS.2007.4621745
22. E. Park, Q. Jenkins, X. Jiang, Measuring trust of human operators in new generation rescue robots. Proc. JFPS Int. Symp. Fluid Power **2008**(7–2), 489–492 (2008). https://doi.org/10.5739/isfp.2008.489
23. E. De Visser, R. Parasuraman, A. Freedy, E. Freedy, G. Weltman, A comprehensive methodology for assessing human-robot team performance for use in training and simulation. Proc. Hum. Factors Ergon. Soc. Annu. Meet. **50**(25), 2639–2643 (2006). https://doi.org/10.1177/154193120605002507

24. A. Steinfeld et al., Common metrics for human-robot interaction, in *Proceedings of the 1st ACM SIGCHI/SIGART Conference on Human-Robot Interaction* (ACM, Salt Lake City, Utah, USA, 2006), pp. 33–40. https://doi.org/10.1145/1121241.1121249
25. J. Garreau, Bots on the ground: In the field of battle (or even above it), robots are a soldier's best friend. Wash. Post **6**(2007), D01 (2007)
26. S. Ogreten, S. Lackey, D. Nicholson, Recommended roles for uninhabited team members within mixed-initiative combat teams, in *2010 International Symposium on Collaborative Technologies and Systems* (IEEE, Chicago, IL, USA, 2010), pp. 531–536.https://doi.org/10.1109/CTS.2010.5478468
27. L. Riazuelo et al., RoboEarth semantic mapping: a cloud enabled knowledge-based approach. IEEE Trans. Autom. Sci. Eng. **12**(2), 432–443 (2015). https://doi.org/10.1109/TASE.2014.2377791
28. A. Frank, Y.S. Khamis Al Aamri, A. Zayegh, IoT based smart traffic density control using image processing, in *2019 4th MEC International Conference on Big Data and Smart City (ICBDSC)* (IEEE, Muscat, Oman, 2019), pp. 1–4. https://doi.org/10.1109/ICBDSC.2019.8645568
29. T.S. Gomides, R.E. De Grande, A.M.D. Souza, F.S.H. Souza, L.A. Villas, D.L. Guidoni, An adaptive and distributed traffic management system using vehicular ad-hoc networks. Comput. Commun. **159**, 317–330 (2020). https://doi.org/10.1016/j.comcom.2020.05.027
30. M. Sarrab, S. Pulparambil, M. Awadalla, Development of an IoT based real-time traffic monitoring system for city governance. Glob. Transit. **2**, 230–245 (2020). https://doi.org/10.1016/j.glt.2020.09.004
31. "Janahan: IoT based smart traffic signal monitoring... - Google Scholar." Accessed: 31 Oct 2024. [Online]. Available: https://scholar.google.com/scholar_lookup?title=IoT%20based%20smart%20traffic%20signal%20monitoring%20system%20using%20vehicles%20counts&publication_year=2018&author=S.K.%20Janahan&author=M.R.M.%20Veeramanickam&author=S.%20Arun&author=K.%20Narayanan&author=R.%20Anandan&author=S.J.%20Parvez
32. P. Agarwal, P. Matta, S. Sharma, Analysis based traffic flow control decision using IoT sensors. Mater. Today Proc. **46**, 10707–10711 (2021). https://doi.org/10.1016/j.matpr.2021.01.540
33. M. Ponnusamy, A. Alagarsamy, Traffic monitoring in smart cities using internet of things assisted robotics. Mater. Today Proc. **81**, 290–294 (2023). https://doi.org/10.1016/j.matpr.2021.03.192
34. "Sahu: Traffic management system using IoT - Google Scholar." Accessed: 31 Oct 2024. [Online]. Available: https://scholar.google.com/scholar_lookup?title=Traffic%20managem ent%20system%20using%20IoT&publication_year=2021&author=S.K.%20Sahu&author=A.%20Basant&author=T.%20Vasudev&author=K.%20Khati&author=N.%20Lawrence
35. P.N. Chowdhury, T. Chandra Ray, J. Uddin, A vehicle detection technique for traffic management using image processing, in *2018 International Conference on Computer, Communication, Chemical, Material and Electronic Engineering (IC4ME2)* (IEEE, Rajshahi, 2018), pp. 1–4. https://doi.org/10.1109/IC4ME2.2018.8465599
36. "Greeshma: Traffic control using computer vision - Google Scholar." Accessed: 31 Oct 2024. [Online]. Available: https://scholar.google.com/scholar_lookup?title=Traffic%20control%20using%20computer%20vision&publication_year=2019&author=C.A.%20Greeshma&author=K.R.%20Nidhindas&author=P.S.%20Sreejith
37. "Bhuiyan: Computer vision based traffic monitoring... - Google Scholar." Accessed: 31 Oct 2024. [Online]. Available: https://scholar.google.com/scholar_lookup?title=Computer%20vision%20based%20traffic%20monitoring%20and%20analyzing%20from%20on-road%20videos&publication_year=2019&author=M.S.R.%20Sajib&author=T.A.U.H.%20Bhuiyan
38. G. Liu et al., Smart traffic monitoring system using computer vision and edge computing. IEEE Trans. Intell. Transp. Syst. **23**(8), 12027–12038 (2022). https://doi.org/10.1109/TITS.2021.3109481
39. S.E. Navarro, B. Hein, H. Wörn, Capacitive tactile proximity sensing: from signal processing to applications in manipulation and safe human-robot interaction, in *Soft Robotics*, ed. by A.

Verl, A. Albu-Schäffer, O. Brock, A. Raatz. (Springer, Berlin, Heidelberg, 2015), pp. 54–65. https://doi.org/10.1007/978-3-662-44506-8_6
40. K. Gui, H. Liu, D. Zhang, Toward multimodal human-robot interaction to enhance active participation of users in gait rehabilitation. IEEE Trans. Neural Syst. Rehabil. Eng. **25**(11), 2054–2066 (2017). https://doi.org/10.1109/TNSRE.2017.2703586
41. A. Ferlinc, E. Fabiani, T. Velnar, L. Gradisnik, The importance and role of proprioception in the elderly: a short review. Mater. Socio Medica **31**(3), 219 (2019). https://doi.org/10.5455/msm.2019.31.219-221
42. C. Deuerlein, M. Langer, J. Seßner, P. Heß, J. Franke, Human-robot-interaction using cloud-based speech recognition systems. Procedia CIRP **97**, 130–135 (2021). https://doi.org/10.1016/j.procir.2020.05.214
43. T. Groechel et al., Kinesthetic curiosity: towards personalized embodied learning with a robot tutor teaching programming in mixed reality, in *Experimental Robotics*, ed. by B. Siciliano, C. Laschi, O. Khatib. Springer Proceedings in Advanced Robotics, vol. 19 (Springer International Publishing, Cham, 2021), pp. 245–252. https://doi.org/10.1007/978-3-030-71151-1_22
44. S. Lackey, D. Barber, L. Reinerman, N.I. Badler, I. Hudson, Defining next-generation multi-modal communication in human robot interaction. Proc. Hum. Factors Ergon. Soc. Annu. Meet. **55**(1), 461–464 (2011). https://doi.org/10.1177/1071181311551095
45. R.C. Luo, S.-R. Chang, C.-C. Huang, Y.-P. Yang, Human robot interactions using speech synthesis and recognition with lip synchronization, in *IECON 2011—37th Annual Conference of the IEEE Industrial Electronics Society* (IEEE, Melbourne, Australia, 2011), pp. 171–176. https://doi.org/10.1109/IECON.2011.6119307
46. Y. Zhao, H. Wang, Q. Ji, Audio-visual Tibetan speech recognition based on a deep dynamic Bayesian network for natural human robot interaction. Int. J. Adv. Robot. Syst. **9**(6), 258 (2012). https://doi.org/10.5772/54000
47. A. Tsiami, P.P. Filntisis, N. Efthymiou, P. Koutras, G. Potamianos, P. Maragos, Far-field audio-visual scene perception of multi-party human-robot interaction for children and adults, in *2018 IEEE International Conference on Acoustics, Speech and Signal Processing (ICASSP)* (IEEE, Calgary, AB, 2018), pp. 6568–6572. https://doi.org/10.1109/ICASSP.2018.8462425
48. B. House, J. Malkin, J. Bilmes, The VoiceBot: a voice controlled robot arm, in *Proceedings of the SIGCHI Conference on Human Factors in Computing Systems* (ACM, Boston, USA, 2009), pp. 183–192. https://doi.org/10.1145/1518701.1518731
49. T. Belpaeme et al., Guidelines for designing social robots as second language tutors. Int. J. Soc. Robot. **10**(3), 325–341 (2018). https://doi.org/10.1007/s12369-018-0467-6
50. J. Humphry, C. Chesher, Preparing for smart voice assistants: cultural histories and media innovations. New Media Soc. **23**(7), 1971–1988 (2021). https://doi.org/10.1177/1461444820923679
51. M. Hasanuzzaman, T. Zhang, V. Ampornaramveth, H. Gotoda, Y. Shirai, H. Ueno, Adaptive visual gesture recognition for human–robot interaction using a knowledge-based software platform. Robot. Auton. Syst. **55**(8), 643–657 (2007). https://doi.org/10.1016/j.robot.2007.03.002
52. O. Celiktutan, E. Sariyanidi, H. Gunes, Computational analysis of affect, personality, and engagement in human–robot interactions. The research reported in this chapter was completed while O. Celiktutan and E. Sariyanidi were with the computer laboratory, University of Cambridge, United Kingdom, in *Computer Vision for Assistive Healthcare* (Elsevier, 2018), pp. 283–318. https://doi.org/10.1016/B978-0-12-813445-0.00010-1
53. S. Pyo, J. Lee, K. Bae, S. Sim, J. Kim, Recent progress in flexible tactile sensors for human-interactive systems: from sensors to advanced applications. Adv. Mater. **33**(47), 2005902 (2021). https://doi.org/10.1002/adma.202005902
54. G. Hoffman, C. Breazeal, Achieving fluency through perceptual-symbol practice in human-robot collaboration, in *Proceedings of the 3rd ACM/IEEE international conference on Human robot interaction* (ACM, Amsterdam, The Netherlands, 2008), pp. 1–8. https://doi.org/10.1145/1349822.1349824

55. K. Malinovska, I. Farkas, J. Harvanova, M. Hoffmann, A connectionist model of associating proprioceptive and tactile modalities in a humanoid robot, in *2022 IEEE International Conference on Development and Learning (ICDL)* (IEEE, London, UK, 2022), pp. 336–342. https://doi.org/10.1109/ICDL53763.2022.9962195
56. E. Coronado, T. Kiyokawa, G.A.G. Ricardez, I.G. Ramirez-Alpizar, G. Venture, N. Yamanobe, Evaluating quality in human-robot interaction: a systematic search and classification of performance and human-centered factors, measures and metrics towards an industry 5.0. J. Manuf. Syst. **63**, 392–410 (2022). https://doi.org/10.1016/j.jmsy.2022.04.007
57. M. Ferati, A. Babar, K. Carine, A. Hamidi, C. Mörtberg, Participatory design approach to Internet of Things: co-designing a smart shower for and with people with disabilities, in *Universal Access in Human-Computer Interaction. Virtual, Augmented, and Intelligent Environments*, ed. by M. Antona, C. Stephanidis. Lecture Notes in Computer Science, vol. 10908 (Springer International Publishing, Cham, 2018), pp. 246–261. https://doi.org/10.1007/978-3-319-92052-8_19
58. B. Fischer, A. Peine, B. Östlund, The importance of user involvement: a systematic review of involving older users in technology design. Gerontologist **60**(7), e513–e523 (2020). https://doi.org/10.1093/geront/gnz163
59. E.J. Rose, E.A. Björling, Designing for engagement: using participatory design to develop a social robot to measure teen stress, in *Proceedings of the 35th ACM International Conference on the Design of Communication* (ACM, Halifax, Nova Scotia, Canada, 2017), pp. 1–10. https://doi.org/10.1145/3121113.3121212
60. D. Deuff et al., Together alone, Yōkobo, a sensible presence robject for the home of newly retired couples, in *Designing Interactive Systems Conference* (ACM, Virtual Event Australia, 2022), pp. 1773–1787. https://doi.org/10.1145/3532106.3533485
61. S. Capy et al., Yōkobo: a robot to strengthen links amongst users with non-verbal behaviours. Machines **10**(8), 708 (2022). https://doi.org/10.3390/machines10080708
62. S. Capy, E. Coronado, P. Osorio, S. Hagane, D. Deuff, G. Venture, Integration of a presence robot in a smart home, in *2023 3rd International Conference on Computer, Control and Robotics (ICCCR)* (IEEE, Shanghai, China, 2023), pp. 192–197. https://doi.org/10.1109/ICCCR56747.2023.10194211
63. S. Bartoloni et al., Towards designing society 5.0 solutions: The new Quintuple Helix—design thinking approach to technology. Technovation **113**, 102413 (2022). https://doi.org/10.1016/j.technovation.2021.102413
64. E. Coronado et al., Towards a modular and distributed end-user development framework for human-robot interaction. IEEE Access **9**, 12675–12692 (2021). https://doi.org/10.1109/ACCESS.2021.3051605
65. H.R. Lee et al. Steps toward participatory design of social robots: mutual learning with older adults with depression, in *Proceedings of the 2017 ACM/IEEE International Conference on Human-Robot Interaction* (ACM, Vienna, Austria, 2017), pp. 244–253. https://doi.org/10.1145/2909824.3020237

Fuzzy Frontiers: Advances in Fuzzy Systems for Real-World Applications

V. Esther Jyothi, Ramakrishna Regulagadda, A. Manjula, G. V. K. Murthy, Puvvada Nagesh, and G. Siva Nageswara Rao

Abstract This paper explores the fuzzy frontiers of computational systems, highlighting recent advances in fuzzy logic and fuzzy systems for real-world applications. Focused on bridging the gap between theoretical developments and practical implementations, the study reviews innovative methodologies and algorithms within the realm of fuzzy systems. Emphasizing the adaptability and uncertainty management capabilities of fuzzy logic, the paper showcases applications in diverse fields such as control systems, decision-making, and pattern recognition. By providing insights into the evolving landscape of fuzzy systems, this work aims to contribute to the effective deployment of fuzzy logic in addressing complex challenges in real-world scenarios.

Keywords Fuzzy Frontiers · Fuzzy systems · Fuzzy logic · Real-world applications · Advances · Methodologies · Algorithms · Adaptability · Uncertainty management · Control systems · Decision-making · Pattern recognition · Evolving landscape · Complex challenges

V. Esther Jyothi
Department of Computer Applications, V R Siddhartha Engineering College, SAHE-A Deemed to be University, Vijayawada, India

R. Regulagadda (✉) · P. Nagesh
Department of CSE, Koneru Lakshmaiah Education Foundation, Guntur, Andhra Pradesh, India
e-mail: rramakrishna@kluniversity.in

A. Manjula
Department of CSE, Jyothishmathi Institute of Technology and Science, Karimnagar, India

G. V. K. Murthy
Department of EEE, PACE Institution of Technology & Sciences, Ongole, Andhra Pradesh, India

G. Siva Nageswara Rao
Department of Computer Science and Engineering, Koneru Lakshmaiah Education Foundation, Vaddesaram, Guntur, Andhra Pradesh, India

R. Sharma and G. Jeon (eds.), *Data Analytics for Smart Robotics and Its Applications*, Intelligent Systems Reference Library 272,
https://doi.org/10.1007/978-3-031-87697-4_7

1 Introduction

1.1 Understanding Fuzzy Logic and Fuzzy Systems

A development of classical logic, fuzzy logic was first presented by Lotfi Zadeh in 1965 and deals with the idea of partial truth—truth values that fall between "completely true" and "completely false". Unlike binary logic, where variables may only take on the values of 0 or 1, fuzzy logic variables may have a truth value that ranges between 0 and 1. Because of this quality, fuzzy logic is especially well-suited for handling ambiguity and imprecision, more precisely representing situations that arise in the actual world. Fuzzy systems leverage fuzzy logic to model complex systems and processes that are not easily described using traditional binary logic. These systems consist of fuzzy sets, rules, and inference mechanisms that can process inputs and generate outputs based on fuzzy principles. This chapter provides a foundational understanding of fuzzy logic and systems, setting the stage for more detailed discussions in subsequent chapters.

1.2 Importance of Fuzzy Systems in Modern Technology

Fuzzy systems are crucial in modern technology due to their ability to handle ambiguity and approximate reasoning, which are common in real-world applications. They are employed in many different domains, such as data analysis, artificial intelligence, and control systems. The adaptability and flexibility of fuzzy systems allow them to be integrated into various technological frameworks, enhancing their capability to deal with complex, nonlinear, and dynamic environments [1].

The chapter will explore the significance of fuzzy systems, highlighting their contributions to advancements in technology and their role in enhancing the robustness and efficiency of intelligent systems. The chapter will show how fuzzy systems have been successfully used to tackle real-world issues through examples and case studies, highlighting their significance and applicability in today's technology environments [2].

2 Foundations of Fuzzy Logic

2.1 Historical Background and Development

The development of fuzzy logic can be traced back to Lotfi Zadeh's seminal work in the mid-twentieth century. This section explores the historical background of fuzzy logic, looking at significant turning points and contributions that have influenced its development. A thorough grasp of fuzzy logic's underlying principles will be provided by discussing the early usage of fuzzy logic in a variety of engineering and scientific areas [3].

1. Key Milestones in Fuzzy Logic

Following Zadeh's introduction of fuzzy sets, the field experienced rapid development and adoption across various disciplines. The creation of fuzzy control systems in the 1970s was one of the first significant turning points. These systems utilized fuzzy logic to control processes that were difficult to model mathematically, such as industrial automation and climate control systems.

In the 1980s, fuzzy logic began to gain traction in Japan, largely due to its application in consumer electronics. Japanese engineers successfully applied fuzzy logic to develop intelligent washing machines, air conditioners, and automotive systems. The success of these products demonstrated the practical utility of fuzzy logic and spurred further interest and research in the field.

2. Advancements in Fuzzy Theory and Applications

The 1990s and 2000s saw significant theoretical advancements in fuzzy logic. New kinds of fuzzy sets, as interval type-2 fuzzy sets, were created by researchers to provide a more adaptable way to describe uncertainty. These developments made it possible to create increasingly complicated fuzzy systems that can deal with increasing degrees of ambiguity and complexity. During this period, fuzzy logic also found applications in areas such as decision-making, pattern recognition, and artificial intelligence. Fuzzy logic was integrated with neural networks and evolutionary algorithms, among other computer techniques, to create hybrid systems that merged the advantages of many approaches [4].

3. Influential Research Papers and Projects

Several influential research papers and projects have shaped the development of fuzzy logic. The foundation for fuzzy control systems was established by Mamdani and Assilian's 1975 work, "An Experiment in Linguistic Synthesis with a Fuzzy Logic Controller," in addition to Zadeh's initial study. Their approach demonstrated how linguistic rules could be used to design controllers for complex systems. Fuzzy inference systems research by Sugeno and Ebrahim Mamdani in the 1990s advanced the field. Fuzzy inference systems of the Sugeno and Mamdani types have evolved into industry standards for fuzzy controller and decision-making system design. These frameworks provided systematic methods for incorporating expert knowledge into fuzzy systems [5].

The contributions of notable figures in the field of fuzzy logic cannot be overstated. Apart from Lotfi Zadeh, other key figures include Ebrahim Mamdani, whose work on fuzzy controllers was foundational, and Michio Sugeno, who developed the Sugeno-type fuzzy inference system. Their work has had a long-lasting influence on fuzzy logic's theoretical advancement and real-world applications [6].

In more recent years, researchers like Jerry Mendel have advanced the field through their work on type-2 fuzzy logic systems, which provide enhanced capabilities for handling uncertainty and imprecision. Their research has opened new avenues for applying fuzzy logic to increasingly complex and dynamic problems.

4. Modern Developments and Future Directions

Fuzzy logic has been developing over the last ten years, and new uses in domains including big data analytics, the Internet of Things (IoT), and autonomous systems have emerged. Fuzzy logic has been combined with deep learning and machine learning approaches to produce strong hybrid systems that can handle a variety of real-world problems. In terms of the future, fuzzy logic's flexibility and capacity to work with new technology will determine its success. Research is ongoing in areas such as explainable AI, where fuzzy logic's capacity for interpretability and transparency offers significant advantages. Additionally, advancements in computational power and algorithmic efficiency are expected to further enhance the capabilities and applications of fuzzy systems.

2.2 Key Concepts and Terminologies

The fundamental ideas and vocabulary of fuzzy logic, including fuzzy sets, membership functions, and linguistic variables, are introduced in this section. It explains how these elements interact within a fuzzy system to process information and make decisions. By breaking down these fundamental components, readers will gain a clear understanding of the building blocks of fuzzy systems.

1. Fuzzy Sets

The idea of a fuzzy set is central to fuzzy logic. A fuzzy set permits partial membership, in contrast to classical sets where an element can only be in the set or out of it (binary membership). A fuzzy set element may have a membership value between 0 and 1, where 1 denotes full membership and 0 denotes no membership. This allows for the representation of uncertainty and gradations of truth.

Example: If the cutoff is 6′0″, a person who is 5′11″ might not be included in a traditional group of "Tall people." In a fuzzy set, this person could have a membership value of 0.8, indicating they are somewhat tall.

2. Membership Functions

Every point in the input space is mapped to a membership value between 0 and 1, according to a membership function. Triangular, trapezoidal, and Gaussian membership functions are examples of common function types. In fuzzy systems, these functions are used to represent linguistic variables.

Example: For the fuzzy set "Medium height," a triangle membership function might be employed, with high membership values for heights approximately 5′9″ and lower values for heights much shorter or taller.

3. Linguistic Variables

Linguistic variables are those whose values are natural language words or sentences rather than numerical values. Every language variable has a fuzzy set assigned to it.

Example: In a temperature control system, "Temperature" might be a linguistic variable with fuzzy sets such as "Cold," "Warm," and "Hot."

4. Fuzzy Rules

Conditional statements known as fuzzy rules specify how to carry out control actions in response to the values of linguistic variables. They are usually written in the form of "IF–THEN" statements.

For example, "IF temperature is cold, then set heater to high" could be a fuzzy rule for a heating system.

5. Fuzzy Inference Systems (FIS)

A Fuzzy Inference System (FIS) is a framework for mapping inputs to outputs using fuzzy logic. It involves several steps: fuzzification (converting inputs into fuzzy values), applying fuzzy rules, aggregation (combining the results of all rules), and defuzzification (converting the fuzzy output back into a crisp value) [7].

6. Types of FIS:

Mamdani FIS: produces outcomes that are easy to understand and analyze by using fuzzy sets for both input and output.

Sugeno FIS: offers computational efficiency by using fuzzy sets as input and linear functions as output.

7. Fuzzification and Defuzzification

The process of converting precise input values into membership degrees for fuzzy sets is known as "fuzzification." The opposite procedure, known as defuzzification, transforms fuzzy outputs into precise values that can be utilized as real system inputs or choices [8].

Example: In a temperature control system, the crisp input temperature is converted into degrees of membership in fuzzy sets like "Cold" and "Warm." After processing through fuzzy rules, the fuzzy output is converted back to a specific heater setting.

8. Fuzzy Clustering

A type of clustering known as fuzzy clustering allows each data point to have a possible membership in several groups at different levels. For this, fuzzy c-means (FCM) is a popular algorithm.

Example: In image segmentation, fuzzy clustering can assign pixels to multiple regions with different degrees of membership, allowing for smooth transitions between segments.

9. Fuzzy Control Systems

Fuzzy logic is used by fuzzy control systems to monitor and regulate processes. They are particularly useful in systems where mathematical models are difficult to derive.

Example: An anti-lock braking system (ABS) in vehicles uses fuzzy logic to control brake pressure, ensuring optimal braking performance under various road conditions [9].

10. Fuzzy Decision Making

Making decisions in a hazy and unclear context is known as fuzzy decision making. Fuzzy logic is used to assess several factors and arrive at the optimal choice.

Example: In project management, fuzzy decision making can help in selecting the best project based on fuzzy criteria like cost, risk, and benefit.

11. Fuzzy Pattern Recognition

Fuzzy logic is used in fuzzy pattern recognition to find patterns and classify information based on imprecise data.

Example: In handwriting recognition, fuzzy pattern recognition can classify handwritten characters even when they vary significantly from the standard form.

12. Advanced Concepts

Type-2 Fuzzy Sets

By permitting fuzzy membership values, type-2 fuzzy sets expand on conventional (type-1) fuzzy sets. This gives the modeling of uncertainty an extra degree of freedom.

Example: In decision making under high uncertainty, type-2 fuzzy sets can provide more robust modeling of imprecise criteria.

13. Fuzzy Neural Networks

Fuzzy neural networks blend the interpretability of fuzzy logic with the learning power of neural networks. These hybrid models have the ability to learn from data and express the information as fuzzy rules [10].

Example: A fuzzy neural network might be used in a medical diagnosis system, learning from patient data to generate interpretable diagnostic rules.

14. Fuzzy Genetic Algorithms

Fuzzy genetic algorithms use principles of natural evolution (selection, crossover, mutation) combined with fuzzy logic to solve optimization problems. They are effective in complex search spaces where traditional methods struggle.

Example: In engineering design, fuzzy genetic algorithms can optimize design parameters under uncertain constraints. Fuzzy Bayesian networks integrate fuzzy logic with Bayesian probability theory to model uncertain and imprecise knowledge. They are useful in scenarios where uncertainty is both stochastic and fuzzy.

Example: In risk assessment, fuzzy Bayesian networks can model the probability of different risk levels considering imprecise input data.

2.3 Mathematical Foundations of Fuzzy Logic

Fuzzy logic is grounded in mathematical principles that allow it to handle imprecision and uncertainty. This section covers the mathematical underpinnings of fuzzy logic, including set theory, fuzzy set operations, and fuzzy inference mechanisms. Through mathematical expressions and examples, the section elucidates how fuzzy logic operates at a computational level.

1. Fuzzy Sets and Functions of Membership

Membership functions are used in fuzzy logic to express sets. These functions give each element in the discourse universe a degree of membership. To reflect different kinds of uncertainty, these membership functions can be triangular, trapezoidal, or Gaussian. A fuzzy set A in a discourse universe X is mathematically characterized by its membership function $\mu(x)$, which associates a degree of membership between 0 and 1 with each element x in X.

2. Operations on Fuzzy Sets

Fuzzy logic extends traditional set operations to fuzzy sets. The membership functions of the relevant fuzzy sets are used to define the union, intersection, and complement operations. For instance, the maximum of each fuzzy set's membership function defines the union of two fuzzy sets, A and B: $(A - B)(x) = \max(\mu A(x), \mu B(x))$.

3. Fuzzy Logic Connectives

Fuzzy logic introduces connectives to combine and manipulate fuzzy propositions. The connectives AND ($\wedge$), OR ($\vee$), and NOT ($\neg$) are the most often utilized ones. T-norms, T-conorms, and negations are used, respectively, to define these connectives. For instance, the minimum of two fuzzy propositions A and B's membership functions is the fuzzy AND operation of those two propositions: $(A \wedge B)(x) = \min(\mu A(x), \mu B(x))$.

4. Fuzzy Rules and Inference

Language variables are used to represent rules in fuzzy logic systems as IF-THEN expressions. These guidelines are combined in fuzzy inference to produce a clear output. The Mamdani model, which use fuzzy logic to determine each rule's degree of satisfaction and then aggregates these degrees of satisfaction to get a fuzzy output, is the most often used technique for fuzzy inference.

5. Defuzzification

The process of turning a fuzzy output into a crisp value for use in control or decision-making is known as defuzzification. There are several defuzzification techniques, including the maximum membership principle, the centroid method, and the weighted average method. These methods calculate a single crisp value from the fuzzy output, representing the "best" or "most representative" value.

6. Fuzzy Logic Control

Fuzzy logic is a control technology called fuzzy logic control (FLC), which is used to model and control nonlinear systems. FLC systems consist of four main components: fuzzification, rule evaluation, aggregation, and defuzzification. FLC systems are particularly useful in applications where precise mathematical models are difficult to obtain or where systems exhibit nonlinear behavior.

7. Fuzzy Mathematics

Traditional mathematical ideas are expanded to allow for fuzzy sets and fuzzy reasoning in fuzzy mathematics. Fuzzy differential equations, fuzzy algebra, fuzzy calculus, and fuzzy arithmetic are all included. These mathematical tools provide formal frameworks for reasoning and decision-making under uncertainty.

8. Applications of Fuzzy Logic

Numerous domains, such as artificial intelligence, pattern recognition, decision support systems, expert systems, control systems, and decision support systems, have found use for fuzzy logic. Its ability to model and reason with uncertainty makes it particularly well-suited for applications where traditional binary logic and crisp sets are inadequate.

9. Future Directions

The mathematical foundations of fuzzy logic continue to evolve, with ongoing research focused on extending fuzzy set theory to handle more complex and dynamic systems. Further developments in fuzzy logic include the creation of sophisticated algorithms, the incorporation of fuzzy logic into other computing paradigms (such neural networks and evolutionary algorithms), and the use of fuzzy logic in newly developing fields like big data analytics and Internet of Things systems.

3 Methodologies and Algorithms in Fuzzy Systems

3.1 Fuzzy Rule-Based Systems

Fuzzy logic applications are based on fuzzy rule-based systems. The construction and operation of these systems, including the development of fuzzy rules, the layout of rule bases, and the application of inference processes, are examined in this section. Examples of rule-based systems in real-world applications will be provided to illustrate their practical utility.

3.2 *Fuzzy Inference Systems (FIS)*

For the purpose of using fuzzy logic principles in decision-making, fuzzy inference systems, or FIS, are indispensable. This section discusses different types of FIS, such as Mamdani and Sugeno models, and their respective methodologies. The section will provide step-by-step explanations of the fuzzy inference process, including fuzzification, rule evaluation, aggregation, and defuzzification.

3.3 *Hybrid Fuzzy Systems*

To improve performance and adaptability, hybrid fuzzy systems integrate fuzzy logic with other computational methods like neural networks and evolutionary algorithms. This section examines various hybrid methodologies, highlighting their advantages and applications. Case studies will be presented to demonstrate how hybrid systems can solve complex problems more effectively than traditional approaches.

4 Applications F Fuzzy Systems

4.1 *Control Systems*

Process control and industrial automation both make extensive use of fuzzy control systems. This section examines the use of fuzzy logic in the design of controllers for situations that are uncertain and dynamic. Examples include temperature control, robotic navigation, and automotive systems, showcasing the versatility and effectiveness of fuzzy control.

1. Components of Control Systems

Plant: The physical system or process being controlled is known as the plant. It might include anything from straightforward mechanical systems like pumps and motors to intricate ones like chemical reactors and aeroplanes.

Controller: The controller is the brain of the control system, responsible for processing information from sensors, computing control actions, and sending commands to actuators. Controllers can be implemented using various techniques, including classical control methods, modern control theory, and intelligent control approaches.

Sensors: Sensors are tools that measure the output of the system or pertinent variables and report back to the controller. Common types of sensors include temperature sensors, pressure sensors, accelerometers, and encoders.

Actuators: Actuators are devices that act upon the system to produce control actions based on commands from the controller. Examples of actuators include motors, valves, solenoids, and hydraulic cylinders.

Feedback Loop: Feedback is an essential feature of control systems, enabling them to continuously adjust the system's behavior based on measured output. Feedback can be negative (to stabilize the system) or positive (to amplify or reinforce certain behaviors).

2. Types of Control Systems

Open-loop Control Systems: In open-loop control systems, the input command alone determines the control action, disregarding the output of the system. These systems are usually employed for simple procedures where stability and precision are not as important, and they do not include feedback.

Closed-loop (Feedback) Control Systems: Closed-loop control systems modify the control action based on input from the system's output. The accuracy, stability, and resilience of the system against uncertainties and disruptions are all enhanced by this feedback loop.

3. Control System Design Methods

Classical Control Methods: Control systems are often designed using classical control techniques including frequency response analysis, root locus analysis, and PID (Proportional-Integral-Derivative) control. These methods are based on linear models and algebraic analysis.

Modern Control Theory: Modern control theory encompasses state-space representation, optimal control, and robust control techniques. It offers more advanced tools for designing control systems, particularly for nonlinear and time-varying systems.

4. Intelligent Control Approaches

Intelligent control approaches that leverage machine learning and artificial intelligence to manage complex and uncertain systems include adaptive control, neural network control, and fuzzy logic control. These methods excel in situations where traditional control approaches may struggle.

5. Applications of Control Systems

Industrial Automation: To maximize production efficiency, guarantee product quality, and improve safety, control systems are widely utilized in manufacturing processes, robotics, and industrial machinery.

Aerospace and Defense: Control systems play a critical role in aircraft flight control, missile guidance systems, unmanned aerial vehicles (UAVs), and spacecraft navigation, ensuring stable and precise operation.

Automobile Systems: Engine control, electronic stability control (ESC), adaptive cruise control (ACC), anti-lock braking systems (ABS), and other control systems are essential to a number of automobile applications that enhance vehicle performance, safety, and fuel economy.

Home and Building Automation: Control systems are increasingly used in smart homes and building automation systems to regulate temperature, lighting, security, and energy consumption, providing comfort, convenience, and efficiency.

6. Emerging Trends in Control Systems

Networked Control Systems (NCS): NCS integrate control systems with communication networks, enabling distributed control, remote monitoring, and collaborative decision-making over large-scale systems.

Cyber-Physical Systems (CPS): CPS combine physical processes with computational and communication capabilities, leading to autonomous and interconnected systems with applications in smart cities, healthcare, and industrial IoT.

Machine Learning for Control: Control systems are increasingly using machine learning techniques, such as deep learning, reinforcement learning, and predictive analytics, for anomaly detection, predictive maintenance, and adaptive and autonomous control.

4.2 Decision-Making Systems

Fuzzy logic enhances decision-making processes by accommodating uncertainty and subjective judgment. This section discusses the application of fuzzy systems in decision support systems, risk assessment, and resource allocation. Through practical examples, the section illustrates how fuzzy logic can improve the quality and reliability of decisions in various domains.

1. Types of Decision-Making Systems

Operational Decision Support Systems (ODSS): ODSS provide decision support for routine or operational decisions in day-to-day business operations. They focus on transactional data, real-time monitoring, and operational efficiency, helping users make decisions quickly and effectively.

Tactical Decision Support Systems (TDSS): TDSS support mid-level managers and planners in making tactical decisions that impact medium-term goals and objectives. These systems analyze historical data, forecast future trends, and optimize resource allocation to achieve organizational objectives.

Strategic Decision Support Systems (SDSS): SDSS assist senior executives and strategic planners in making high-level strategic decisions that shape the long-term direction of an organization. These systems analyze external market trends, competitive intelligence, and scenario planning to guide strategic planning and decision-making.

2. Applications of Decision-Making Systems

Business and Finance: Decision-making systems are widely used in business and finance for portfolio management, risk analysis, investment decisions, supply chain optimization, pricing strategies, and customer relationship management (CRM).

Healthcare: In healthcare, decision-making systems support clinical decision-making, diagnosis, treatment planning, patient monitoring, and healthcare management. To enhance patient outcomes and operational efficiency, these systems make use of clinical decision support systems (CDSS), electronic health records (EHRs), and medical imaging.

Engineering and Operations: Decision-making systems are essential in engineering and operations for process optimization, predictive maintenance, quality control, logistics planning, and resource allocation. These solutions assist businesses in increasing productivity, cutting expenses, and streamlining processes.

3. Emerging Trends in Decision-Making Systems

Explainable AI (XAI): XAI techniques aim to make AI and machine learning models more transparent and interpretable, enabling users to understand how decisions are made and trust the recommendations provided by decision-making systems.

Ethical Decision-Making: With increasing concerns about fairness, accountability, and bias in AI systems, ethical considerations are becoming integral to decision-making systems design. To make sure that decisions are in line with moral standards and society norms, frameworks and algorithms for ethical decision-making are being developed.

Human-AI Collaboration: Decision-making systems are evolving to facilitate collaboration between humans and AI, combining the strengths of both to improve decision-making outcomes. These systems leverage human expertise, intuition, and creativity, while AI algorithms provide data-driven insights and recommendations.

4.3 Pattern Recognition and Image Processing

Fuzzy systems play a significant role in pattern recognition and image processing, where ambiguity and noise are prevalent. This section covers methodologies for applying fuzzy logic to tasks such as image segmentation, feature extraction, and object recognition. Real-world applications in medical imaging, facial recognition, and remote sensing will be presented to highlight the impact of fuzzy systems in this field.

1. Components of Pattern Recognition and Image Processing

Feature extraction is the process of locating and obtaining pertinent features or attributes from unprocessed data, like signals or images. These features may include edges, corners, textures, colors, shapes, or statistical descriptors, which are used to represent and differentiate objects or patterns in the data.

Following their extraction, features must be expressed in an appropriate manner in order to be used for additional analysis or categorization. This step involves encoding the extracted features into a feature vector or matrix representation. The input data is then categorized or labeled using classification techniques, such as support vector machines (SVM), k-nearest neighbors (KNN), decision trees, and deep neural networks, depending on these properties.

Image Enhancement and Restoration: Image processing techniques are employed to enhance the quality of digital images, remove noise, correct distortions, and improve visual clarity. Histogram equalization, denoising, sharpening, and contrast modification are common image enhancing techniques. With the use of modeling and compensation for image degradation processes, image restoration approaches seek to restore the original image from distorted or degraded versions.

Segmentation and Object Detection: Segmentation is the process of dividing an image according to specific parameters, including color, texture, or intensity, into meaningful areas or objects. Object detection algorithms identify and localize specific objects or regions of interest within an image, enabling machines to understand and interact with their surroundings. Techniques such as thresholding, edge detection, region growing, and deep learning-based object detection are commonly used for these tasks.

Pattern Recognition and Image Processing Applications

Medical Imaging: In medical imaging applications such as CT (computed tomography) scanning, MRI (magnetic resonance imaging), X-ray analysis, and histopathology, pattern recognition and image processing are essential. These techniques assist in disease diagnosis, treatment planning, surgical guidance, and medical research.

Surveillance and Security: Pattern recognition and image processing are used in surveillance systems for monitoring, detecting, and tracking objects or activities in real-time video streams. These systems are deployed in public spaces, airports, transportation hubs, and border security to enhance safety and security.

Autonomous Vehicles: Image processing and pattern recognition are essential for autonomous vehicles (AVs) to perceive and interpret their environment. AVs use cameras, LiDAR (light detection and ranging), and radar sensors to detect obstacles, recognize traffic signs and signals, and navigate safely in complex traffic scenarios.

Biometrics: Pattern recognition techniques are widely used in biometric systems for identity verification and authentication. Pattern recognition algorithms are used by biometric modalities including voice, facial, iris, and fingerprint recognition to compare biometric traits to templates that have been saved and confirm individual identities.

2. Emerging Trends in Pattern Recognition and Image Processing

Deep Learning: Convolutional neural networks (CNNs), in particular, are deep learning algorithms that have revolutionized image processing and pattern recognition tasks. They have achieved state-of-the-art performance in a variety of applications, including semantic segmentation, object detection, and picture classification.

GANs, or Generative Adversarial Networks: By learning from actual data distributions, a type of deep learning models called GANs may produce convincing synthetic images. GANs have applications in image synthesis, style transfer, image-to-image translation, and data augmentation.

Explainable AI (XAI): The need for explainable AI techniques is developing as deep learning models get more intricate and opaque, offering insights into model decisions and improving the transparency and interpretability of image processing and pattern recognition systems.

5 Advancements in Fuzzy Systems

5.1 Recent Developments in Fuzzy Theory

This section reviews the latest theoretical advancements in fuzzy logic, including new mathematical models and computational techniques. We'll talk about innovations like type-2 fuzzy logic and intuitionistic fuzzy sets and offer some insights into their possible uses and advantages.

Applications for type-2 fuzzy logic can be found in a number of areas, such as pattern recognition, control systems, and decision-making, where accurate modeling of uncertainty is essential for reliable performance. For example, in control systems, type-2 fuzzy logic controllers have been shown to outperform their type-1 counterparts in handling nonlinearities and uncertainties in dynamic environments. Similarly, in decision-making scenarios, type-2 fuzzy sets provide a richer representation of uncertainty, enabling more informed and confident decisions in complex and uncertain situations.

The advent of intuitionistic fuzzy sets, which enhance conventional fuzzy sets with an extra degree of uncertainty known as non-membership, is another new breakthrough in fuzzy theory. The reluctance or ambiguity around components that do not belong to the set is represented by a third parameter in intuitionistic fuzzy sets, which is different from classical fuzzy sets, which only take membership and non-membership degrees into account. This extension enhances the expressiveness of fuzzy logic models, particularly in domains where decision-makers exhibit hesitancy or lack of confidence in their judgments.

In many domains where ambiguity and vagueness are inherent, such as image processing, medical diagnosis, and decision-making, intuitionistic fuzzy sets have been used. An much more thorough framework for modeling and reasoning under uncertainty is provided by intuitionistic fuzzy sets, which capture not only the degree of membership but also the degree of non-membership and reluctance. For example, in medical diagnosis, where decisions are often based on incomplete or uncertain information, intuitionistic fuzzy sets enable more accurate and interpretable diagnoses by accounting for both positive and negative evidence.

5.2 Algorithmic Innovations

Advancements in algorithms are crucial for enhancing the performance of fuzzy systems. This section explores recent developments in fuzzy algorithms, including optimization techniques, learning algorithms, and real-time processing methods. Examples of cutting-edge algorithms and their applications will be provided to demonstrate their effectiveness.

One notable area of advancement is in optimization techniques tailored for fuzzy systems. Fuzzy logic controllers, rule-based systems, and fuzzy inference systems have all benefited from the adaptation and extension of conventional optimization techniques including genetic algorithms, particle swarm optimization, and simulated annealing. In order to enhance system performance and adaptability in dynamic situations, these optimization techniques allow the automatic tweaking of fuzzy system characteristics, such as membership functions, rule weights, and rule structures.

Fuzzy systems have also been transformed by advances in learning algorithms, which allow them to learn from data and adjust to changing circumstances. Fuzzy logic has been combined with machine learning techniques like neural networks, support vector machines, and deep learning models to develop hybrid intelligent systems that can recognize membership functions and fuzzy rules in data. These hybrid approaches improve the modeling capabilities and predictive accuracy of fuzzy systems in a variety of applications, such as pattern recognition, control systems, and decision-making, by utilizing the advantages of both fuzzy logic and machine learning.

Moreover, real-time processing methods have been developed to enable efficient and responsive operation of fuzzy systems in dynamic environments. Real-time fuzzy control algorithms leverage parallel processing, distributed computing, and hardware acceleration techniques to achieve low-latency and high-throughput execution of fuzzy inference tasks. These algorithms are particularly relevant in applications requiring rapid decision-making, such as autonomous vehicles, robotics, and industrial automation systems, where timely responses are critical for ensuring safety and efficiency.

Adaptive neuro-fuzzy inference systems (ANFIS), which integrate fuzzy logic with neural networks for data-driven modeling and control, and online learning algorithms for dynamic fuzzy systems that can adjust in real-time to changing environmental conditions are two examples of very advanced algorithms. These algorithms have been successfully applied in various domains, including intelligent transportation systems, energy management, and fault diagnosis, demonstrating their effectiveness in addressing complex real-world problems.

5.3 Integration with Emerging Technologies

The integration of fuzzy systems with cutting-edge technologies like big data analytics, blockchain, and the Internet of Things (IoT) is becoming more common. This section examines the synergies between fuzzy logic and these technologies, highlighting innovative applications and future research directions.

1. Synergies Between Fuzzy Logic and Emerging Technologies

Internet of Things (IoT):

Fuzzy logic plays a vital role in IoT systems by enabling intelligent decision-making and control in interconnected devices and networks.

IoT devices often operate in dynamic and uncertain environments, where traditional control methods may fall short. IoT systems may handle uncertainty with flexibility and adaptability by using fuzzy logic, which enables them to make context-aware judgments based on imprecise or partial data.

Applications include smart home automation, industrial IoT (IIoT) for predictive maintenance and process optimization, and smart healthcare systems for remote patient monitoring and personalized treatment.

2. Blockchain:

Fuzzy logic allows for sophisticated data processing and decision making, which can improve the efficiency and dependability of blockchain-based systems.

Fuzzy logic can be used to interpret and reason about complicated situations and events in blockchain-based smart contracts and decentralized apps (DApps), which increases the expressiveness and flexibility of smart contract logic.

Fuzzy logic can also be applied in blockchain-based supply chain management systems for optimizing logistics, managing uncertainties in supply chain operations, and ensuring transparency and trust among stakeholders.

3. Big Data Analytics:

Fuzzy logic techniques are well-suited for handling the uncertainty and imprecision inherent in big data analytics tasks, such as data preprocessing, clustering, and classification.

Fuzzy clustering algorithms can partition large datasets into meaningful groups based on similarity and proximity, enabling insights and patterns to be extracted from complex and heterogeneous data sources.

Fuzzy inference systems can be integrated with big data analytics platforms to perform intelligent data processing tasks, such as anomaly detection, sentiment analysis, and recommendation systems, by combining fuzzy logic reasoning with statistical and machine learning techniques.

4. Innovative Applications and Future Directions

Smart Cities: Integration of fuzzy logic with IoT technologies for smart city applications, such as intelligent transportation systems, energy management, and environmental monitoring, to optimize resource utilization, enhance sustainability, and improve quality of life for residents.

Healthcare: Leveraging fuzzy logic and big data analytics for personalized medicine and healthcare decision support systems, enabling tailored treatment plans and predictive analytics for disease prevention and management.

Finance: Utilizing fuzzy logic and blockchain technology for algorithmic trading, fraud detection, and risk management in financial markets, enhancing decision-making processes and mitigating financial risks.

Cybersecurity: Integrating fuzzy logic with blockchain-based cybersecurity solutions for threat detection, intrusion detection, and anomaly detection in network security, providing adaptive defense mechanisms against evolving cyber threats.

6 Conclusion

To sum up, this work offers a thorough investigation of the hazy boundaries found in computational systems. Through the presentation of current achievements in fuzzy logic and fuzzy systems, the study has helped close the gap between theoretical breakthroughs and real-world applications. Through a review of innovative methodologies and algorithms, it has demonstrated the adaptability and uncertainty management capabilities of fuzzy logic. Furthermore, by demonstrating applications in a variety of domains, including pattern recognition, decision-making, and control systems, this study has shown the adaptability and potency of fuzzy systems in handling challenging real-world problems. Moving forward, the insights provided in this paper aim to facilitate the effective deployment of fuzzy logic, contributing to the advancement of computational systems and their application in various domains.

References

1. L.A. Zadeh, Fuzzy sets. Inf. Control **8**(3), 338–353 (1965)
2. J. Bezdek, *Pattern Recognition with Fuzzy Objective Function Algorithms* (Plenum Press, New York, 1981)
3. T. Takagi, M. Sugeno, Fuzzy identification of systems and its applications to modeling and control. IEEE Trans. Syst. Man Cybern. **15**(1), 116–132 (1985)
4. J. Yen, R. Langari, Fuzzy Logic: Intelligence, Control, and Information. Prentice Hall (1999)
5. P.P. Wang, H.H. Wang, Fuzzy control systems: design, analysis, and performance evaluation. IEEE Trans. Fuzzy Syst. **4**(2), 103–114 (1996)
6. K. Tanaka, T. Ikeda, H. Wang, Fuzzy regulators and fuzzy observers: relaxed stability conditions and LMI-based designs. IEEE Trans. Fuzzy Syst. **6**(2), 250–265 (1998)
7. G.J. Klir, B. Yuan, Fuzzy Sets and Fuzzy Logic: Theory and Applications. Prentice Hall (1995)
8. E.H. Mamdani, Application of fuzzy algorithms for control of simple dynamic plant. Proc. Inst. Electrical Eng. **121**(12), 1585–1588 (1974)
9. R.R. Yager, D.P. Filev, Essentials of Fuzzy Modeling and Control. Wiley (1994)
10. W. Pedrycz, F. Gomide, An Introduction to Fuzzy Sets: Analysis and Design. MIT Press (1998)

Application of the Internet of Things and Multimedia Data Towards a Smart City

Risa Triassanti, Mohammed Ayad Alkhafaji, and Ahmad Al Yakin

Abstract Smart Cities provide a comprehensive framework for smart city projects. The urban network helps cities gain economic power, develop unique technological solutions, and build efficient systems. As part of the smart economy and information technology, smart cities are an advanced development product. A wide range of networking technologies are combined in the system, including wireless networks, broadcast networks, mesh networks, telecommunication networks, and a sensor network that connects everything in the system. Besides integrating the many sensors into everyday objects, the Internet of Things also uses specific protocols to exchange information and communications for tracking, managing, monitoring, and recognizing intelligent achievements. Smart cities are not only supported, but also interconnected. An IoT-based machine learning approach to developing a smart city is explored in this chapter.

Keywords Internet of Things · Smart city · Machine learning · Intelligent system · Intelligent transport system

1 Introduction

According to a Cisco report released in 2008, things related to the Web have exceeded the breaking point for individuals [1]. Additionally, it states that by 2020, these things will reach the limit of fifty billion, propelling us towards advanced civilization. In the Web of Things (IoT), these devices collaborate and communicate via the internet. As

R. Triassanti (✉)
Universitas PGRI Ronggolawe Tuban, Tuban, Jawa Timur, Indonesia
e-mail: triassantirisa@gmail.com

M. A. Alkhafaji
College of Technical Engineering, National University of Science and Technology, Dhi Qar, Nasiriyah, Iraq

A. Al Yakin
Universitas Al Asyariah Mandar, Polewali Mandar, Indonesia

R. Sharma and G. Jeon (eds.), *Data Analytics for Smart Robotics and Its Applications*, Intelligent Systems Reference Library 272,
https://doi.org/10.1007/978-3-031-87697-4_8

a result of the IoT infringement, people are not able to expect everyday comforts in a useful manner. In addition to medicinal services [2], mechanization, transportation, and crisis reaction to various scenarios when humans find themselves incapable of making decisions, it also covers the field of mechanization. It is unlikely that the Internet will ever be defined as a network of computers in its traditional sense again.

There are currently no specific technologies or sets of technologies that make up Smart Cities. Contrary to popular belief, these technologies are mostly related to the generation of data about the city and allowing interaction within it. The term "smart" refers to the integration of various types of information into decision-making procedures through rules and algorithms of various complexity. The development of sensor networks in smart cities requires a wide range of technologies, both on a hardware level (transducers, signal conditioning, communications links, energy harvesting, etc.) as well as a software level (such as data presentation, relocation, and web integration).

Smart cities are those that apply advanced technology to improve traditional networks and services for their residents' benefits [3]. Urban urbanization and the growth of the urban population are driving the implementation of smart cities to meet the needs of the expanding population and improve living conditions [4]. The goal of a smart city is not just to conserve resources and reduce pollution; it involves a more responsive and interactive government system, better water and waste disposal facilities, more energy-efficient lighting and heating systems, and safer public spaces as well. The IoT is therefore used in smart cities to collect measurements (time-series data) regularly (sensors, embedded systems, smart meters) and analyze them to improve services and infrastructure [5].

A major strength of deep learning is the ability to extract intricate patterns from complicated datasets [6], which is particularly useful when dealing with smart cities' time-series data. In this way, algorithms are capable of capturing multiple time series simultaneously, harnessing their interactions, and providing more accurate forecasts as a result [7]. The application of deep learning techniques has been found to be beneficial in a wide variety of time-series forecasting scenarios. These include retail markets [8], healthcare markets [9], biology markets [10], medicine markets [11], as well as aviation markets [12], energy markets [13], climate markets [14], the automotive industry [15], and finance markets [16].

2 Related Work

The Internet of Things has developed interestingly over the last few years [17]. Especially in vehicular environments, IoT is being adopted. Autonomous vehicles are raising global concerns in the automotive industry. In the future, IoT and its frameworks will bring about a new era of automotive technology. Similarly, the development of automobiles is also anticipated to be the next frontier of IoT [18]. Self-driving vehicles can be controlled via a distributed IoT system [19]. Using a distributed IoT system, we investigate how delay-accuracy affects the real-time performance. As a result of the authors' assumptions, self-driving vehicles and roadside infrastructure could play a key role in improving intersection efficiency and reducing fuel consumption's environmental impact. An emerging area of research is introduced within the Quality of IoT-experience theme in automated vehicles [20, 21]. Aside from the end-users affected by the evaluation, the ecosystem itself must be considered when evaluating the quality. In this work, a novel architecture for IoT in autonomous vehicles is presented that considers data quality, network quality, and contextual quality. As a result of the convergence of IoT and AI, vehicles will become fully autonomous for Industry 4.0, which bridges the gap between the physical and digital worlds [22]. Compared to this [23], examines the challenges associated with integrating connected vehicles, cloud computing, and IoT technology. By integrating ITS, we can solve issues such as fuel costs, high carbon emissions, and traffic congestion [24]. Automation, and particularly the IoT, are important tools for cities to meet the requirements of the citizenry while managing costs in the current environment. IoT is indeed capable of addressing or ameliorating many city challenges. A graphic depiction of some of these applications can be found in Fig. 1.

While there is no formal and universal definition of the concept of a "Smart City," [25], explains that a smart city is one in which all infrastructure and services are integrated into a cohesive whole and monitored and controlled with intelligent devices, in order to operate sustainably and intelligently. It is, therefore, evident that

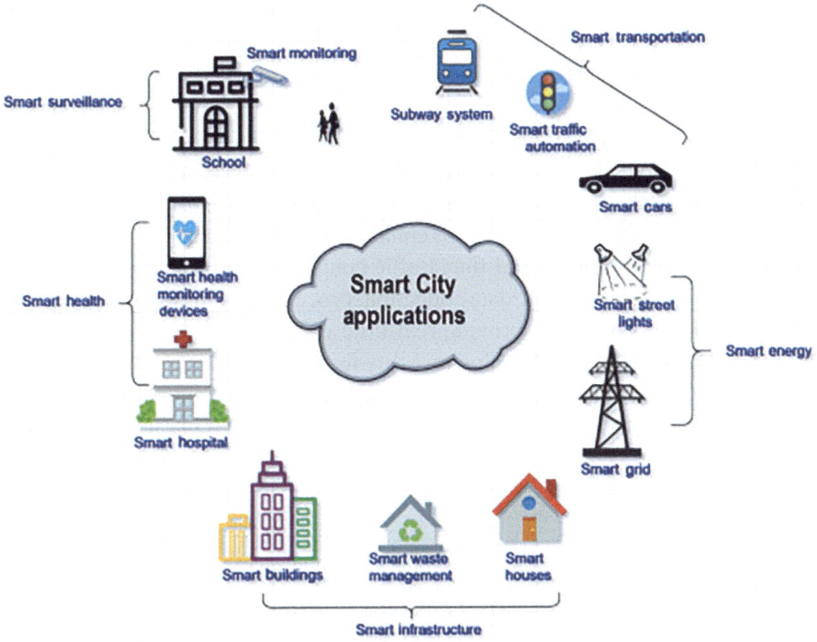

Fig. 1 IoT applications for smart cities illustrated in graphics

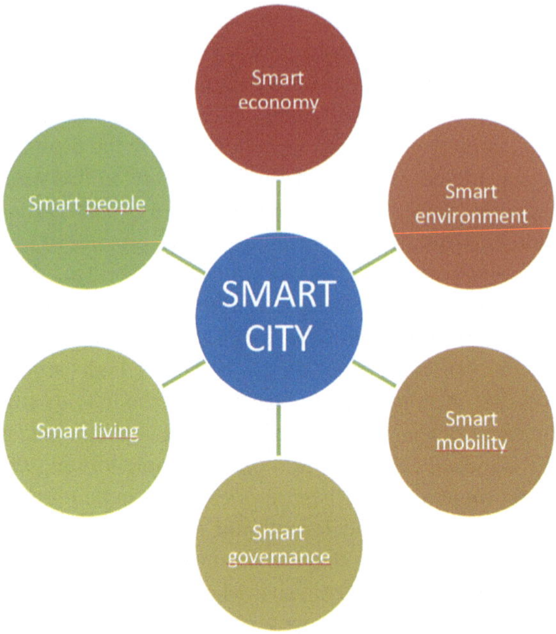

Fig. 2 Concept of smart cities

a Smart City (in Fig. 2) should be interoperable across the various IoT deployments, which, at present, are mainly closed and vertically integrated.

Researchers propose these kinds of solutions using a variety of IoT sensors. This paper proposes a cluster-based approach for monitoring road conditions using real-time traffic data obtained with IR sensors [26, 27]. With the help of an Android device, the algorithm uses a clustering technique to perform algorithmic computations. Through Bluetooth technology, drivers can receive the data processing results through a mobile application. To enhance mobility, the system uses roadside messaging devices to provide real-time traffic congestion information.

We propose a real-time traffic data collection, processing, and storage model based on IoT [28]. This system can provide real-time information about traffic congestion and unusual incidents, enhancing mobility throughout the region. An integrated magnetic sensor node measures road conditions in real time and stores them in a MongoDB database for further analysis. Using Wi-Fi-enabled platforms, the microcontrollers collect data, which is then sent to IoT platforms. During rush hour, the message unit can display early warning messages that will help commuters and drivers avoid traffic jams.

3 Proposed Methodology

As IoT technology advances rapidly and is applied more deeply in modern science and technology, ITS and IoT technology are becoming closer together [25]. As IoT technology is introduced, ITS functions will be enhanced, and transportation services will be more convenient and comfortable [29]. In this section, ITS service models based on IoT are implemented by analyzing the needs of different travellers as well as ITS features, and a platform for traffic control and decision support systems is developed. By using IoT, transport facilities will be more efficient. A diagram of ITS's control center functions and services can be found in Fig. 3 [30].

Using this construction, traffic data can be collected in real-time and traffic analysis can be generated. It is possible to predict traffic congestion based on road conditions and traffic flow, and travelers can receive real-time travel information if needed. Using the application services, cargo transportation tracking can be paired with real-time transportation safety services. The construction and maintenance management process includes regular collection and processing of data about static facilities and environmental conditions. Operations control is one of the most important components of ITS, as it provides a mechanism for controlling traffic and planning routes by analyzing and controlling information [31].

In IoT-based intelligent transportation applications, perception layers consist of vehicles, personnel, environments, and other infrastructure. Sensors on road signs and beacons collect road information. Sharing and utilizing data are key components of the network layer. A network layer includes satellites, computer networks, wireless mobile communication technologies, and other methods of transmitting traffic data. Application-layer services can include traffic flow-based signal light adjustments, real-time traffic information inquiry, and traffic management and control. ITS based on IoT allows traffic information to be collected, transmitted, controlled, and applied. The service level of ITS will continue to improve as a result of IoT technology. According to Fig. 4, IoT-based ITS modules include the following aspects.

Having collected, transmitted, processed, and analyzed the data, it can offer traffic travelers a variety of information security services. Information can be fully utilized, traffic congestion situation can be released timely, best routes can be mapped out,

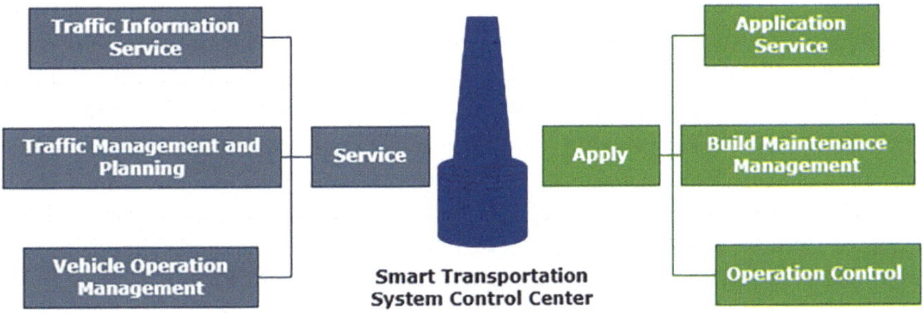

Fig. 3 A functional model for ITS services

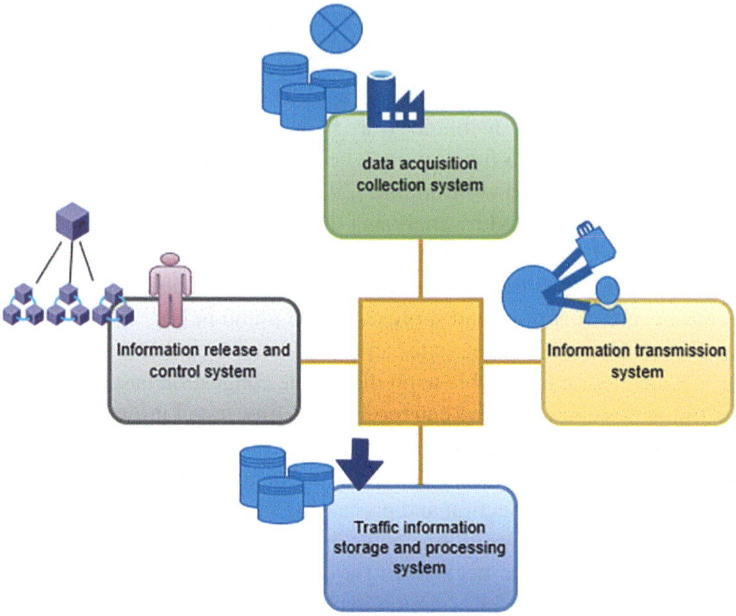

Fig. 4 ITS module based on IoT

and travelers can use it as a reference. Reduce traffic congestion by controlling traffic lights' length reasonably based on main road traffic flow information. Information about road conditions for vehicle faults can be released to accident handling agencies in a timely fashion [32].

3.1 AI Inspection Robot

Robotic inspections are proposed as an alternative to human inspections in complex and hazardous highway tunnel environments. It can continuously inspect tunnel equipment, vehicle movement, and road conditions, as well as perform long-term, real-time inspections and monitoring. In this way, it becomes a valuable automated inspection method for tunnels, detecting various anomalies promptly. Maintaining, repairing, and managing tunnel safety hazards quickly reduces the occurrence of accidents and minimizes property losses. Furthermore, inspection robots are capable of responding promptly to accidents that occur inside tunnels, reducing losses associated with accidents. Inspecting results and real-time monitoring images will be uploaded from local monitoring centers to the remote control center. Several robot units can be controlled simultaneously in multiple sections and tunnels by the remote control center backend to inspect the interior conditions of each section and tunnel

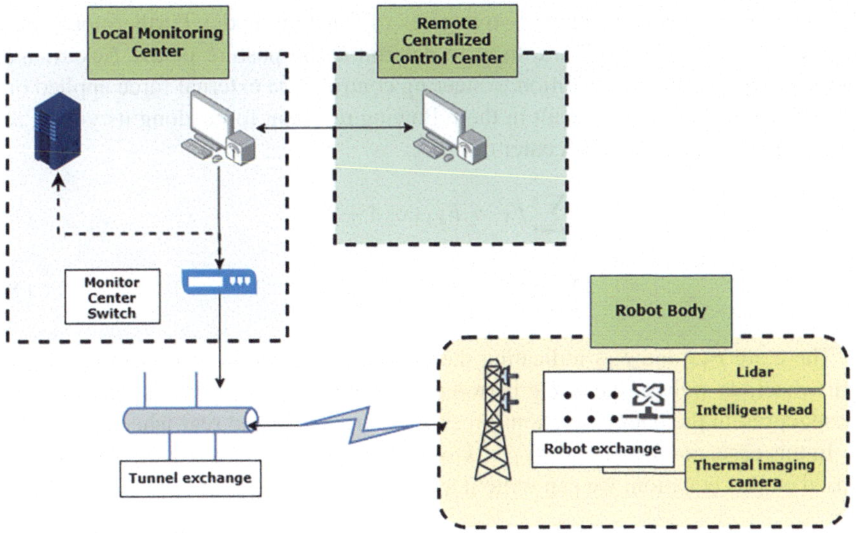

Fig. 5 An architectural diagram of the system for inspecting robots

simultaneously. It is shown in Fig. 5 how the tunnel inspection robot system is designed.

An intuitive user interface provides real-time information about the robot's status and the condition of the tunnel through the human–computer interaction component. Controlling the process of regular inspections, unique inspection tasks, and the resumption of inspection tasks are the main responsibilities of inspection settings. A monitoring platform is used to deliver instructions to the inspection robot in accordance with the requirements of highway tunnel inspections on a daily basis. Inspection targets are set, automatic inspection tasks are assigned, and targets are identified based on the requirements of highway tunnel inspections on a daily basis. Robot's body movement, PTZ camera, navigation system, related devices, and PTZ camera are controlled via remote control. In order to identify tunnel problems, this module processes and analyzes images obtained from internal tunnel devices. Moreover, this module analyzes and processes thermal image data.

3.2 *Model of Ackermann Chassis Kinematics*

Dynamically constrained, the Ackermann chassis lacks homogeneity. To plan a path, the dynamic constraints of the robot chassis need to be taken into consideration, which is why the four-wheel forward rotation rear-drive mode is used. When designing steering control strategies for an Ackermann chassis, steering characteristics must be considered [20]. There are various topics discussed in this section, such as the linear two-degrees-of-freedom model of the Ackermann chassis, and the model that reflects the linear two-degrees-of-freedom model. Furthermore, two-degrees-of-freedom models are used to determine the posture of the Ackermann chassis during motion in addition to steering control. The external force applied on Ackermann's chassis will result in the following resultant force along its y-axis, as well as a moment around its center of mass.

$$\sum F_Y = F_{Y1} \cos\delta + F_{Y2} \tag{1}$$

$$\sum M_Z = aF_{Y1} \cos\delta + bF_{Y2} \tag{2}$$

There are F_{Y1} and F_{Y2} indicating the distance between the entrance wheel and exit wheel, as well as the angle between the front wheel and the entrance wheel, these represent the lateral reaction forces at the front axle and rear wheel.

In this case, $\cos\delta \approx 1$ and $F_{Y1} = k_1\alpha_1 + k_2\alpha_2, F_{Y2} = K_2\alpha_2$ are the values of δ. Based on this equation, we can write it as follows:

$$\sum F_Y = F_{Y1}\alpha_1 + F_{Y2}\alpha_2 \tag{3}$$

$$\sum M_Z = aF_{Y1}\alpha_1 + bF_{Y2}\alpha_2 \tag{4}$$

α_1 and α_2 represent the sideslip angle of the front and rear axles, respectively. A relative motion equation follows from the principle of relativity:

$$u_1 \sin\zeta = v + a.w_r = utab\zeta \tag{5}$$

u_1 represents the linear speed at the midpoint of the front axle. It represents the angle between the axis and u_1; X and Y axes are represented by u and v, respectively; w_r represents the rotational angular velocity in the z axis of Ackermann chassis. $\beta = v/u$ where β is the center of mass, and

$$\zeta = \frac{v + aw_r}{u} = \beta + \frac{aw_r}{u} \tag{6}$$

Using the Ackermann chassis coordinate system, the side slip angles are measured at the front and rear wheels:

$$\alpha_1 = -(\delta - \zeta) = \beta + \frac{aw_r}{u} - \delta \tag{7}$$

$$\alpha_2 = \frac{v + bw_r}{u} = \beta + \frac{bw_r}{u} \tag{8}$$

The differential equation of motion for the Ackermann chassis with two degrees of freedom can be found by combining Eqs. (3), (4), (7), and (8):

$$(k_1 + k_2)\beta + \frac{1}{u}(ak_1 - bk_2)w_r - k_1\delta = m(v + uw_r) \tag{9}$$

$$(ak_1 - bk_2)\beta + \frac{1}{u}\left(a^2k_1 + b^2k_2\right)w_r - ak_1\delta = I_z w_r \tag{10}$$

Based on rotational inertia and angular acceleration measurements Z axis are expressed by I_z and w_r, respectively; m represents the chassis' mass.

Ackermann's chassis is therefore demonstrated and analyzed as a model scheme for using the tunnel inspection robot's autonomous mobile chassis as its mobile chassis.

3.3 Simultaneous Localization and mapping Algorithm

Originally, it was used to position robots in robotics. Early SLAM algorithms did not use visual cameras. In robotics, SLAM is commonly performed with a laser rangefinder. This process involves the simultaneous location of and building of maps in real time, also known as concurrent mapping and positioning (CML). Basically, the problem is to put a robot in new surroundings and see if it can map out the environment by gradually drawing a map while deciding what direction to take. It's typical for SLAM to be used for tasks such as sweeping robots. It is possible to move around the room with no obstacles if you use the so-called consistent map. Based on $p(x_{1:t'}m|z_{1:t'}u_{1:t})$ of map m and robot movement $x_{1:t'}$, the SLAM algorithm has been improved for several rounds for the inspection robot. In SLAM, Markov chains are obeyed. A probability distribution such as the one above can be calculated using the odometer's measured value $u_{1:t}$ and the laser's observed value $z_{1:t'}$. As a result of decomposing the joint probability distribution, we get the following:

$$p(x_{1:t'}m|z_{1:t'}u_{1:t}) = (m|z_{1:t'}u_{1:t}) * p(p(x_{1:t'}|z_{1:t'}u_{1:t}) \tag{11}$$

The decomposed equation above can be used to estimate the robot trajectory, and then a map can be created using the robot trajectory and laser measurements. An updated map is generated using the odometer, laser, and SIR filter data. Steps to follow are as follows:

1. **Initialization**: Initialization occurs at $t = 0$. x_t represents the current robot pose, w_t represents the weight, Map m represents the map, and Node Tree represents the trajectory
2. **Sampling**: With the new generation particle set $\left\{x_{t-1}^i\right\}$, the odometer motion sampling model has been modified to incorporate Gaussian noise as part of the previous generation particle set $\left\{x_t^i\right\}$.
3. **Weight estimation**: Based on laser data, map m, and particle pose X_t, the likelihood field distance search model finds the optimal pose for particles and estimates their weight.

4. **Resampling**: In resampling, particles are weighed. There may be many samples of high-quality particles, and many samples of low-quality particles. Particles are resampled if the reciprocal of the sum of squares of their normalized weights is less than a certain threshold N_{res}

$$N_{res} = \frac{1}{\sum_{i=1}^{N} \left(\widetilde{w_i}\right)^2}. \tag{12}$$

3.4 The Best Example Has Been Chosen for the Map Update

As a result of updated laser data $z_{(1:t)}$ and its optimal route tree $x_{(1:t)}$, the grid map is updated based on [33]. A built-in check robot is used to model the gesture of the odometer. During time $t-1$ to t, the robot experiences rotational δ_{rot1}, translational δ_{trans}, and rotational δ_{rot1}. There is both rotational and translational Gaussian noise [34].

$$x_t = \begin{pmatrix} x_{t-1} \\ y_{t-1} \\ \theta_{t-1} \end{pmatrix} + \begin{pmatrix} \delta_{trans}\cos(\theta + \delta_{rot1}) \\ \delta_{trans}\sin(\theta + \delta_{rot2}) \\ \theta + \delta_{rot1} + \delta_{rot2} \end{pmatrix} \tag{13}$$

$p(x_t|x_{t-1}, u_t)$ is the probability distribution underlying the motion sampling algorithm. The control quantity $u_t = (x_{t-1}, x_t)^T$ represents pose estimation based on odometer feedback for two groups of robots. The new species groups $\left\{x_{t-1}^i\right\}$ can be calculated based on the particle set $\left\{x_t^i\right\}$ from the previous generation.

A movement model error occurs when the tire slips during the actual operation of the inspection robot. It is generally possible to correct small errors with the SIR filter. Nevertheless, when the model is too large, robots will not be able to model it. This is especially true for large map models [35]. Due to these issues, this exploration incorporates a compass onto the robot chassis, using EKF to dynamically weigh and calculate an angle of attitude [36].

$$u_t = \overline{u_t} + K_t(z_t - h(\overline{u_t})) \tag{14}$$

Kalman gain $h(\overline{u_t})$ is represented by K_t, whereas observation equation is represented by u_t.

4 Result Analysis and Discussion

Figure 6 shows a comparison between IoT-only and IoT-aided approaches as measured by Data Delivery Ratio (DDR). It is evident from the comparison that the proposed method maintains and minimizes data loss more effectively and reliably than other methods, demonstrating its superior effectiveness and reliability. To overcome the limitations often associated with IoT-based solutions, data transmission protocols could be optimized, error correction could be enhanced, or resource management strategies may be designed to be more efficient. A major benefit of the proposed method is that it effectively tackles network congestion, interference, and limited bandwidth issues, enabling it to deliver data more consistently and effectively, proving its potential as a more robust solution for data-intensive applications.

An analysis of alternative approaches and the proposed method is depicted in Fig. 7. In this method, fewer hops are required, which means that data travels across the network more directly. The reduced hop count means that data packets encounter fewer intermediate nodes, which leads to a lower network latency and reduced energy consumption. Optimally routing data to its destination minimizes the use of multiple relay points. Improvements in routing protocols or better resource allocation may facilitate this efficiency, allowing it to be used for applications requiring speed and efficiency.

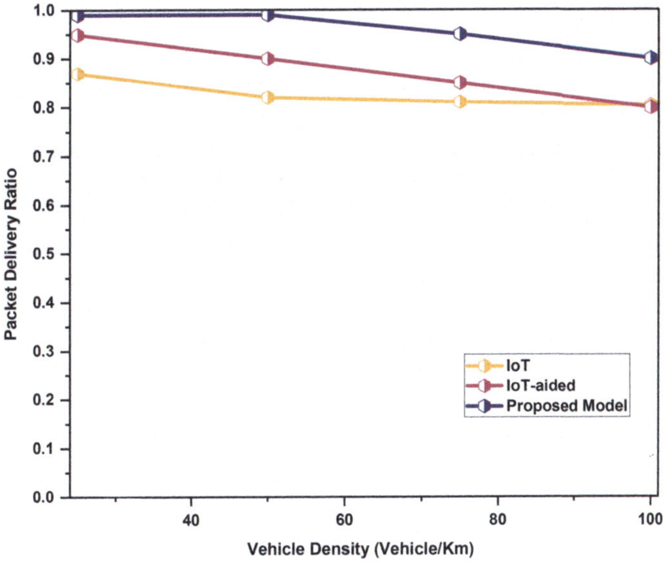

Fig. 6 The ratio of packets delivered versus the density of vehicles

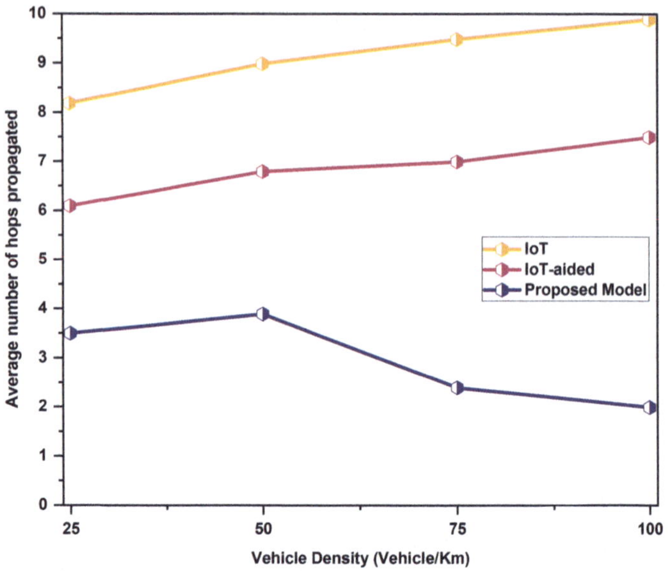

Fig. 7 Percentage of vehicles propagating versus hops propagated

The proposed method has lower overhead than existing methods, especially for non-critical messages, as shown in Fig. 8. The reduced overhead for non-critical messages suggests that the proposed method prioritizes critical data, reducing unnecessary transmissions and optimizing network resources.

It can be seen from Fig. 9 that there is significantly less delay in transmission using the proposed method compared to IoT-only integrations and IoT-aided integrations. As a result of this reduction in average delay, the proposed method appears to improve the speed and responsiveness of data transmission. This may be the result of improved routing efficiency, reduced congestion on the network, or a more direct channel of communication. This method reduces delay times in real-time communication, thereby enhancing overall network performance and user experience.

A comparison of the average number of hops required for different types of messages is shown in Fig. 10, with non-critical messages propagating faster than high-critical messages. As a result of this reduction in hops for non-critical messages, it may be that the proposed method prioritizes critical messages, ensuring their successful delivery by routing them through more reliable or redundant paths. By optimizing network efficiency and conserving resources, the method ensures critical messages receive the most priority by routing non-critical messages along shorter, less resource-intensive routes.

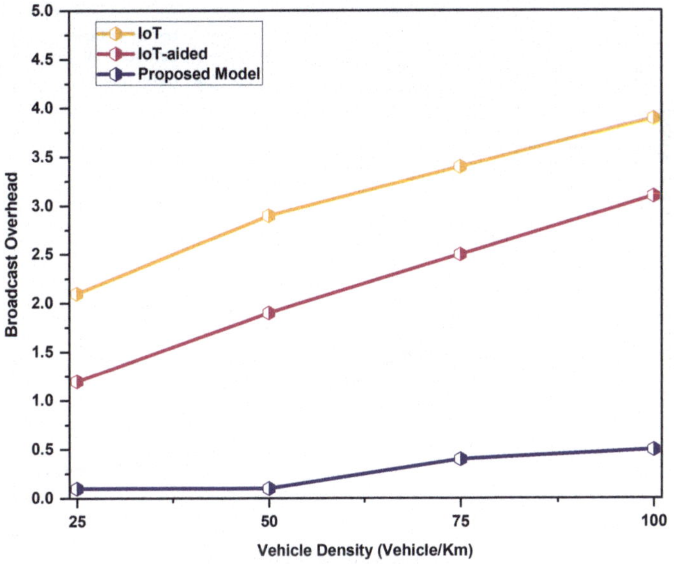

Fig. 8 Density of vehicles versus overhead broadcasts

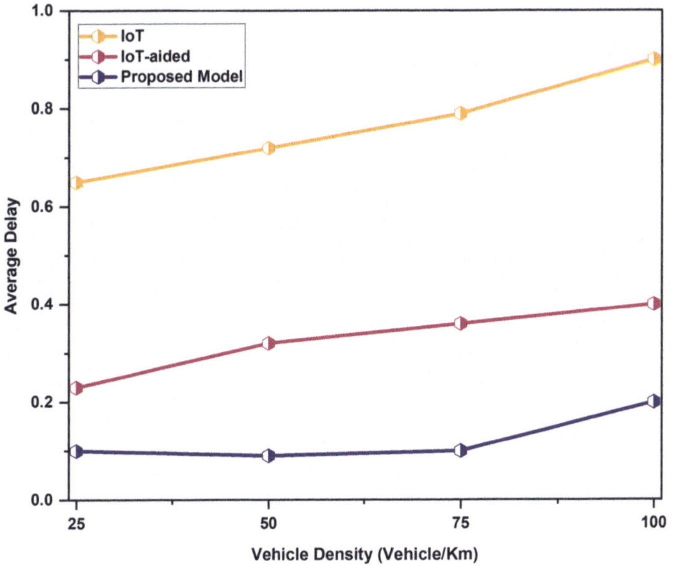

Fig. 9 The average delay versus the density of vehicles

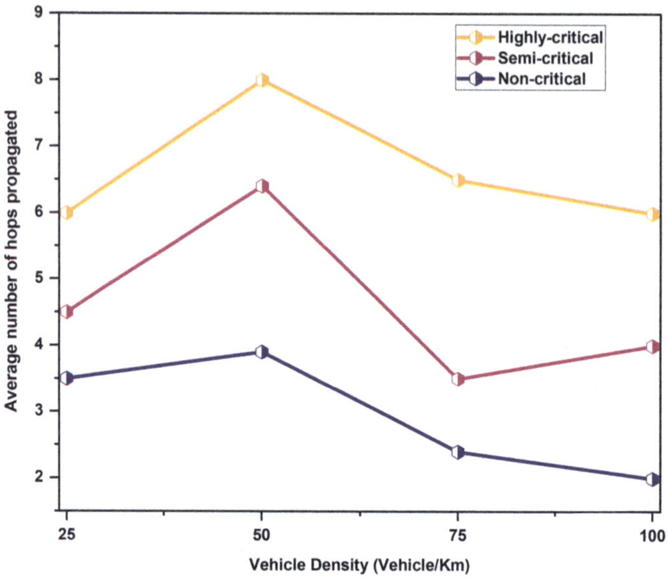

Fig. 10 Percentage of vehicles propagating versus hops propagated

5 Conclusion

As a result of smart cities and urban planning, society has gained the ability to make more informed decisions in a timely manner, thereby accelerating national development. Smart Cities (SC) are cities that incorporate Information and Communication Technologies (ICTs) in order to improve resident quality of life and streamline city operations. Using advanced networks and sensors to create Smart Cities, as presented in this paper, shows how machine learning-based IoT approaches can improve urban environments significantly. As a result of this interconnected framework, not only are individual Smart Cities able to carry out their functions, but also are able to interact with other Smart Cities, fostering efficiency and collaboration among them. Using IoT and machine learning, urban infrastructure can be tracked, managed, and intelligently managed in real-time. By integrating smart cities, they can serve the needs of dynamic urban environments, support sustainable economic growth, and improve the quality of life for residents. Smart city initiatives play a crucial role in transforming urban living in the future, as highlighted in this study.

References

1. G. Saranya, G. Kanimozhi, Effect of internet of things IoT application and smart cities. Sci. Technol. **2**(07) (2020). Accessed: 30 Oct 2024. [Online]. Available: https://ijadst.com/ajradmin/certificates/116/IJADST_2020111.pdf
2. A. Ahmad, A. Paul, M.M. Rathore, H. Chang, Smart cyber society: Integration of capillary devices with high usability based on cyber-physical system. Future Gener. Comput. Syst. **56**, 493–503 (2016). https://doi.org/10.1016/j.future.2015.08.004
3. O. Pribyl, M. Svitek, L. Rothkrantz, Intelligent mobility in smart cities. Appl. Sci. **12**(7), 3440 (2022). https://doi.org/10.3390/app12073440
4. B.N. Silva, M. Khan, K. Han, Towards sustainable smart cities: a review of trends, architectures, components, and open challenges in smart cities. Sustain. Cities Soc. **38**, 697–713 (2018). https://doi.org/10.1016/j.scs.2018.01.053
5. T. Panagiotakopoulos, Y. Kiouvrekis, L.-M. Misthos, C. Kappas, RF-EMF exposure assessments in Greek schools to support ubiquitous IoT-based monitoring in smart cities. IEEE Access **11**, 7145–7156 (2023). https://doi.org/10.1109/ACCESS.2023.3237970
6. Y. LeCun, Y. Bengio, G. Hinton, Deep learning. Nature, **521**(7553), 436–444 (2015)
7. B. Lim, S. Zohren, Time-series forecasting with deep learning: a survey. Philos. Trans. R. Soc. Math. Phys. Eng. Sci. **379**(2194), 20200209 (2021). https://doi.org/10.1098/rsta.2020.0209
8. J.-H. Böse et al., Probabilistic demand forecasting at scale. Proc. VLDB Endow. **10**(12), 1694–1705 (2017). https://doi.org/10.14778/3137765.3137775
9. S. Kaushik et al., AI in healthcare: time-series forecasting using statistical, neural, and ensemble architectures. Front. Big Data **3**, 4 (2020). https://doi.org/10.3389/fdata.2020.00004
10. T.L. Leise, Analysis of nonstationary time series for biological rhythms research. J. Biol. Rhythms **32**(3), 187–194 (2017). https://doi.org/10.1177/0748730417709105
11. L.A. Lynn, Artificial intelligence systems for complex decision-making in acute care medicine: a review. Patient Saf. Surg. **13**(1), 6 (2019). https://doi.org/10.1186/s13037-019-0188-2
12. G. Vonitsanos, T. Panagiotakopoulos, A. Kanavos, A. Tsakalidis, Forecasting air flight delays and enabling smart airport services in apache spark, in *Artificial Intelligence Applications and Innovations. AIAI 2021 IFIP WG 12.5 International Workshops*, ed. by I. Maglogiannis, J. Macintyre, L. Iliadis. IFIP Advances in Information and Communication Technology, vol. 628 (Springer International Publishing, Cham, 2021), pp. 407–417. https://doi.org/10.1007/978-3-030-79157-5_33
13. F. Martínez-Álvarez, A. Troncoso, G. Asencio-Cortés, J. Riquelme, A survey on data mining techniques applied to electricity-related time series forecasting. Energies **8**(11), 13162–13193 (2015). https://doi.org/10.3390/en81112361
14. M. Mudelsee, Trend analysis of climate time series: a review of methods. Earth-Sci. Rev. **190**, 310–322 (2019). https://doi.org/10.1016/j.earscirev.2018.12.005
15. S. Nousias, E.-V. Pikoulis, C. Mavrokefalidis, A.S. Lalos, Accelerating deep neural networks for efficient scene understanding in multi-modal automotive applications. IEEE Access **11**, 28208–28221 (2023). https://doi.org/10.1109/ACCESS.2023.3258400
16. O.B. Sezer, M.U. Gudelek, A.M. Ozbayoglu, Financial time series forecasting with deep learning: a systematic literature review: 2005–2019. Appl. Soft Comput. **90**, 106181 (2020). https://doi.org/10.1016/j.asoc.2020.106181
17. N. Kumar, P. Rani, V. Kumar, P.K. Verma, D. Koundal, Teeech: Three-tier extended energy efficient clustering hierarchy protocol for heterogeneous wireless sensor network. Expert Syst. Appl. **216**, 119448 (2023)
18. X. Krasniqi, E. Hajrizi, Use of IoT technology to drive the automotive industry from connected to full autonomous vehicles. IFAC-Pap. **49**(29), 269–274 (2016). https://doi.org/10.1016/j.ifacol.2016.11.078
19. B.V. Philip, T. Alpcan, J. Jin, M. Palaniswami, Distributed real-time IoT for autonomous vehicles. IEEE Trans. Ind. Inform. **15**(2), 1131–1140 (2019). https://doi.org/10.1109/TII.2018.2877217

20. D. Minovski, C. Ahlund, K. Mitra, Modeling quality of IoT experience in autonomous vehicles. IEEE Internet Things J. **7**(5), 3833–3849 (2020). https://doi.org/10.1109/JIOT.2020.2975418
21. N. Hussain, P. Rani, H. Chouhan, U.S. Gaur, Cyber security and privacy of connected and automated vehicles (CAVs)-based federated learning: challenges, opportunities, and open issues. Fed. Learn. IoT Appl. 169–183 (2022)
22. H. Khayyam, B. Javadi, M. Jalili, R.N. Jazar, Artificial intelligence and Internet of Things for autonomous vehicles, in *Nonlinear Approaches in Engineering Applications*, ed. by R.N. Jazar, L. Dai (Springer International Publishing, Cham, 2020), pp. 39–68. https://doi.org/10.1007/978-3-030-18963-1_2
23. J.A. Guerrero-ibanez, S. Zeadally, J. Contreras-Castillo, Integration challenges of intelligent transportation systems with connected vehicle, cloud computing, and internet of things technologies. IEEE Wirel. Commun. **22**(6), 122–128 (2015). https://doi.org/10.1109/MWC.2015.7368833
24. P. Rani, N. Hussain, R.A.H. Khan, Y. Sharma, P.K. Shukla, Vehicular intelligence system: time-based vehicle next location prediction in software-defined internet of vehicles (SDN-IOV) for the smart cities. Intell. Things AI-IoT Based Crit.-Appl. Innov. 35–54 (2021)
25. N. Hussain, P. Rani, Comparative studied based on attack resilient and efficient protocol with intrusion detection system based on deep neural network for vehicular system security, in *Distributed Artificial Intelligence* (CRC Press, 2020), pp. 217–236
26. S. Kumar Janahan, M.R.M. Veeramanickam, S. Arun, K. Narayanan, R. Anandan, S. Javed Parvez, IoT based smart traffic signal monitoring system using vehicles counts. Int. J. Eng. Technol. **7**(2.21), 309 (2018). https://doi.org/10.14419/ijet.v7i2.21.12388
27. N. Hussain, P. Rani, N. Kumar, M.G. Chaudhary, A deep comprehensive research architecture, characteristics, challenges, issues, and benefits of routing protocol for vehicular ad-hoc networks. Int. J. Distrib. Syst. Technol. IJDST **13**(8), 1–23 (2022)
28. M. Sarrab, S. Pulparambil, M. Awadalla, Development of an IoT based real-time traffic monitoring system for city governance. Glob. Transit. **2**, 230–245 (2020). https://doi.org/10.1016/j.glt.2020.09.004
29. M. Babar, F. Arif, Real-time data processing scheme using big data analytics in internet of things based smart transportation environment. J. Ambient. Intell. Humaniz. Comput. **10**(10), 4167–4177 (2019). https://doi.org/10.1007/s12652-018-0820-5
30. A. Rudskoy, I. Ilin, A. Prokhorov, Digital twins in the intelligent transport systems. Transp. Res. Procedia **54**, 927–935 (2021)
31. A.Y. Awad, S. Mohan, Internet of things for a smart transportation system. Int. J. Interdiscip. Telecommun. Netw. IJITN **11**(1), 57–70 (2019)
32. G. Lei, Y. Zheng, Research on cooperative trajectory planning algorithm based on tractor-trailer wheeled robot. IEEE Access **10**, 64209–64221 (2021)
33. T. Liu, J. Kuang, W. Ge, P. Zhang, X. Niu, A simple positioning system for large-scale indoor patrol inspection using foot-mounted INS, QR code control points, and smartphone. IEEE Sens. J. **21**(4), 4938–4948 (2020)
34. J. Szrek, J. Jakubiak, R. Zimroz, A mobile robot-based system for automatic inspection of belt conveyors in mining industry. Energies **15**(1), 327 (2022)
35. T. Pathmakumar, M. Kalimuthu, M.R. Elara, B. Ramalingam, An autonomous robot-aided auditing scheme for floor cleaning. Sensors **21**(13), 4332 (2021)
36. N.B. Hollinger, Examining autonomous inspection of geologic repositories. ESARDA Bull. 48 (2022)

Machine Learning Based Solutions for Securing Robotics and IoT Systems

Naman Oli, Bharat Bhushan, Hari Mohan Rai, and Omar S. Saleh

Abstract In recent years, computer applications have significantly changed from traditional data processing to machine learning, due to the increased accessibility and availability of abundance of data gathered from the network and other resources. Machine learning automates human assistance by automating learning methods. In IoT integrated with Robotics, machine learning is applied to secure interconnected devices against cyber threats and insecurities. IoT is a network of interconnected gadgets and systems that exchange data over a network to perform various tasks using machines and robots. These machines range up to industrial machines and healthcare equipment which can be vulnerable to security threats. Machine learning is a prominent way to detect anomalies in real time. It enables predictive threat detection, continuous learning, and automated responses to improve the protection of IoT networks. In this paper, we explore the applications of various machine learning and its algorithms to strengthen the security of IoT devices, robotic machines and network. The paper presents the predictive ability of ML algorithms and their roles in improving threat detection and identification, reviewing their effectiveness against different types of challenges and vulnerabilities in IoT systems. Finally, the paper comprehensively reviews the current machine learning algorithms and their effectiveness in securing IoT network, devices, and robot while also offering some valuable insights and recommendations to create more adaptive and effective security solutions.

Keywords IoT · Smart city · Robotics · ML Algo

N. Oli (✉) · B. Bhushan
Department of Computer Science and Engineering, School of Engineering and Technology, Sharda University, Greater Noida, India
e-mail: namanoli83@gmail.com

H. Mohan Rai
School of Computing, Gachon University, Seongnam, South Korea
e-mail: drhmrai@gachon.ac.kr

O. S. Saleh
Directorate of Studies, Planning and Follow Up, Ministry of Higher Education and Scientific Research, Baghdad, Iraq

R. Sharma and G. Jeon (eds.), *Data Analytics for Smart Robotics and Its Applications*, Intelligent Systems Reference Library 272,
https://doi.org/10.1007/978-3-031-87697-4_9

1 Introduction

The quick expansion of the Internet of Things (IoT) has greatly transformed current ICT (Information and communication technologies) landscape, also a vast range of cellular IoT and robotics devices have projected to be deployed in the coming years. IoT devices and sensors have become integral with various aspects of our daily lives including home environments, healthcare, and transportation. Powered by analytics and cloud computing these devices can individually communicate and exchange data, making widespread adoption possible. These devices are expected to be able to communicate autonomously, and seamlessly exchanging relevant contextual data in between without the need of any human interaction [1]. The capability to independently process and share data is one of the major causes of rapid growth and widespread adoption of IoT technology. As these advantages further develop, IoT is actively being integrated into various industries and is transforming the way technical and informational systems operate by enhancing their efficiency, responsiveness, and decision-making. Although IoT has shown a rapid growth which led to the advancements in data analysis, privacy, security, and energy efficiency, the existing solutions remain primarily focused on communication. To effectively tackle these challenges, IoT-based smart systems are require to have more intelligent and adaptable integrated approaches. IoT's expansion introduces several security risks, i.e. the vulnerabilities in the IoT devices may get exploited by attackers, leading to security and personal breaches and financial losses. Ensuring privacy, security, and business continuity is essential as IoT becomes more integrated into daily life and handles sensitive personal data. An attack on these devices can result in the leakage of personal information and may disrupt workflows, thereby compromising the product quality. Conventional network security solutions often struggle to address the distinct characteristics of IoT devices, which are generally limited and decentralized due to constraints in terms of computational potential and power resources. These limitations often obstruct the deployment of strong security protocols which are mainly used in larger systems, leaving vulnerabilities in IoT devices to various cyber threats. These weaknesses may get exploited by unauthorised users to gather unauthorized control, steal sensitive information, or disrupt services [2]. Absence of these extensive, built-in security features further heightens the vulnerabilities, highlighting the need of more advanced and specialized security measures integrated to the IoT ecosystem.

Mainly IoT devices are obstruct by limited storage capacity and processing power, which may hamper their ability to deploy robust security solutions and maintain extensive databases of threat signatures and malware for protection. Additionally, anomaly-based traditional detection methods have their shortcomings, as they might classify some rare or unusual user activities as threats, leading to inaccuracy in threat detection and compromising overall security. In this scope, machine learning presents a valuable alternative, machine learning algorithms leverages a greater volume of data to continuously learn and adapt, facilitating the creation of more advanced defensive mechanism. These techniques can greatly address the increased complexity of cybersecurity challenges in IoT environments, to help even resource constrained devices to effectively detect potential threats [3]. The expected outcome of this work is to examine the various Machine Learning techniques to enhance IoT security. This involves the analyses of the predictive capabilities and how these algorithms can enhance threat detection, anomaly identification, and evaluate their effectiveness to overcome the specific challenges and vulnerabilities present in IoT environments. This paper reviews the current Machine Learning techniques and their implications for IoT security, to provide meaningful insights and suggestions on developing more adaptive and robust security solutions.

The remainder of this work is structured as follows. Section 2 presents the algorithms of Machine Learning classified under supervised, unsupervised, reinforcement and semi-supervised. Section 3 comprises of IoT its characteristics, key components, working and applications in robotics. Section 4 addresses the cyber threats related to IoT and robotics while presenting its ML related solutions. Section 5 regards the recent advancements.

2 Machine Learning

Machine Learning is a branch of AI that structure and design statistic models and algorithms which enable the computers to imitate human intelligence, learn from and make decisions or predictions based on the available information. Figure 1 depicts the various machine learning models.

2.1 Supervised Learning

Supervised learning is a technique used in machine learning in which the models on some data are trained with labelled inputs and their corresponding outputs. In this model, the algorithm is provided with some case in point of correct decisions, which allows it to know patterns and make predictions about new, unseen data [4]. This model is set to minimising the error between actual labels and the predictions in the training data. To measure performance of the supervised learning algorithm we follow measures like precision, accuracy, recall, and mean squared error, which depends on the nature or the type of problem. It is applied to both classification (categorizing inputs into predefined groups) and regression (predicting continuous values). Here are several widely used supervised learning algorithms.

- **Linear Regression**: Linear regression is enlisted so it predicts a continuous outcome that is based on a single or multiple inputs, modelling the relationship between the inputs and outputs as a linear equation.

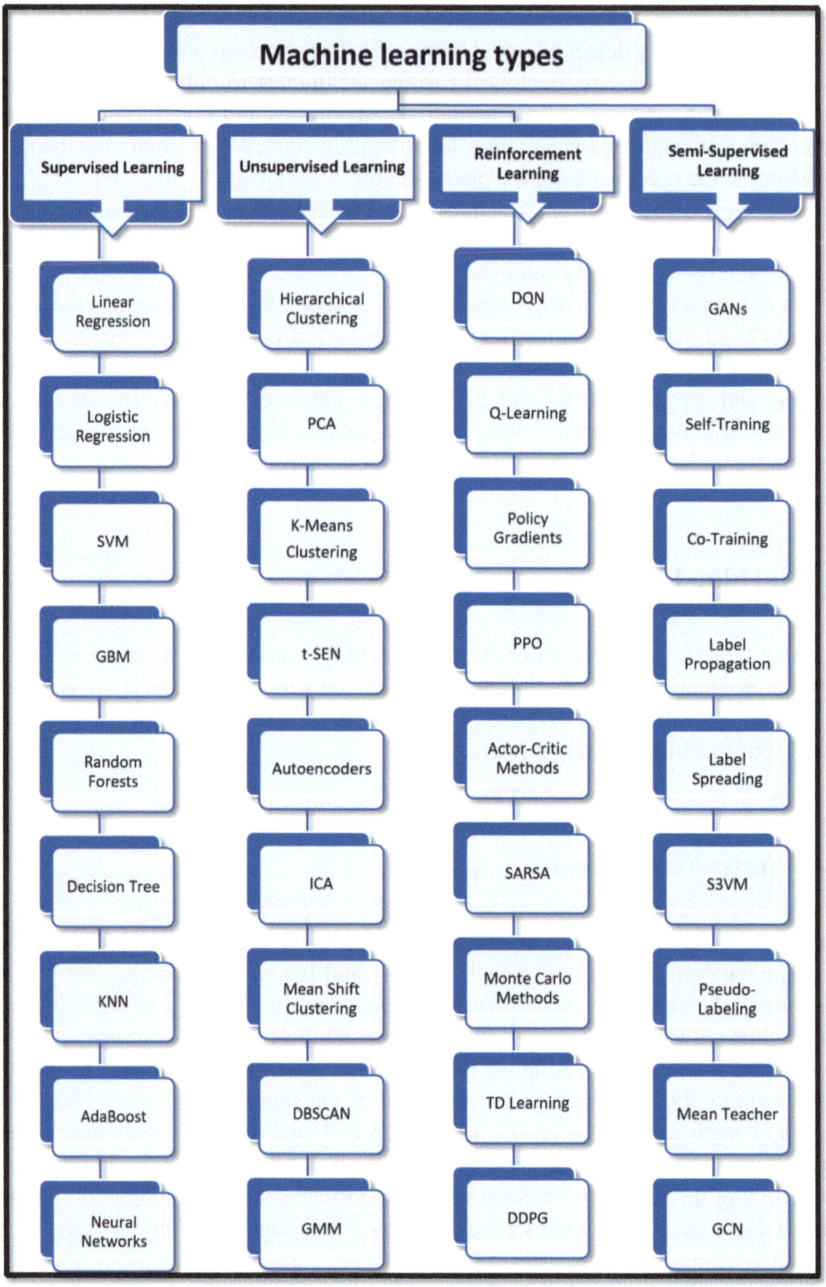

Fig. 1 Machine Learning Types

- **Logistic Regression**: It a technique of the binary classification of input, which presents the probability of a specific input i.e. it comes under a particular class by using the logistic function.
- **Support Vector Machines (SVM)**: SVM identifies the hyperplane separating the different classes within the featured space most effectively and is capable to handle both linear and non-linear classification task.
- **Gradient Boosting Machines (GBM)**: GBM is a composite method that construct trees in a sequential mode, where the errors of the previous ones is corrected by each new tree. Its well-known implementations include XG-Boost and Light-GBM.
- **Random Forest**: Random forest is the ensemble technique which construct multiple decision tree and then merges the predictions, enhancing accuracy and reducing overfitting.
- **Decision Trees**: Decision tree branch off the data into branches based on feature values, and then makes a decision at each node. This process is continued until the leaves represent a final prediction.
- **K-Nearest Neighbours (KNN)**: KNN classifieds data point as it identifies the major class from its k-nearest neighbours. It's a straightforward yet effective approach, particularly with smaller datasets.
- **AdaBoost**: AdaBoost creates models in sequence, modifying misclassified instances that enhances the performance of subsequent models.
- **Neural Networks**: Neural networks are considered as a collection of algorithms modelled to simulate the human intelligence to recognize patterns. They consist of interconnected nodes (neurons) in layers that can capture and model intricate relationships.

2.2 Unsupervised Learning

It is an ML approach in which the algorithm learn from the unlabelled data, unlike supervised learning relying on labelled data for training the model. Unsupervised learning focuses on revealing patterns, structures and relationship within the data by itself. Unsupervised algorithm doesn't have a pre-defined target variable. Its aims to uncover underlying, structures, patterns or anomalies within the data [5]. Few of its main algorithm are listed as follows.

- **Hierarchical Clustering**: A tree-like cluster structure is generated by a hierarchical clustering method, where each cluster is contained within a large cluster. A tree-like structure of clusters is created by a hierarchical clustering technique, where each cluster is enclosed within a bigger cluster.
- **Principal Component Analysis (PCA)**: PCA is a technique that simplifies datasets by reducing the variables and keeping all the required data.
- **K-Means Clustering**: It is a clustering technique which groups data into K clusters on the basis of their distance with the centre of every cluster.

- **t-SEN**: t-SNE is a method of mapping high-specification data into a low dimensional one while preserving the similarities and differences between data sets.
- **Autoencoders**: Autoencoders are composed of neural networks which can compress and decompress the data and can find efficient ways to represent the information in a smaller format.
- **ICA**: ICA is a technique that can break complex signals into simpler independent components, unlike focusing on finding the most important patterns like PCA.
- **Mean Shift Clustering**: The mean Shift clustering technique searches group of data points, as it moves them towards the area with highest density, without any need to know the number of groups.
- **DBSCAN**: A clustering algorithm based on density, BSCAN identifies the clusters of densely packed data while marking those that are in isolation. There's no requirement to signify the number of clusters and will find the cluster with irregular shapes.
- **GMM**: GMM suppose that data is a blend of several normal distributions. It makes use of the EM algorithm to know the best-fitting parameters for these distributions.

2.3 Reinforcement Learning

Reinforcement Learning (RL) is an approach in machine learning where the algorithm is set to learn by interacting with an environment to make its own decisions. Its main aim is to multiply the cumulative rewards with time by learning policies that maps states to actions. This is often achieved through the method of trial and run to check errors, so the machine can benefit from its experiences continuously and then adjusts its policy accordingly to provide better outputs [6]. Some of its major algorithms are as follows.

- **DQN**: DQN is considered as reinforcement learning algorithm which utilizes the advanced neural networking to find Q-valued function, which makes it more suitable to handle large or continuous state spaces.
- **Q-Learning**: Q-Learning, is a method of reinforcement learning algorithm which learns the best actions to take in a particular situation through estimating expected rewards for each action.
- **Policy Gradients**: Policy Gradient is a method which targets to find the most optimal policies by clearly adjusting the direction of its parameters to increases the expected reward.
- **PPO**: PPO is taken as a policy gradient technique that ensures stable learning which limits the size of the policy updates. This helps in maintaining a balance between exploiting known good ones and exploring new actions.
- **Actor-Critic Methods**: Actor-Critic methods compile the plus points of the policy and value based approaches. It uses the actor which update policies and critics to evaluate it.

- **SARSA**: SARSA is a reinforcement learning algorithm which can learn from its own experiences, using the operation it takes in the next state instead of the best possible action.
- **Monte Carlo Methods**: Monte Carlo methods approximate the value of any policy by averaging its total rewards collected through multiple simulations of the policy. They don't require to know how its environment works and simply sample the results of different policy executions.
- **TD Learning**: TD Learning combines elements of the method of Monte Carlo and dynamic programming. rather than waiting for a result, it adjusts the value estimates based on multiple subsequent estimates.
- **DDPG**: DDPG combining deep Q-learning with policy gradients, to effectively handle continuous action spaces.

2.4 Semi-Supervised Learning

A machine learning approach which comes in between the supervised learning and unsupervised learning is called semisupervised learning, effectively leveraging both types of data, labelled and unlabelled. It is considered valuable while labelling of the data which is either costly or very time taking process, as it can use a lot of unlabelled data to enhance a model's performance. Applications for semi-supervised learning can exist in various domains, like bioinformatics, computer vision, and natural language processing [7].

Some of its main algorithms are explained below.

- **GANs**: GANs is a type of generative model with a generator network that can create some data as fake and then a network (discriminator) which is set out so it can differentiate between original and fake samples.
- **Self-Training**: A semi-supervised learning algorithm where a model can be trained in a limited amount or sections of unlabelled data at the initial level. It then uses this knowledge so that for unlabelled data points, it can predict the labels.
- **Co-Training**: The semi-supervised learning technique co-training divides the data into two separate but sufficient sets of features. It takes two models to train them on different sets and both the models label the data for the other one.
- **Label Propagation**: Label Propagation is a technique under semi-supervised learning that uses graph representation of datasets in which the node is the data points, with the edges representing their relation. Labels from labelled nodes are passed on to unlabelled nodes through graph until the labels are stabilized.
- **Label Spreading:** Label Spreading is a technique in semi-supervised learning based on graphs, that uses regularization for modifying the label propagation process. This helps in preventing overfitting and ensures similarity between the label data points are consistent.

- **S3VM:** S3VMs further extend the SVM algorithm including the unlabelled data by finding a decision boundary to maximise the margin between labelled data points, and also considering the distribution of unlabelled data to avoid overfitting.
- **Pseudo-Labeling**: Pseudo-labeling technique is used to train the models in labelled data to use that in predicting labels for unlabelled data. These labels that are predicted are known as pseudo-labels and are taken as the true labels, used to retrain the model.
- **Mean Teacher**: Mean Teacher is a method of semi-supervised learning which considers two models in which one is taken as a student model which trains on both labelled data and the unlabelled data, and another as a teacher model which is basically an unstill average of the weight of the student model.
- **GCN**: GCNs are neural network which are specifically designed for operating on graph-structured data. In semisupervised learning, the GCNs are considered effective for classifying nodes in graphs when only few of the nodes have known labels.

3 Internet of Things

IoT network has been expanding rapidly and holds the potential of revolutionising human life. It's a major advancement in internet and technological world. The core idea of IoT is to equip everyday objects with embedded equipment like sensors or miniature devices and transforming them into smart devices. Few of its key characteristics include wireless connections, embedded inbuilt sensors, mobility, a variety of technological applicants, and support system for different devices. IoT plays a necessary role in transmission of information, data collection and analysis and remotely controlling devices via a network. The motivating factor supporting the substantial growth of IoT includes majority of the software industries, service provider and manufacturing industries [8].

3.1 Definition

IoT is an interconnected web or network of smart physical gadgets which are embedded with electronics, various software, sensors and other type of technologies that are connected with the internet so that they can communicate to each other and collect data or exchange it. These objects with embedded technology can sense and interact with their environment. IoT has enabled us to monitor the physical world and control it digitally. They come in a wide range, including household devices like smart thermostats and wearables, as well as industrial machines and urban infrastructure designed to improve automation, efficiency and real-time data analytics [9]. IoT is further advancing with innovations like enhanced security protocols, smart cities, deeper integration with voice-activated assistants among various other developments.

3.2 Characteristics of IoT

IoT comprises of 7 key characteristics highlighting the fundamental aspects of the system and are crucial for better understanding the opportunities and challenges in IoT. The characteristics are as follows:

- **Sensing**: IoT devices have sensors that can capture and measure data from their real-world environment like motion, light, humidity, heat, or sound. This real-time collection of data allows IoT devices to monitor its surroundings with high precision applications in smart homes, healthcare, and environmental monitoring. The data collected forms the foundation for further processing and decision-making, which enhances the overall functionality of IoT systems and their ability to adjust to the changes in their surroundings [10].
- **Connectivity**: IoT devices are dependent on a variety of communication networks, like 5G, Wi-Fi network, Bluetooth system, Zigbee, and LoRa, to stay inter-connected and share data across systems. These networks make sure that the data collected from the sensor is transmitted to centralized platforms and other connecting devices for further analysis or action. Seamless connectivity plays a crucial role in ensuring IoT systems operate efficiently, and different protocols are often employed depending on the needs of applications, such as consumption of power, range, or bandwidth [11].
- **Automation**: IoT systems can provide devices the ability to perform tasks autonomously without any human intervention. These devices by analysing the collected data can make decisions based on predefined rules or machine learning methods. This is particularly helpful in smart homes, industrial automation, and agricultural system where efficiency needs to be significantly improved. Automation also helps reduce human errors and operational costs, while increasing responsiveness and system reliability [12].
- **Scalability**: As IoT systems have grown, scalability has become a critical factor in managing vast deployments of interconnected devices. These systems have to be capable of supporting a large number of devices across different environments, while processing huge volumes of data efficiently. Scalability ensures that IoT infrastructures can adapt to the increasing demands in smart cities, healthcare systems, or industrial environments, allowing them to grow further without compromising with performance [13].
- **Real-time Data Processing**: Many IoT requirements like those in healthcare, industrial automation, and smart oriented cities requires a simultaneous data processing where decisions are made instantly. IoT systems with realtime data processing can trigger immediate actions or send alerts such as monitoring a patient or controlling machinery in a factory. This ability to act swiftly based on incoming data is crucial for mission-critical applications, enhancing safety, efficiency, and overall effectiveness [14].
- **Heterogeneity**: IoT system consist of a variety of devices, platforms, sensors, and communication protocols creating a highly competent network. This variety

of components in the system presents challenges for standardization and compatibility. Ensuring that all devices of systems within the network can communicate and interact with ease is essential for optimizing performance. Achieving this interoperability is crucial to maintain flexibility, reducing complexity, and fostering innovation in IoT applications [15].

- **Security and Privacy**: IoT gadgets manage a lot of sensitive information, like personal information and even operational details of critical infrastructures. This makes securing essential for IoT systems to ensure privacy. Robust encryption, secure authentication methods, and privacy-preserving techniques are used to protect data against cyber threats like hacking, unauthorized access, and data breaches. Ensuring strong security protocols protects user data, and builds trust in the adoption and expansion of IoT technologies across industries [16].

3.3 IoT: How It Works

IOT devices functions by collecting data through sensors, then processing it and using that information to trigger certain actions or provide insights. This process involves:

- **Data Collection**: IoT devices are already embedded with sensors which can measure physical quantities like temperature, atmosphere humidity, motion, light, or any sound depending on the application. These sensors constantly monitor their environment and collect real-time data. This process forms the most fundamental layer of IoT functionality, enabling systems to gather detailed information from the real world. This data collected is crucial for further strengthening and decision-making of processes in smart cities, homes, health sector, and industrial automation [17].
- **Data Processing**: The data collected is further analysed and then processed by the IoT embedded systems and cloud-based platforms. This data processing may involve filtering, aggregating, or transforming raw data into actionable data. Localized processing helps to reduce latency in time sensitive applications, while cloud-based processing has higher computational power for complex analytics. This stage is crucial in making the data sensible and extracting meaningful insights that will allow automated actions or decision-making [18].
- **Data Transmission**: After processing, it is necessary to transmit the data to the central server or the cloudbased platform for storage or other actions. Wireless networks like Wi-Fi, 5G, Zigbee, or LoRa enables the transmission of data across large distances or among multiple devices. Efficient data transmission is crucial to ensure that the IoT system operates efficiently, especially in large-scale applications like smart cities or industrial monitoring where real-time communication is required [19].
- **Decision-making**: On the basis of analysed data, specific algorithms or predefined rules are applied to make decisions. These decisions may trigger automatic actions like turning on light bulbs, adjusting HVAC systems, controlling machinery, or

even sending notifications to user. IoT system with decision making enables real-time responses or interventions without the need for human involvement. This ability to autonomously make decisions and take actions accordingly is a key aspect of IoT and its potential in areas like smart house, health sector, and industrial automation [20].

3.4 Key Components of IoT

The Internet of Things (IoT) comprises of several essential components that works together to enable different connected devices and systems. These components include:

- **Sensors**: Sensors are a key component of IoT systems which are responsible for measuring various physical quantities like temperature, motion, humidity, light, or any sound. Sensors collect the data from the real world and transfer it to IoT devices and systems for further processing. This data is crucial for monitoring and controlling environments in real-time. Sensors are used for various applications, including smart houses, industrial motorization, health sector, and for environmental monitoring, where precision and real-time feedback are essential for efficient system performance.
- **Actuators**: Actuators are devices that controls the physical process in IoT systems by receiving commands from a central system or other devices. With these commands, the actuators perform specific actions like opening or closing valves, turning lights on or off, adjusting temperature settings, or moving mechanical parts in a factory. Actuators shallows the separation in between the digital and physical worlds, allowing IoT systems to interact with their environment on the basis of data they receive from the sensors [21].
- **Connectivity**: IoT devices to function effectively and efficiently, a connection to the internet or a network is required to communicate or exchange data among devices or system. It's done using various wireless communication techs like Wi-Fi connection, 5G networks, Bluetooth, and low-powered wide-area networks (LPWAN). Connectivity is essential to ensure timely data transfer, remote monitoring, or device interoperability in large scale IoT deployments.
- **Embedded Systems**: Embedded systems consist of small computers or micro-controllers that are integrated within IoT devices to manage the core operations. These systems process the data collected from sensors, control actuators, and handle communication with other devices or platforms. Embedded systems are needed to have efficiency allowing IoT devices to perform complex tasks with minimal power consumption. It has an important role to conduct in-time decision-making, data processing, and automation particularly in applications that demand low latency.
- **Cloud Platforms**: Cloud platforms provide scalable, flexible infrastructures to store information, processing it, and analysing vast amount of the IoT datasets. They offer various services including data storage, machine learning, analytics,

and application development which helps institutes in fetching meaningful insights from their IoT systems. By using cloud platforms, IoT devices can offload complex computations and data storage, ensuring greater scalability and enabling more advanced processing like predictive analytics or real-time monitoring across different industries and logistics [22].

- **Protocols**: IoT protocols are sets of rules that governs the communication and data transfer in between IoT devices ensuring the data will be exchanged efficiently and securely across different networks. Commonly used IoT protocols are HTTP, MQTT (Messaging Queuing Telemetry Transport), LWM2M (Lightweight M2M), and CoAP (Constrained Application Protocol). These protocols are developed to be lightweight to allow IoT devices with limited computational resources to communicate effectively. These protocols guaranty the data is transmitted minimizing delays, improving the overall performance of the IoT system.
- **Security**: Security is a factor which is very important in IoT ecosystems as IoT devices handle private data and operate in highly interconnected environments. Effective security measures are necessary to safeguard IoT devices, the data it collects and the network from cyber-crimes such as hacking, unauthorized access, and sensitive data breaches. These measures typically involve encryption to protect data transmission, authentication protocols to verify device identity, and mechanisms of access control to limit unauthorized use. Ensuring robust security is required to maintaining trust in IoT technologies, especially in applications involving personal data, critical infrastructure, and healthcare systems.

3.5 Robotics in Specific IoT Domains

Table 1 depicts the application, advantages and challenges of Robotics in various IoT domains.

3.5.1 Industrial Internet of Things (IIoT)

The Industrial Internet of Things comprises integrating the IoT technology in industrial workspaces like manufacturing, energy, and logistics. It includes the use of inter-connected machines, sensors and analytics platforms to gather realtime data from processes. This data can further be analysed for the optimization of operations, to increase efficiency, reduce downtime, and the enhancement of predictive maintenance. IIoT has a major role in smart factories where machine-to-machine communication and automation helps to improve productivity and lower the operational costs. The use of IIoT in industries have led to many innovations like autonomous systems and monitoring of critical infrastructure in real-time [23]. Robotics can be integrated with IoT technology which can provide more efficient, smarter industrial process which can be highly automated. It impacts the key areas like smart manufacturing, predictive maintenance and energy efficiency [24].

Table 1 Advantages and limitations of IoT domains

Domain	Robotics Application	Advantages	Challenges
(IIoT) Industrial Internet of Things	• Machine-to-Machine Communication • Collaborative Robots • Predictive Maintenance • Automation	• Increased productivity and efficiency • Reduced downtime and maintenance costs • Improved quality control • Enhanced safety and security	• Data Security • Legacy • Infrastructure • Interoperability • Scalability
(HIoT) Healthcare IoT	• Surgical Robotics • Medication Delivery bots • Smart Hospitals • Telepresence Robots	• Improved patient outcomes and quality of care • Reduced healthcare costs • Enhanced patient engagement. Better disease management	• Data Privacy and Security • Interoperability • Regulatory • Compliance • Technical • Challenges
Smart Cities	• Waste Management Bots • Intelligent Transportation Systems • Smart Grids • Surveillance Drones	• Improved quality of life for citizens • Reduced energy consumption and environmental impact • Enhanced safety and security • Efficient resource management	• Infrastructure Costs • Public • Acceptance • Data Privacy and Security
Smart Agriculture	• Soil Analysis bots • Automated Irrigation • Autonomous • Tractors, Drones • Livestock Monitoring	• Increased crop yields and efficiency • Reduced water consumption and pesticide use • Improved food safety and quality • Enhanced sustainability	• Cost • Data • Connectivity • Weather • Variability
Smart Homes	• Cleaning Robots • Security Robots • Home Automation • Entertainment Robots	• Reducing energy consumption and costs • Enhancing home security with smart devices • Creating a personalized and comfortable living environment	• Complexity • Initial Cost • Security Risks • Interoperability

3.5.2 Healthcare IoT (HIoT)

Healthcare IoT refers to the use of medical devices and sensors which are interconnected, to collect patient data, monitoring health conditions and healthcare process. It may include devices like wearables, smart medical equipment, and remotely patient monitoring system. HIoT makes continuous tracking of vital signs easier which allows healthcare professionals to make valid decisions and offer more personalized care. It supports telemedicine which can reduce the need to visit hospitals and a better access to healthcare. HIoT enhances the outcomes of patients by ensuring consistent time interventions and enables data sharing between patients and doctors in real-time [25]. Robotics applications in IoT have shown major developments in healthcare sector by improving machine operated surgeries, better patient care, and enhancing the services of medicine [26].

3.5.3 Smart Cities

IoT technologies are used in smart cities to improve the standards of urban life by strengthening services like energy management, waste management, transportation and public safety. Sensors can collect different data on factors like water quality, energy consumption, traffic and infrastructure conditions. This dataset can help city officials to make better decisions on the optimisation of resources and improve public services. IoT embedded Smart city also include domains like real-time traffic management, smart street lighting, and efficient waste management systems. The main goal of smart cities is to create sustainable urban environments that are highly responsive to the needs of citizens [27]. Robotics in smart cities has transformed the urban life by providing better public services, systemize city management, and infrastructure efficiency. It provides autonomous solutions for transportation, waste disposal, safety etc [28].

3.5.4 Smart Agriculture

Smart agriculture, also known as precision farming integrates IoT technology to make agriculture practices more efficient and productive. Sensors can monitor the soil condition, weather patterns, and crop health to provide farmers a real-time data to make better decisions. Its key applications may include Automation in irrigation systems, livestock tracking devices and drones for crop monitoring. Smart agriculture helps to reduce water wastage, optimize crop yields, and helps reducing the environmental impact of farming. By using these data-driven insights, farmers can improve the quality and quantity of their produce while making farming more sustainable [29]. Robotics with IoT have revolutionized agriculture and farming by improving the productivity, better resource management and more. It has been found helpful in solving problems like shortage of labour, need of sustained farming, and variability in weather [30].

3.5.5 Smart Homes

IoT devices are used in Smart homes to remotely control household functions like lighting, security, and appliance use. These systems depend on the interconnected devices like voice-activated assistants, smart thermostats and security cameras which can be controlled via mobile applications or voice commands. Smart homes provide convenience, efficiency and security. The owners can easily monitor their homes remotely, alerts about unusual activities can be received, and everyday tasks can also be automated such as turning off lights, or other appliances based on the preference [31]. Powered by IoT, Robotics can be used to enhance domestic life by proving convenience, household tasks automation, and better security. It makes the home environment more smart, responsive, and adaptive [32].

4 Security Threats in Robotics and IoT

There are various challenges present in Internet of Things (IoT) related to its security due to multiple number of devices interconnected through a network, limited storage and processing power, the vulnerabilities in their software and hardware. These threats and risks can broadly be categorised into active, passive and miscellaneous threats on the basis of the method of attack.

4.1 Active Threats

Active threats involve a direct interference with IoT system. Usually the attacker aims to disrupt, control or alter the network and devices. These attacks typically include modification of data, services, or device control [33].

- **Botnets**: Botnets is an interconnected network of IoT machines that are controlled remotely to initiate attacks like distributed denial of service (DDoS) or other unfair activities. It's one of the most prevalent multi-stage cyber-attacks in IoT environment that initially starts with scanning activity and ends at the DDoS (Distributed Denial of Service) attack. The corrupted devices are connected through installed bot program with a central server or a network which sends the further commands to the infected devices to send spams, flooding DDoS and perform other malicious activities. Few of the prominent examples are Mirai, Q-bot, Reaper.
- **Man-in-the-Middle (MITM) Attacks**: The communication of the IoT gadgets or other systems is intercepted, allowing attackers to eavesdrop on data and steal it. MITM attack makes Data vulnerable to tampering which exhibit two noticeable characteristics, a detectable delay in the session and unusual travel times when compared to previous normal transactions, as the attacker typically manipulates or reroutes the data flow which will lead to increased latency.

- **Device Cloning and Counterfeiting**: IoT devices are vulnerable to counterfeiting and cloning, often without security protections same as the authentic versions. Attackers generally use cloned device to access the network and manipulate data transmissions, and the counterfeit devices can introduce vulnerabilities in the network such as hidden malware or weaker encryption, making them easier targets for cyberattacks which will compromise the data integrity and will weaken the reliability of system and user trust. The risks of device cloning and counterfeiting shows the need for secure device provisioning, robust authentication mechanisms, and ceaseless monitoring which will ensure the integrity of the IoT network. Counterfeit devices can be used to create vulnerabilities in the network or infect malicious code.
- **Insecure Interfaces**: Security flaws may exist in the APIs, mobile apps, or web-based interface that are interconnected with IoT devices. They can provide entry points for attackers to make use of vulnerabilities in IoT devices and their networks, which can lead to unauthorized device control, data breaches or manipulation of the functionalities of IoT. It Can be a caused due to various problems like weak authentication and authorization, insufficient encryption, exposed sensitive information, inadequate input validation, lack of regular updates etc.
- **Physical Security Risks**: IoT devices have potential risk of physical attacks, such as theft or tampering, when placed in unprotected area. It can cause an attacker to directly gain the access of IoT devices physically which will further compromise in the entire system's security. IoT tools are often available for different environments including remote areas and unsecured places, making the tools vulnerable to various physical threats. These risks may include device tampering, theft and lost, side-channel attacks, hardware debugging and exploitation, environmental exploitation and physical destruction.

Physical access od IoT devices can allow attackers to directly manipulate them, extract data, or introduce malicious code.

4.2 Passive Threats

In passive attacks, the main focus is on collecting data or eavesdropping without affecting or altering the functioning of the IoT system. It is more difficult to identify these type of attacks as they don't interfere with device operations [34].

- **Data Privacy and Security Breaches**: Unauthorized access to private data stored or collected through IoT devices or channelled over IoT networks. Internet of Things devices are highly interconnected and may transmit sensitive data across the networks which makes this distributed nature and large scale of IoT systems more vulnerable to various security breaches and privacy violations. It includes various types of threats in which unauthorized access, data eavesdropping, weak authentication and authorization, data tampering and insufficient data encryption are most prominent ones.

- **Eavesdropping**: Tracing or monitoring the exchange of data between IoT devices to gain access to secured information. Also known as packet sniffing or traffic interception, it is a major concern in IoT networks as eavesdropping refers to an unauthorized user secretly listening to the interaction of IoT devices, between devices and central clouds or server they connect to. This type of attack mainly exploits insecure data transmissions which leads to privacy breaches and misuse of data.
- **Traffic analysis**: Observing patterns of data without any actual alteration causing security breaches like revealing sensitive user or device data. In this kind of attack, the attackers monitor the traffic patterns of network without the need of intercepting the actual data. Even with a proper encryption, its possible to gather valuable insights by analysing traffic frequency, volume and message timing between IoT devices. Traffic analysis in industrial IoT environments can expose operational routines which may enable more focused attacks. This threat takes advantage of weaknesses in IoT networks, such as predictable communication flows and insufficient security measures, potentially compromising privacy and system integrity. Counteracting traffic analysis requires strong encryption and network obfuscation techniques to mask traffic patterns and protect sensitive information.
- **Outdated systems and security vulnerabilities**: Mostly the IoT devices have either limited or no automatic update capabilities which may leave them vulnerable to known security threats for extended periods. Outdated systems in IoT present serious security vulnerabilities and risks as numerous devices operate on outdated firmware or software with known issues. These vulnerabilities are easily exploited by attackers to gain unauthorized access or control over the machine. Many IoT devices do not have regular update system so they remain at risk of malware and other threats. Manufacturers may discontinue support for old or discontinued devices, leaving them without essential security patches. This oversight can result in data breaches, privacy infringements, and compromised system integrity.
- **Insufficient Encryption**: The data transmitted through IoT tools and their core system are vulnerable for interceptions because of poor or outdated encryption. Insufficient encryption in IoT systems poses a significant security risk as many IoT devices do not utilize robust encryptions for data transmission and storage due to which sensitive information like personal data, health records, and operational commands can be intercepted by attackers during the transmission or accessed through compromised devices. Weak encryption algorithms or incorrect implementations makes it easier to be decrypted and get exploit by attacker. The limited processing power of many IoT devices often results in the adoption of outdated or less secure encryption methods. Unencrypted data is vulnerable to interception, which may lead to potential data theft, tampering, or eavesdropping.

4.3 Miscellaneous Threats

Miscellaneous threats cover the indirect or broader risks that arise from the shortcomings or flaws of IoT ecosystem or weak infrastructure. These threats may not classify under active or passive but can significantly cause damage to the system or exploitation of data [35].

- **Firmware Vulnerabilities**: There can be some weakness in the firmware which can be exploited (i.e. software embedded in hardware) in IoT devices. Firmware vulnerabilities are a major security threat in IoT devices, as many devices depend on outdated or inadequately designed firmware that may contain known flaws. These vulnerabilities are put to use to gain unauthorized access, inject malware, or disturb the functionality of devices by attackers. Unlike conventional software, firmware typically does not receive regular updates which leaves devices vulnerable over time. Manufacturers may not offer long-term support for older devices which heightens the risk. These vulnerabilities can result in data breaches, privacy infringements, and compromised system integrity.
- **Supply Chain Attacks**: Security can be compromised during the manufacturing or supply process of IoT tools. Supply chain attacks involve exploiting vulnerabilities in the supply of IoT devices that compromises with the security. Attackers can infiltrate the supply chain at various stages including the manufacturing, design, or distribution of the devices. Malicious code may be embedded in firmware or hardware components before the devices reach consumers to enable unauthorized access once the devices are operational. These attacks take advantage of the trust among manufacturers, suppliers, and customers making them very challenging to detect. It may result in data leaks, unauthorized access to sensitive systems, and may even cause disruption to the critical infrastructure.
- **Privacy Invasion**: Personal information like health condition, location or password are collected by some IoT devices. If these devices lack in security, it will lead to unauthorized access of data that can be misused. Privacy invasion is a major threat associated with IoT devices, which collect large volumes of personal data from users regularly. Various smart devices like voice assistants, cameras, and health monitors continuously gather sensitive information such as location, daily chores, and even conversations. If this data is not properly secured, unauthorized individuals can access it resulting in significant privacy breaches.
- **Insecure Firmware**: Vulnerable or outdated firmware on IoT devices can potentially leave them vulnerable to attackers and security breaches. Insecure firmware poses a major threat in the IoT ecosystem, as numerous devices depend on firmware that may contain vulnerabilities or outdated security measures. Firmware serves as the foundational software that manages hardware functions but if it is poorly designed or not regularly updated, it may leave devices open for attackers.
- **Lack of Standardization**: Due to inconsistent security standards across different platforms and devices make comprehensive security more difficult, increasing the potential of breaches. The absence of standardization in the IoT devices presents various threat to security and privacy. With a wide variety of manufacturers,

protocols, and device types, consistency in security practices and guidelines are often lacking throughout the industry. This can result in various problems such as vulnerabilities in communication protocols, incompatible devices and uneven security measures.

These threats can be detected or prevented using Machine Learning (ML) and its algorithms. ML algorithms can identify those patterns and anomalies that indicates malicious activities, such as unusual network traffic, device behaviour deviations, or suspicious sensor readings. ML models, by continuously learning from new data can adapt themselves for evolving threats and provide real-time alerts. We will explore more of the specific applications and benefits of ML for IoT securities.

5 ML Methods for IoT Securities

Machine learning methods Comprises of supervised and unsupervised approaches which can further be divided into SVM, KNN, PCA, ICA and other methods. These methods present high accuracy in detecting various known threats and can be fine-tuned for the detection of specific attack types. Supervised and Unsupervised Methods can be combined to follow a hybrid approach for enhanced IoT security. It allows for a flexible and adaptive system that can detect known threats and new, unknown anomalies [36]. In this parameter, security across different layers of IoT systems is conveyed using machine learning (ML) algorithms. The layers are categorized into several key areas like ML for perception layer security, ML for network layer security, and ML for application layer security. Each layer plays a crucial role in the overall security of IoT systems. For example, in perception layer ML techniques are used to secure data collection process from sensors and devices, while in the network layer, it helps to detect threats such as data breaching or unauthorized access. In application layer, ML can enhance security for data processing [37].

5.1 ML at Physical/Perception Layer

The physical layer in IoT includes the use of real-world objects like sensing gadgets and actuators that can collect and process information. IoT devices like motion, heat, and acceleration sensors are adopted to perform different sensing tasks. As the number of IoT devices is growing, so does the amount of data they produce which reflects a correlation between IoT device expansion and generation of big data. Machine learning (ML) in term of analysing data plays a vital role and can provide better abilities of making decisions and enhanced security in IoT environments.

In this context, perception-layer authentication is a promising application of ML methods. ML-based authentication offers more adaptable solutions rather than the traditional techniques by learning patterns of device communication, detect anomalies in real time, thus providing better defence against threats like spoofing in wireless networks. From the perspective of physical layer, security threats are investigated since the physical layer security techniques may achieve low-computational complexity, perfect secrecy and improvised channel changes.

On the basis of modified physical authentication flow, soft information PDA algorithm, and hash map-based authentication key randomization, a physical layer authentication model is suggested.

Wang et al. [38] provided a solution that mitigate the threats related to security are explored in the perspective of physical layer, as technologies related to physical layer security offer benefits such as to achieve a perfect secrecy, lowering the computational complexity, minimal consumption of resources, and effective adaptation to channel variations. Wang et al. [39] presented a physical layer authentication approach to solve issues related to unavailability of authentication infrastructure. It is structured on the modified physical authentication flow, a PDA algorithm (modified), around soft information, and authentication key randomization on the basis of hash mapping. Benaya et al. [40] proposed Intelligent Omni-Surfaces (IOS) as an alternative to Intelligent Reflecting Surfaces (IRS) along with Unmanned Aerial Vehicles (UAV) is proposed to ensure secure communication in an Internet-of-Things (IoT) system. Huang et al. [41] presented an authentication method using radio frequency fingerprinting (RFF) based on deep learning is proposed. This approach can be implemented on the base station side without adding complexity to the IoT devices. Sharma et al. [42] proposed a lightweight misbehaviour detection method, based on behaviour rule specification, is proposed for IoT-embedded cyber-physical systems (BRIoT). It models the system in such a way that misbehaviour in an IoT devices caused by attacks exploiting vulnerabilities can be detected through automatic model checking. Rajkumar et al [43] proposed a security model at physical layer for the security of wireless communication. It basically generates artificial noise that confuses the attackers using methods like eavesdropping. Neupane et al [44] presented a survey around three dimensions i.e. legal and ethical concerns, attack surfaces and human-robot interaction security. Explained the holistic understanding to the developers and other readers about these areas to enhance the security of AI-Robotics system.

5.2 ML at Network Layer

In IoT, the network layer has an important role in safeguarding transmission of data between devices and gateways while also handling communication protocols and routing data across complex networks. The use of machine learning at this layer has become increasingly important to enhance security and maintain smooth data flow.

A primary ML application at the network layer is intrusion detection, where the analysis of network traffic in real-time helps in identifying unusual patterns or anomalies that might represent malicious activities like Distributed Denial of Service (DDoS) attacks or any other unauthorized access attempts. ML has another key application in security threat prediction where it examines the past incidents to predict potential attack patterns which allows a network to take proactive security measures. Through incorporating ML into network layer, IoT systems are able to detect threats more quickly, dynamically adapt defences, and optimize communication greatly, improving overall network security. Zahar et al. [45] presented an innovative technique for detecting jamming in IoT systems, with a focus on transmission and environmental dynamics within any industry. In an industrial environment, it first analyse the conditions for communication and jamming and then collects the quality of service (QoS) data parameters of connection during the normal communication, and blocking the attacks on production lines within a server-client architecture using the IoT technology. Saleem et al. [46] provided a survey on network layer security using 5G in IoT applications, outlining the vulnerabilities in term od security and requirements at the network layer in networks with wireless sensors, IoT, and 5G-enabled IoT systems. Lee et al. [47] presented a formulation on IoT device energy consumption model and have figured out the benefits of saving energy if the IoT device are embedded with the e AP system model, in comparison to the traditional model with AP system. Jahangeer et al. [48] presented a comprehensive review on internal cyberattacks in the network layer. Focused in the literature considering to present ways such that sinkhole attacks can be detected and prevented. Ullah et al. [49] designed an innovative anomaly-based intrusion detection IoT based networks model. It initially uses a convolutional neural network (CNN) to create a classification model that's multiclass. The model proposed is then implemented using 1D, 2D, and 3D CNNs. Siriweera et al. [50] surveyed the architecture of cloud robotics and explained about a design with top-down approach for highly variable and complex challenges by using an architecture with unified framework as the cognitive approach. Raza et al. [51] proposed DTLS as a better approach for security in communication for IoT devices rather than using the standard method the key establishment requires certificates like the X.509 and public keys.

5.3 ML at Application Layer

Application layer in IoT handles data processing, user interactions, and delivers services which makes it a critical area to ensure data security, privacy and integrity. ML ensures the overall enhanced security at this layer by analyzing large volumes of data, detecting threats and providing real-time defence measures against vulnerabilities. Major applications of ML at the application layer are in user authentication and access control, malware detection and data analytics. ML algorithms can uncover certain security patterns and insights that might signal risks or inefficient encryption, providing better decision-making and can lead to more rigid security protocols, and application architectures. Integrating machine learning into application layer enhances IoT security as it enables faster threat detection, effective data protection, and adaptive strategies to counter emerging threats.

Amjad et al. [52] investigated 34 influential research works where the interoperability for IIOT is performed of the application layer protocols. Diego et al. [53] presented a study on the advantages offered by IoT safe, which is a popular security decoupling method and is applied with intersection with the Constrained Application Protocol (CoAP). Hassija et al. [54] provided a review of the challenges in security and potential threats in IoT applications. A discussion based on various existing and emerging technologies which aim to achieve a valuable trust in the IoT applications. Hoang et al. [55] presented the advancements of machine learning assisted physical layer (PHY) security are discussed where security in PHY has been categorized into two different types, PHY authentication and secure PHY transmission. Tailored ML tools are applied to identify greater opportunities and challenges in tracking PHY security challenges. Zhou et al. [56] presented the taxonomy of smart contracts related security issues in IoT technology is invested. To tackle these challenges in terms of security and address the constraints of current approaches, a tree-based machine learning vulnerability detection (TMLVD) method is proposed to analyse the vulnerabilities in smart contracts. Liu et al. [57] conducted an experiment to control a mobile robot remotely and was successful using a 33.6 kb/s modem network, presenting the improvements in robot teleportation and their sensibility. Centeno et al. [58] presented the importance for a lunar surface imagery of Multi-frame Super Resolution method. Evaluated the method using measures like signal to noise ratio over other methods of super resolution.

6 Recent Advancements

Hussain et al. [59] presented a comprehensive review of evolving security requirements, newly emerging attack vectors, and current security solutions for IoT networks. This work presents the latest Machine Learning methods aimed to address diverse security challenges in IoT and discuss potential future research directions that can enhance ML-based IoT security. Kornaros et al. [60] proposed a well explored study of Machine Learning methodologies for IoT framework securities and study of various solutions lookout for hardware-based ML methods for authentication, malware detection, access offloading and access control. Al-Garadi et al. [61] provided a survey of ML algorithms and the current advances that may can be used in enhancing the existing IoT system security methods. ML methods present various IoT security opportunities challenges which can be proven as potential advance research directions. Bagga et al. [62] presented distributing attacks by using data mining approach has been really successful in detection of attacks with low costs and high performance. Intrusion detection system anomaly is evaluated for the use of IoT and one-class SVM in a real smart building scenario. Abdalzaher et al. [63] presented ML benchmarking and introducing a taxonomy that may classify the ML models as linear and non-linear ones on the basis of the type of problem. Kornaros et al. [64] presented hardwarebased Machine Learning ways for IoT controlled access, secure offloading, authentication schemes and malware detection. Illustrates the value of various IoT security approaches to be addressed in flexible, effective and seamless manner and to provide solutions for integrating accelerators trade-off and customize embedded device architecture to effectively use ML-based methods.

In another work, Samriya et al. [65] provided security enhancement cloud-based technique rooted on customer Internet of Things utilization AML techniques for smart health systems. The trust-based encryption cryptographic analysis cloud network security can be used to enhance the security of cloud networks. Zhou et al. [66] presented an analysis for vulnerabilities where smart contracts can be performed with the help of tree-based ML vulnerability detection (TMLVD) method. TMLVD uses the intermediate representation of abstract syntax trees derived smart contracts as input for a treebased learning network to generate a predictive model. IIly et al. [67] provided ML algorithms for the detection of attack affinity of different smart home security attacks and the quality of its features. A software defined networking-based architecture and intrusion prevention mechanisms within home IoT systems and network. Alhogail et al. [68] proposed a study based on high performance optimised model to detect intrusions in IoT networks based on MQTT (Message Queue Telemetry Transport protocol) for smart homes. On the basis of empirical evaluation, the technique with best results is the Generalized Linear Model with random over sampling with an accuracy of 100% and a 100% f-score. Sahu et al. [69] addressed the requirement to integrate open-source platforms for image-guided robotic research to upgrade situational awareness. Presented an ENTRI named open-source toolkit that integrates an AMBF with a 3-D slicer. Chen et al. [70] proposed a pressure control strategy configurated with affordable pneumatic system and by using only 2-state 2-way solenoid valve, is capable to program pressure states (Table 2).

7 Conclusion

Machine Learning has become one of the most prominent technologies for improving the security of Internet of Things and robots. It provides a dynamic and adaptive method to protect robotics integrated IoT system by using specific algorithms like supervised learning, unsupervised learning, semi-supervised learning and reinforcement learning to detect security threats in real-time. ML models can continually learn from various IoT collected data to increase their accuracy and efficiency in a variety of applications like malware protection, intrusion detection, and predictive threat detection. IoT is a network of interconnected devices that collect and exchange data with each other over the internet. This vast interconnected system of IoT and robots presents significant security vulnerabilities as IoT devices generally have limited computational resources and lack robust security measures. Thus, they become a target of cyber-crime or other security threats. These security breaches may occur in different IoT layers, Physical Layer, Network Layer, or Application Layer. Machine learning is a prominent tool to counter these threats by implementing its algorithms and technologies like hash map-based authentication key randomization, intrusion detection, and, malware detection and data analytics.

Table 2 Summary of recent advances

References	Year	Contributions
Hussain et al. [59]	2020	Latest Machine Learning methods addressing diverse security challenges in IoT
Kornaros et al. [60]	2022	ML methodologies for IoT framework securities and various solutions for hardware-based ML methods
Al-Garadi et al. [61]	2020	ML methods to enhance existing IoT system security methods
Bagga et al. [62]	2020	Data mining approach to detect attacks with low cost and high performance
Abdalzaher et al. [63]	2023	Taxonomies to identify the ML models as linear and nonlinear ones
Kornaros et al. [64]	2022	Hardware-based Machine Learning methods to provide solutions for integrating accelerators trade-off
Samriya et al. [65]	2024	Security enhancement for cloud based on customer IoT utilization AML techniques for smart healthcare
Zhou et al. [66]	2022	A tree-based ML vulnerability detection (TMLVD) method for vulnerability analysis to generate a prediction model
IIIy et al. [67]	2022	A networking-based intrusion prevention mechanism for home IoT systems and network
Alhogail et al. [68]	2024	Advanced model for intrusion detection in IoT networks based on MQTT for smart homes
Sahu et al. [69]	2024	An open-source toolkit to integrate AMBF with a 3-D slicer
Chen et al. [70]	2024	Using Pneumatic soft robotics with low cost to program pressure states

References

1. M. Goudarzi, H. Wu, M. Palaniswami, R. Buyya, An application placement technique for concurrent IoT applications in edge and fog computing environments. IEEE Trans. Mobile Comput. **20**(4), 1298–1311 (2021). https://doi.org/10.1109/TMC.2020.2967041
2. C. Xenofontos, I. Zografopoulos, C. Konstantinou, A. Jolfaei, M.K. Khan, K.-K.R. Choo, Consumer, commercial, and industrial IoT (In)security: attack taxonomy and case studies. IEEE Internet Things J. **9**(1), 199–221 (2022). https://doi.org/10.1109/JIOT.2021.3079916
3. N. Neshenko, E. Bou-Harb, J. Crichigno, G. Kaddoum, N. Ghani, Demystifying IoT security: an exhaustive survey on IoT vulnerabilities and a first empirical look on internet-scale IoT exploitations. IEEE Commun. Surv. Tutorials **21**(3), 2702–2733 (2019). https://doi.org/10.1109/COMST.2019.2910750
4. Y.-F. Li, L.-Z. Guo, Z.-H. Zhou, Towards safe weakly supervised learning. IEEE Trans. Pattern Anal. Mach. Intell. **43**(1), 334–346 (2021). https://doi.org/10.1109/TPAMI.2019.2922396
5. M. Usama et al., Unsupervised machine learning for networking: techniques, applications and research challenges. IEEE Access **7**, 65579–65615 (2019). https://doi.org/10.1109/ACCESS.2019.2916648
6. X. Wang et al., Deep reinforcement learning: a survey. IEEE Trans. Neural Netw. Learn. Syst. **35**(4), 5064–5078 (2024). https://doi.org/10.1109/TNNLS.2022.3207346

7. F. Noorbehbahani, M. Saberi, Ransomware detection with semi-supervised learning, in *2020 10th International Conference on Computer and Knowledge Engineering (ICCKE)*, Mashhad, Iran, pp. 024–029 (2020). https://doi.org/10.1109/ICCKE50421.2020.9303689
8. J. Shehu Yalli, M. Hilmi Hasan, A. Abubakar Badawi, Internet of Things (IoT): origins, embedded technologies, smart applications, and its growth in the last decade. IEEE Access **12**, 91357–91382 (2024). https://doi.org/10.1109/ACCESS.2024.3418995
9. N. Mazhar, R. Salleh, M. Zeeshan, M.M. Hameed, Role of device identification and manufacturer usage description in IoT security: a survey. IEEE Access **9**, 41757–41786 (2021). https://doi.org/10.1109/ACCESS.2021.3065123
10. S.C. Mukhopadhyay, S.K.S. Tyagi, N.K. Suryadevara, V. Piuri, F. Scotti, S. Zeadally, Artificial Intelligence-based sensors for next generation IoT applications: a review. IEEE Sens. J. **21**(22), 24920–24932 (2021). https://doi.org/10.1109/JSEN.2021.3055618
11. J. Ding, M. Nemati, C. Ranaweera, J. Choi, IoT connectivity technologies and applications: a survey. IEEE Access **8**, 67646–67673 (2020). https://doi.org/10.1109/ACCESS.2020.2985932
12. T. Sauter, A. Treytl, IoT-enabled sensors in automation systems and their security challenges. IEEE Sens. Lett. **7**(12), 1–4 (2023). Art no. 7500904. https://doi.org/10.1109/LSENS.2023.3332404
13. G. Chen, W. Dong, F. Qiu, G. Guan, Y. Gao, S. Zeng, Scalable and interactive simulation for IoT applications with TinySim. IEEE Internet Things J. **10**(23), 20984–20999 (2023) https://doi.org/10.1109/JIOT.2023.3285244
14. S.-I. Hahm et al., Reliable real-time operating system for IoT devices. IEEE Internet Things J. **8**(5), 3705–3716 (2021). https://doi.org/10.1109/JIOT.2020.3025612
15. E. Pagliari, L. Davoli, G. Ferrari, Harnessing communication heterogeneity: architectural design, analytical modeling, and performance evaluation of an IoT multi-interface gateway. IEEE Internet Things J. **11**(5), 8030–8051 (2024). https://doi.org/10.1109/JIOT.2023.3317672
16. A. Alwarafy, K.A. Al-Thelaya, M. Abdallah, J. Schneider, M. Hamdi, A survey on security and privacy issues in edge-computing-assisted Internet of Things. IEEE Internet Things J. **8**(6), 40044022 (2021). https://doi.org/10.1109/JIOT.2020.3015432
17. I. Ali, I. Ahmedy, A. Gani, M.U. Munir, M.H. Anisi, Data collection in studies on Internet of Things (IoT), Wireless Sensor Networks (WSNs), and Sensor Cloud (SC): similarities and differences. IEEE Access **10**, 33909–33931 (2022). https://doi.org/10.1109/ACCESS.2022.3161929
18. M. Zhang, H. Zhang, D. Yuan, M. Zhang, Learning-based sparse data reconstruction for compressed data aggregation in IoT networks. IEEE Internet Things J. **8**(14), 11732–11742 (2021). https://doi.org/10.1109/JIOT.2021.3059735
19. J. Han, G.H. Lee, J. Lee, T.Y. Kim, J.K. Choi, A novel deep-learning-based robust data transmission period control framework in IoT edge computing system. IEEE Internet Things J. **9**(23), 23486–23505 (2022). https://doi.org/10.1109/JIOT.2022.3203156
20. N.A. Nabeeh, M. Abdel-Basset, H.A. El-Ghareeb, A. Aboelfetouh, Neutrosophic multi-criteria decision making approach for IoT-based enterprises. IEEE Access **7**, 59559–59574 (2019). https://doi.org/10.1109/ACCESS.2019.2908919
21. N. Chaabane, S. Mahfoudhi, K. Belkadhi, Interpolation-based IoT sensors selection. IEEE Sens. J. **24**(21), 36143–36147 (2024). https://doi.org/10.1109/JSEN.2024.3461833
22. M. Aazam, S. Islam, S.T. Lone, A. Abbas, Cloud of Things (CoT): Cloud-Fog-IoT task offloading for sustainable Internet of Things. IEEE Trans. Sustain. Comput. **7**(1), 87–98 (2022). https://doi.org/10.1109/TSUSC.2020.3028615
23. Z.-H. Sun, Z. Chen, S. Cao, X. Ming, Potential requirements and opportunities of blockchain-based industrial IoT in supply chain: a survey. IEEE Trans. Comput. Soc. Syst. **9**(5), 1469–1483 (2022). https://doi.org/10.1109/TCSS.2021.3129259
24. T. Czimmermann et al., An autonomous robotic platform for manipulation and inspection of metallic surfaces in industry 4.0. IEEE Trans. Autom. Sci. Eng. **19**(3), 1691–1706 (2022). https://doi.org/10.1109/TASE.2021.3122820
25. M.N. Bhuiyan, M.M. Rahman, M.M. Billah, D. Saha, Internet of Things (IoT): a review of its enabling technologies in healthcare applications, standards protocols, security, and market

opportunities. IEEE Internet Things J. **8**(13), 10474–10498 (2021). https://doi.org/10.1109/JIOT.2021.3062630
26. A. Ingole, S. Sonwane, Robotics and automation in healthcare, in *2024 International Conference on Healthcare Innovations, Software and Engineering Technologies (HISET)*, Karad, India, pp. 78–81 (2024). https://doi.org/10.1109/HISET61796.2024.00036
27. E.H. Houssein, M.A. Othman, W.M. Mohamed, M. Younan, Internet of Things in smart cities: comprehensive review, open issues, and challenges. IEEE Internet Things J. **11**(21), 34941–34952 (2024). https://doi.org/10.1109/JIOT.2024.3449753
28. J. Song, A. Song, F. Zhang, RoboMaze: swarm robotics and coordinated navigation in smart city, in *2021 IEEE 23rd International Conference on High Performance Computing & Communications; 7th International Conference on Data Science & Systems; 19th International Conference on Smart City; 7th International Conference on Dependability in Sensor, Cloud & Big Data Systems & Application (HPCC/DSS/SmartCity/DependSys)*, Haikou, Hainan, China, pp. 1881–1885 (2021). https://doi.org/10.1109/HPCC-DSSSmartCity-DependSys53884.2021.00279
29. A.A. AlZubi, K. Galyna, Artificial Intelligence and Internet of Things for sustainable farming and smart agriculture. IEEE Access **11**, 78686–78692 (2023). https://doi.org/10.1109/ACCESS.2023.3298215
30. X. Han, L. Xu, Y. Peng, Z. Wang, Trend of intelligent robot application based on intelligent agriculture system, in *2021 3rd International Conference on Artificial Intelligence and Advanced Manufacture (AIAM)*, Manchester, United Kingdom, pp. 205–209 (2021). https://doi.org/10.1109/AIAM54119.2021.00050
31. X. Wang, X. Qian, Information-centric IoT-based smart home control and monitoring system. IEEE Sens. J. **24**(21), 35722–35729 (2024). https://doi.org/10.1109/JSEN.2024.3462929
32. B. Chen, S. Qin, W. Wang, Y. Ge, Design of smart home control system based on STM32, in *2021 4th International Conference on Robotics, Control and Automation Engineering (RCAE)*, Wuhan, China, p. 208212 (2021). https://doi.org/10.1109/RCAE53607.2021.9638880
33. K. Lounis, M. Zulkernine, Attacks and defenses in short-range wireless technologies for IoT. IEEE Access **8**, 88892–88932 (2020). https://doi.org/10.1109/ACCESS.2020.2993553
34. Y.R. Siwakoti, M. Bhurtel, D.B. Rawat, A. Oest, R.C. Johnson, Advances in IoT security: vulnerabilities, enabled criminal services, attacks, and countermeasures. IEEE Internet Things J. **10**(13), 11224–11239 (2023). https://doi.org/10.1109/JIOT.2023.3252594
35. Y. Yang, J. Wang, P. Liu, A. Fu, Y. Zhang, Uncovering access token security flaws in multiuser scenario of smart home platforms. IEEE Internet Things J. **11**(22), 36841–36857 (2024). https://doi.org/10.1109/JIOT.2024.3429417
36. W.G. Hatcher, C. Qian, F. Liang, W. Liao, E.P. Blasch, W. Yu, Secure IoT search engine: survey, challenges issues, case study, and future research direction. IEEE Internet Things J. **9**(18), 16807–16823 (2022). https://doi.org/10.1109/JIOT.2022.3153171
37. A.K. Yerrapragada, T. Eisman, B. Kelley, Physical layer security for beyond 5G: ultra secure low latency communications. IEEE Open J. Commun. Soc. **2**, 2232–2242 (2021). https://doi.org/10.1109/OJCOMS.2021.3105185
38. D. Wang, B. Bai, K. Lei, W. Zhao, Y. Yang, Z. Han, Enhancing information security via physical layer approaches in heterogeneous IoT with multiple access mobile edge computing in smart city. IEEE Access **7**, 54508–54521 (2019). https://doi.org/10.1109/ACCESS.2019.2913438
39. Y. Wang, J. Jin, Y. Li, C. Choi, A reliable physical layer authentication algorithm for massive IoT systems. IEEE Access **8**, 80684–80690 (2020). https://doi.org/10.1109/ACCESS.2020.2989395
40. A.M. Benaya, M.H. Ismail, A.S. Ibrahim, A.A. Salem, Physical layer security enhancement via intelligent omni-surfaces and UAV-friendly jamming. IEEE Access **11**, 2531–2544 (2023). https://doi.org/10.1109/ACCESS.2023.3233947
41. D. Huang, A. Al-Hourani, Physical layer spoof detection and authentication for IoT devices using deep learning methods. IEEE Trans. Mach. Learn. Commun. Netw. **2**, 841854 (2024). https://doi.org/10.1109/TMLCN.2024.3417806

42. V. Sharma, I. You, K. Yim, I.-R. Chen, J.-H. Cho, BRIoT: behavior rule specification-based misbehavior detection for IoT-embedded cyber-physical systems. IEEE Access **7**, 118556–118580 (2019). https://doi.org/10.1109/ACCESS.2019.2917135
43. S. Rajkumar, D.N.K. Jayakody, R. Alkanhel, A. Muthanna, Physical layer security in N-pair NOMA-PLNC wireless networks. IEEE Access **10**, 91356–91371 (2022). https://doi.org/10.1109/ACCESS.2022.3198978
44. S. Neupane et al., Security considerations in AI-robotics: a survey of current methods, challenges, and opportunities. IEEE Access **12**, 22072–22097 (2024). https://doi.org/10.1109/ACCESS.2024.3363657
45. F.T. Zahra, Y.S. Bostanci, M. Soyturk, LSTM-based jamming detection and forecasting model using transport and application layer parameters in Wi-Fi based IoT systems. IEEE Access **12**, 3294432958 (2024). https://doi.org/10.1109/ACCESS.2024.3371673
46. K. Saleem, G.M. Alabduljabbar, N. Alrowais, J. Al-Muhtadi, M. Imran, J.J.P.C. Rodrigues, Bio-inspired network security for 5G-enabled IoT applications. IEEE Access **8**, 229152–229160 (2020). https://doi.org/10.1109/ACCESS.2020.3046325
47. S. Lee, H. Choi, T. Kim, H.-S. Park, J.K. Choi, A novel energy-conscious access point (eAP) system with cross-layer design in Wi-Fi networks for reliable IoT services. IEEE Access **10**, 6122861248 (2022). https://doi.org/10.1109/ACCESS.2022.3181304
48. A. Jahangeer, S.U. Bazai, S. Aslam, S. Marjan, M. Anas, S.H. Hashemi, A review on the security of IoT networks: from network layer's perspective. IEEE Access **11**, 71073–71087 (2023). https://doi.org/10.1109/ACCESS.2023.3246180
49. I. Ullah, Q.H. Mahmoud, Design and development of a deep learning-based model for anomaly detection in IoT networks. IEEE Access **9**, 103906–103926 (2021). https://doi.org/10.1109/ACCESS.2021.3094024
50. A. Siriweera, K. Naruse, Survey on cloud robotics architecture and model-driven reference architecture for decentralized multicloud heterogeneous-robotics platform. IEEE Access **9**, 40521–40539 (2021). https://doi.org/10.1109/ACCESS.2021.3064192
51. S. Raza, L. Seitz, D. Sitenkov, G. Selander, S3K: scalable security with symmetric keys—DTLS key establishment for the Internet of Things. IEEE Trans. Autom. Sci. Eng. **13**(3), 1270–1280 (2016). https://doi.org/10.1109/TASE.2015.2511301
52. A. Amjad, F. Azam, M.W. Anwar, W.H. Butt, A systematic review on the data interoperability of application layer protocols in industrial IoT. IEEE Access **9**, 96528–96545 (2021). https://doi.org/10.1109/ACCESS.2021.3094763
53. J.D. de Hoz Diego, J. Saldana, J. Fernández-Navajas, J. Ruiz-Mas, Decoupling security from applications in CoAP-based IoT devices. IEEE Internet Things J. **7**(1), 467–476 (2020). https://doi.org/10.1109/JIOT.2019.2951306
54. V. Hassija, V. Chamola, V. Saxena, D. Jain, P. Goyal, B. Sikdar, A survey on IoT security: application areas, security threats, and solution architectures. IEEE Access **7**, 82721–82743 (2019). https://doi.org/10.1109/ACCESS.2019.2924045
55. T.M. Hoang, A. Vahid, H.D. Tuan, L. Hanzo, Physical layer authentication and security design in the machine learning era. IEEE Commun. Surv. Tutorials **26**(3), 1830–1860 (2024). https://doi.org/10.1109/COMST.2024.3363639
56. Q. Zhou, K. Zheng, K. Zhang, L. Hou, X. Wang, Vulnerability analysis of smart contract for blockchain based IoT applications: a machine learning approach. IEEE Internet Things J. **9**(24), 24695–24707 (2022). https://doi.org/10.1109/JIOT.2022.3196269
57. J.N.K. Liu, M. Wang, B. Feng,iBotGuard: an Internet-based intelligent robot security system using invariant face recognition against intruder. IEEE Trans. Syst. Man Cybern. Part C (Appl. Rev.) **35**(1), 97–105 (2005). https://doi.org/10.1109/TSMCC.2004.840051
58. J.I. Delgado-Centeno, P.J. Sanchez-Cuevas, C. Martinez, M.A. Olivares-Mendez, Enhancing lunar reconnaissance orbiter images via multi-frame super resolution for future robotic space missions. IEEE Robot. Autom. Lett. **6**(4), 7721–7727 (2021). https://doi.org/10.1109/LRA.2021.3097510
59. F. Hussain, R. Hussain, S.A. Hassan, E. Hossain, Machine learning in loT security: current solutions and future challenges. IEEE Commun. Surv. Tutorials **22**(3), 1686–1721 (2020). https://doi.org/10.1109/COMST.2020.2986444

60. G. Kornaros, Hardware-assisted machine learning in resource-constrained IoT environments for security: review and future prospective. IEEE Access **10**, 58603–58622 (2022). https://doi.org/10.1109/ACCESS.2022.3179047
61. M.A. Al-Garadi, A. Mohamed, A.K. Al-Ali, X. Du, I. Ali, M. Guizani, A survey of machine and deep learning methods for Internet of Things (IoT) security. IEEE Commun. Surv. Tutorials **22**(3), 1646–1685 (2020). https://doi.org/10.1109/COMST.2020.2988293
62. M. Bagaa, T. Taleb, J.B. Bernabe, A. Skarmeta, A machine learning security framework for Iot systems. IEEE Access **8**, 114066–114077 (2020). https://doi.org/10.1109/ACCESS.2020.2996214
63. M.S. Abdalzaher, M.M. Fouda, H.A. Elsayed, M.M. Salim, Toward secured IoT-based smart systems using machine learning. IEEE Access **11**, 20827–20841 (2023). https://doi.org/10.1109/ACCESS.2023.3250235
64. G. Kornaros, Hardware-assisted machine learning in resource-constrained IoT environments for security: review and future prospective. IEEE Access **10**, 58603–58622 (2022). https://doi.org/10.1109/ACCESS.2022.3179047
65. J.K. Samriya, C. Chakraborty, A. Sharma, M. Kumar, S.K. Ramakuri, Adversarial ML-based secured cloud architecture for consumer internet of things of smart healthcare. IEEE Trans. Consum. Electron. **70**(1), 2058–2065 (2024). https://doi.org/10.1109/TCE.2023.3341696
66. Q. Zhou, K. Zheng, K. Zhang, L. Hou, X. Wang, Vulnerability analysis of smart contract for blockchain based IoT applications: a machine learning approach. IEEE Internet Things J. **9**(24), 24695–24707 (2022). https://doi.org/10.1109/JIOT.2022.3196269
67. P. Illy, G. Kaddoum, K. Kaur, S. Garg, ML-based IDPS enhancement with complementary features for home IoT networks. IEEE Trans. Netw. Serv. Manage. **19**(2), 772–783 (2022). https://doi.org/10.1109/TNSM.2022.3141942
68. R. Alasmari, A.A. Alhogail, Protecting smart-home IoT devices from MQTT attacks: an empirical study of ML-based IDS. IEEE Access **12**, 25993–26004 (2024). https://doi.org/10.1109/ACCESS.2024.3367113
69. M. Sahu et al., ENTRI: enhanced navigational toolkit for robotic interventions. IEEE Trans. Med. Robot. Bionics **6**(4), 1405–1408 (2024). https://doi.org/10.1109/TMRB.2024.3475827
70. P. Chen, Q. Ding, Y. Liu, Z. Deng, J. Huang, Programmable pressure control in pneumatic soft robots with 2-way 2-state solenoid valves. IEEE Robot. Autom. Lett. **9**(7), 6448–6455 (2024). https://doi.org/10.1109/LRA.2024.3406201

Internet of Robotic Things in Smart Domains: Applications and Challenges

Satish Khatak and Rajeev Sharma

Abstract Incorporating the Internet of Robotic Things (IoRT) into different smart domains has enabled robotic systems to be intelligent and connected, enabling advancements in automation and smart environments as well. IoRT identifies robots and the Internet of Things (IoT) as key components for improving human–robot interaction (HRI) and machine-to-machine interaction (M2M). This facilitates smart cities, healthcare, agriculture, and security applications. With IoRT systems, data can be processed in real-time and autonomous decisions can be made in a variety of environments through the use of cloud-based resources. In this paper, we discuss networking connectivity, robotic autonomy, cyber security, and interoperability of systems as key components of IoRT. IoRT architectures and frameworks are described to demonstrate their potential to revolutionize industry and society through the development of secure, efficient, and responsive systems.

Keywords Application · Internet of robotic things (IORT) · Internet of thing (IOT) · Cyber security · Robotic · Smart domain

1 Introduction

It has been a decade since robotic systems helped to facilitate various technological advancements. A combination of robotics and networking technologies enabled robots to perform more functions during the 1990s [1]. Through the integration of local and distributed data, IoRT determines how sensors from different sources can be incorporated and explained. This enables data to be used for monitoring and verifying physical phenomena [2, 3]. Robots that are networked are described as devices

S. Khatak (✉) · R. Sharma
Department of ECE, The Technological Institute of Textile & Sciences, Bhiwani, Haryana, India
e-mail: khatak1977@yahoo.co.in

R. Sharma
e-mail: rajeevsharma78@yahoo.com

R. Sharma and G. Jeon (eds.), *Data Analytics for Smart Robotics and Its Applications*,
Intelligent Systems Reference Library 272,
https://doi.org/10.1007/978-3-031-87697-4_10

that use standard networking protocols like TCP, UDP, or IEEE 802.11 to communicate over the LAN. Today, robots are widely used in engineering and manufacturing. For humanity to grow in the digital age, robotic systems are critical components. A number of robot systems have evolved into IoRT applications, including artificial intelligence, robots, sophisticated networks, and electric mobility [4]. By using various paths, networks, and services, robots can now be connected to anything, anywhere, and anytime. With the advancement of intelligent networking. Networked robotic devices may form the pillar of future IoRT applications, which includes edge nodes [5]. Several goals are being pursued with IoT and robotic technologies: (1) to detect, sense, and track, and (2) Creating the ability to move and interact. These two objectives together have improved the development of IoRT. In the words of the author [4], IoRT refers to a global active network framework that is dynamic, adaptable, and self-adjusting. Data communication protocols (the rules that govern data exchange over a network) and interoperability protocols (the rules that govern data exchange between multiple systems) define the characteristics of data communications. Through the use of intelligent interfaces, cognition, and connectivity, both virtual and physical robots are capable of making decisions and acting according to various rules. Sensors and actuators are integrated into the Internet of Things so that people, devices, and processes can collaborate together.

Robotic interaction services (ROI) and human–robot interactions (HRI) are some of the functions performed by IoRT. Various equipment is needed for robotic systems, including microphones, cameras, LIDARs, RADARs, and even sensors for interaction and reaction [6]. A number of facilities, including assisted living facilities, hotels, and hospitals, use robots with HRI. A number of robots can be monitored continuously through IoRT and HRI. It is possible for data leakage to occur in IoRT communications. The privacy of customers is compromised when data is leaked. IoRT systems can pose a security risk if data collected by sensing units or sensors is transmitted to them, for instance. The IoRT system detects physical environments through sensing units, while sensors provide information to devices [7, 8]. Robots have long been used to solve problems as a result of artificial intelligence built into them, making them intelligent and autonomous. The IoRT is the fusion of two technologies. Several robots need to be aware of each other in their environment in order to work together to achieve more complex tasks, which is the reason for multiple robot systems (MRS), in which require (M2M communication. Another option is the Internet of Things, which is a network of smart, singularly addressable devices (things) interconnected via the Internet and capable of sharing and processing data. Everywhere and everything is connected through the Internet of Things. A wide range of industries have benefited from IoT, including healthcare, military, industrial processes, businesses, and smart cities.

There has been some overlap between the technologies in different domains. The use of robots has expanded across a variety of sectors, including human assistance, healthcare, industrial processes, military support, and emergency response. They have also performed important functions in monitoring, maintaining, and regulating environmental and equipment conditions [2, 6, 9, 10]. The Internet of Things has a variety of applications, including healthcare, smart environments, and even homes, factories, and offices. Thus, IoRT, which is a combination of these two, provides a smarter, more autonomous system that can be applied across these domains. While IoRT promises to deliver both worlds, it also poses new challenges that are necessary to achieve its full potential and expand its scope and complexity.

Since the advent of robotic systems, various socioeconomic aspects of human society have changed greatly [11]. Across all industries, robotic manipulators are widely used for repetitive, tedious, dangerous, or critical tasks, including assembling products, painting cars, wrapping boxes, and fabricating shields. Preprogrammed robots have proven to be highly accurate, precise, durable, and fast in various structured industrial applications. As robot technology was incorporated with existing networks in the 1990s, unstructured environments became more accessible to robots, resulting in networked robotics [12].

There has been a lot of attention paid to the Internet of Things in the last few decades due to its exponential growth. It uses smart technology to create a network of real-world objects, called 'Things,' that exchange information online [13, 14]. Throughout human history, this revolution has shaped every aspect of life. A wide variety of organizations are adopting the Internet of Things concept, such as military, robotics, healthcare, nanotechnology, or space. Gartner estimates that 127 IoT devices connect to the Web every second by 2025, and this number will grow to 75 billion by then. The IoT has reached 75 billion units by 2025, according to Gartner, which estimates 127 new devices will connect to the web every second. The global IoT industry will spend 1.29 trillion dollars by 2020, and 93% of businesses will adopt it.

When robots are first applied in these applications, they are merely a single machine with limited computational capability and hardware components. A Networked Robotic System was created to resolve these issues through the connection of robots over wires and wireless networks [15]. In addition to network latency, memory limitations, and computation and learning limitations, their inherent resource constraints resulted in high costs. Real-life scenarios require sophisticated data analysis and high computing power to complete complex tasks. Cloud Robotics [16] has recently been proposed as a solution to these limitations, allowing access to resources on demand and supporting operations through cloud infrastructure.

A fertile ground for Internet of Things technology can be found in robotics. In recent decades, robotics has revolutionized several aspects of human society with its modern and rapidly evolving technology. The science and engineering of robotics involves conceiving, designing, manufacturing and operating machines. In disaster areas or on extraterrestrial missions, for instance, it can perform tedious, repetitive, and hazardous tasks humans cannot afford, such as rescuing people in disasters.

2 Related Work

A general overview of IoT is presented here. It is first presented what IoT is and how it works. Cloud Robotics is included in the IoRT, which later becomes IoT. As a result of Kevin Ashton's work at the Auto-ID Center in 1999, the term "IoT" was coined. Internet of Things sensors and actuators play a significant role. Communication, collaboration, and practical analysis can be integrated with IoT to enable real-time decision-making. There is a variety of smart objects which are controlled by radio tags, sensors, actuators, and smart devices within this concept, as well as the ubiquitous presence of radio tags, sensors, actuators, and smart devices within humans and their socioeconomic cultures. In addition to implementing unique addressing schemes, securing communication channels, and using standard architectures, these platforms facilitate interactions with neighbors and encourage cooperation [17]. A virtual personality, physical characteristics, and identity are all possible for physical and digital things. These "things" integrate seamlessly into information networks and use intelligent interfaces. Communication protocols are interoperable and standard, allowing the infrastructure to adjust itself dynamically. Users and their environments often communicate data associated with it. The purpose of this paper is to use The International Telecommunication Union (ITU) definition of Internet of Things given in [18], will be used as it encompasses all the essential elements of IoT, I believe:

ITU Definition for IoT: A modern information society's infrastructure consists of interconnected (physically and virtually) systems based on interoperable, existing and evolving technologies for information and communication.

Internet of Robotic Things Definition: Interoperability describes the set of information and communication technologies that are interoperable both now and in the future, which connect robotic objects in an information society. Cloud computing, cloud storage, and other Internet technologies can be made available to robots through converged cloud infrastructures and shared services, improving their computing power, and communication capitals available and connecting data centres with cloud-based middleware to enhance independence, remove maintenance and update overheads, and enhance connectivity with modern data centers, offloading tasks to the cloud, robot mobility may be reduced and operating time may be reduced because of reduced real-time deadlines (cloud data transfer rates).

Robotics and IoT are merged in IoRT. According to Dan Kara, in a report published by ABI The IoT is a category of intelligent devices that gather data from many sensors and use both local and distributed intelligence for controlling objects and determining their best course of action. Robotic things are interconnected through the IoRT, enabling advanced robotic services. By using the IP protocol, along with its IPv6 version, robotic things can benefit from modern communication technologies and interoperability capabilities to process data, store memories, compute overhead, and be more secure. This mitigates robotic issues and allows them to be resolved. IoRT also deploys the IoT technology, which enables it, beyond networked robots and cloud robots.

IoRT's multidisciplinary nature provides advanced robotic capabilities, making it suited for interdisciplinary solutions spanning a wide range of fields. With IoRT, ordinary robots are enhanced in their ability to perceive, move, manipulate, and make decisions, while at the higher level (decision making, autonomy, cognitive interaction) and at the system-level (configurability, adaptability, dependability). These challenges are addressed by combining IoT features with multi-radio access, artificial intelligence, and cognitive technologies. Integrated into Industrial 4.0, the Internet of Things (IoT) and smart services have become indispensable components of manufacturing [19]. Industrial revolutions 1 to 3 saw the invention of mechanical, electrical, and electronic technologies leading to an increase in productivity. With the increasing need for an improved quality of life, more and more personal and digital services are being produced [20]. Integrated vertically, horizontally, and end-to-end comprise Industry 4.0's paradigm. A horizontal integration involves integrating data networks and resources across the value chain so that industries can cooperate effectively and services can be delivered in real time. An example of vertical integration is the development of smart factories with their networked manufacturing systems, which allow custom manufacturing to replace conventional manufacturing processes. Lastly, end-to-end integration entails digitizing both raw materials and manufacturing processes throughout a product life cycle.

A new generation of technologies is emerging as a result of Industry 4.0, and a new set of terms is emerging as well. An IoRT connects smart technologies with robotic systems, enabling robots to share, network, and collect information from humans and machines in order to perform elaborate tasks. CPSs, or Cyber-Physical Systems, are among the factors associated with Industry 4.0. With new technologies advancing, CPS applications have become fundamental, allowing all physical devices to be connected to the Internet to enable smarter products and processes [21]. By integrating Industry 4.0 and CPS, smart spaces are created and smart factories are built. Applications and processes can be monitored based on sensor and actuator data in a defined control area. As it relates to industrial manufacturing, Smart Factory involves the use of IoT technology to collect, analyze and exchange data related to manufacturing processes [22].

3 Challenges Facing the Internet of Robotic Things (IORT) Industry

With the fourth industrial revolution, robots, cloud technologies and CPS are blending together to benefit different domains. IoRT uses the cloud to offload robotic applications, access large amounts of data, and allow robots to share information, aiming to gain new skills and knowledge. Compared to traditional robotic applications, this can be advantageous.

In the field of IoRT, many applications have been developed and implemented. It is important to note that, even though the collected literature covers a variety of different domains, it is common to find overlap between methodological approaches and technologies.

- **Manufacturing**

IoT technology has become increasingly embedded in the manufacturing process, and industrial robots have continued to evolve over the last few years and continue to change. Industry 4.0 is embraced by manufacturing largely due to IoRT, which embeds sensors, automates processes and monitors products and services [23]. Spot welding and spray painting are among the manufacturing applications where Industrial IoRT can be used. Spot welding involves applying a welding tool independently to a specific point on an object [24]. Using spray paint, paint is applied along a predetermined trajectory to a surface [25]. When there is a minor deviation from the intended path, a real-time correction may be required to achieve adequate production and prevent errors. As a result of their hardware limitations, industrial robots face significant difficulties in processing and storing data. Hence, cloud computing is advantageous, since such task executions can be achieved by utilizing cloud technologies, ensuring integration and interoperability [26].

- **Agriculture**

In recent years, smart agricultural systems have gained considerable interest as a means of enhancing crop yields, controlling weeds and gathering field data [27]. Incorporating services, products, and knowledge into systems like IoRT can enhance farm productivity, quality, and profitability.

IoRT is used in precision agriculture to manage crop sizes, weather conditions, and resources using herbicides, fertilizers, and irrigation [28]. Data collection, control, and monitoring can also be supported by WSN. An RF transceiver, microcontroller, various sensors, and power sources combine to make up a wireless sensor network (WSN). A specific area is monitored by such devices. Since the devices have limited autonomy, the deployment of heterogeneous sensor nodes is critical for maintaining the widest coverage and minimizing the number of nodes [29].

- **Surveillance**

In surveillance systems, places and people are monitored in order to ensure their safety and security. In response to the advent of high-tech surveillance, high-risk environments, military borders, public spaces, and homes have increasingly become monitored. A closed circuit television (CCTV) camera can be used indoors or outdoors in this context [30]. While such technology has several benefits, it also entails several issues and limitations, primarily because of potential tampering and blind spots. Nevertheless, such technology has several benefits, but also causes several issues and limitations. A system with more cameras can partially resolve such drawbacks, but will also become more expensive and complex. Robots can be used to detect humans in any environment, such as homes, military environments, industrial factories, retail and wholesale stores, and more generally, any environment where humans exist. While conducting surveillance missions, the robot also transmits real-time data about GPS, magnetic fields, air quality, and environmental conditions [31].

- **Health-care and Education**

Educational and health-care sectors are viewed as major beneficiaries of IoRT systems. There are several applications in which IoRT may be beneficial to health, society, and economy, particularly for patients with specific needs. Robots, sensors, and IoT devices will be incorporated into health care systems so that patients will receive real-time health information and can receive diagnoses in real-time, resulting in fewer human errors such as incorrect medication, dosage, and procedure prescriptions [32]. Also, IoRT technology can be used to track patients, staff, and ambulances, to automatically gather data, and to sense the environment.

Therefore, robots must adopt appropriate and adaptive behavior in order to develop into exploitable homework helpers and teachers by maintaining and achieving appropriate social relationships with humans. Electrodermal activity (EDA) can be measured during human–robot interactions (HRI) since it represents emotional responses like attention, anxiety, and excitement. A child's EDA response may differ from an adult's [33], since they may respond differently to certain impulses. To provide child education through IoRT technologies, data must be collected, processed, and analyzed.

4 Framework and Architecture for the Internet of Robotic Things

In addition to Smart Environments and autonomous agents (robots), IoRT is also an integral part of the IoT. The concept of smart environments includes a wide range of requests, such as smart buildings, smart machines, and smart rooms [34]. An example of a smart environment is a smart room, a smart machine, a smart building, or a smart city. HVAC systems are typically used to maintain reasonable temperatures and air humidity in the space, or simple sensors and actuators are used to monitor these conditions. The air conditioning or windows can be opened or closed with simple drives, and the heating can be adjusted at the appropriate time. In addition, managing energy consumption can also be a task. For example, it may be necessary to turn off the electricity when someone leaves a building, or to begin household appliances, such as washing machines, dishwashers, and lawn waterers, only when there are minimal power overloads or nightly tariffs.

Even though Smart Environment can perform monitoring functions and simple actuator actions, a range of indoor actions cannot be performed by it (such as moving objects or performing certain operations). An agent of this type is a robot, which may be an assistive robot, a manipulator, a service robot, or an autonomous mobile vehicle. Integrated smart environments and robotics combine the capabilities of both through intelligent agents. As robots become more autonomous, their level of sensor data collection and onboard processing should also increase, allowing them to operate independently. Indoor environments contain a variety of types of sensors (RFID, occupancy sensor, surveillance camera, magnetic sensor, infrared/sonic beacon, etc.) as well as computing resources (mobile phones, routers, computers, servers).

Smart environment resources enhance robot functionality, but the Smart Environment also plays an important role in the IoRT concept. The Smart Environment appears to employ agents (robots) to perform complex tasks beyond monitoring and performing simple actions. Through this integration, robots are able to receive tasks from Smart Environments (like Smart Buildings or Smart Factory), which are also able to monitor the execution of tasks. The sensor network of a robot provides information on navigation, collision avoidance, object detection, and human interaction. Basically, IoT refers to an advanced level of Internet of Things, combining cloud computing, wireless sensing, control, guidance, safety, and security into one system. On the Smart Environment side, it includes distributed monitoring and networked control, multi-agent control, planning and control, localization, navigation, swarms, and human–robot interactions.

The Internet of Things Robotics (IoRT) might also be called IoT Robotics, which combines IoT with robots. As a result of the IoT, robots can operate more intelligently than they would ordinarily be in order to accomplish tasks in this context. Integrating robots with smart environments allows them to learn about their environment and interact with things other than other robots. As robots become more knowledgeable and smarter due to context awareness and their autonomous capabilities, they will be better able to perform more complex tasks. IoRT has not yet been defined universally, but some studies have attempted to describe it. In a nutshell, we can say that:

4.1 IoRT Architecture

It has been attempted to provide an architectural overview of IoRT's landscape through studies. Illustration of the IoRT landscape, illustrating robots' and IoT's roles, their main applications, technologies, and domains of use [35]. In IoT robots, components respond to and communicate with each other. In order for robotic systems to function effectively, certain abilities are required. Among the basic abilities of people are perception, which allows them to recognize and locate their surroundings; movement, which lets them move from point to point in a dynamic environment; manipulation ability, which involves processing information and altering the environment as required.

- **Physical**

Physical layers are included in IoRT architectures. A robot, a sensor, or an actuator are all hardware components. As intelligent agents, robots are capable of communicating with each other and setting up more complex systems than a single robot can execute alone, in order to achieve a specific goal. Robots can be represented in a variety of ways, such as vehicles, drones, appliances, and health care equipment. Actuators and sensors are devices used by agents to detect signals from the environment, while actuators are tools used to perform actions. Combine sensors and actuators to enhance and optimize robot performance. The physical layer serves as an abstraction of all technical information and provides services to the network and control layers related to its periphery.

- **Network and Control**

Physical layer components are efficiently integrated through control protocols and communications processes at the Network and Control layer. The robots can communicate with each other and with the cloud in a number of ways. In communications, several different protocols are used depending on how far away sender and receiver are from each other and how much power is required to send data. Communication over the Internet can be accomplished using any protocol. A number of short-range protocols are used at the link layer and physical layer, including IEEE 802.11 (WiFi) and Bluetooth. It is common to use long-range protocols, such as Sigfox and LoRaWAN. IEEE 802.15.4 is not the only protocol based on this standard; 6LowPAN and Zigbee are also based on it. There are two protocols commonly used in IoT networks: 6LowPAN and RPL (Running Protocol for Low Power and Lossy Networks).

- **Service and Application**

An implementation of an IoRT service or application is located at the top of the architecture, and it monitors, processes, and controls end-user operations. Protocols can be selected from a number of options. UDP is one of the transport layer protocols used by CoAP (Constrained Application Protocol). Resources-constrained devices on lossy, low-power networks benefit from it. It is unclear whether CoAP is secure or appropriate for IoT devices. There are several protocols that can be used to transmit data, including MQTT (Message Queue Telemetry Transport), which works on a publish-subscribe basis, and HTTP (Hypertext Transfer Protocol). With MQTT, you can send messages over unreliable networks in a lightweight way and be assured of their reliability. A robot system may benefit from avoiding REST, however, since it adds complexity. In addition to XMPP (eXtensible Message Presence Protocol), there are other common application protocols. With this open standard, messages can be sent in real-time between devices in a decentralized fashion. To communicate from the cloud to an exoskeleton robot, a cloud platform used MQTT for data transmission and HTTP for data communication [36]. In other words, considering IoRT's multi-dimensional capabilities, additional components and layers are required in order to address challenges about connectivity, sensing, security, controls, data analytics, and other API's.

4.2 IoRT Architecture Layer

Robots can perform tasks with high precision by operating in smart environments with the IoRT architecture. There are four layers in the IoRT architecture:

- **Hardware Layer**: As a part of the IoRT architecture, hardware layers are used to connect robots with the world. There are many different types of devices in this category, such as automobiles, robots, homes appliances, medical equipment, mobile phones, sensors, satellites, defense equipment, and field equipment. Autonomous vehicles, robots, home appliances, medical devices, mobile phones, sensors, satellites, defense equipment, and field equipment all belong to this field.
- **Network Connectivity**: In addition to cellular, short-range, and medium-range network connectivity, there are three options at the network layer for connecting sensors and robots.
- **Internet Connectivity**: Connectivity to the Internet is the only means of facilitating communication between devices and accessing information anywhere and everywhere. All types of communication are made possible by Internet connectivity, which is considered to be the heart of the IoRT architecture. IoRT is based on IoT, which enables machines to communicate with machines as well as machines to communicate with humans using IoT standardized protocols [6]. This protocol performs many functions such as sending messages, multicasting, queuing, addressing, packet switching, and datagram-based operations.
- **Robotic Platform Support**: In order to conduct experiments, researchers highly utilize robotic platforms with basic software capabilities. It is possible to implement IoRT robotics with different Robot operating systems (ROS), such as ROS, RT Middleware, RSNP, CANOpen, UNR-PF, ORiN, etc.
- **M2M2A Cloud Platform Support**: For the growth of advanced robotic systems such as the Internet of Things (IoRT), M2M and M2A architectures provide a solid foundation. A network is a system for exchanging information among machines without involving humans. As part of the architecture, data acquisition is also analyzed, user data is managed, and sensor data is also managed by this component.
- **Big Data Services**: As a result of Big Data, we can optimize, analyze, predict, make arithmetical answers, etc.
- **IoT Business Cloud Services**: The Internet of Things can reduce the cost of working overhead for diverse complex business activities and improve the efficiency of organizations and manufacturers at the same time. In the IoRT model, every business transaction can be executed on a cloud computing platform, such as a SaaS, PaaS, or IaaS cloud computing platform. The possibility of performing complex e-Commerce operations is made possible by a variety of packages and APIs. Through Cloud and Robotics, business operations can be improved in accuracy, ease of use, and business decisions in complex activities by using IoRT.
- **IoT Cloud Robotics Infrastructure**: Only services are included in this component. IoRT can perform a variety of services through the use of IoT and Cloud,

includes manipulating images, recognizing objects, recognizing voices, and navigating maps.

- **Application Layer**: Atop the IoRT architecture is the application layer. When robotics and IoT are combined, they can be incorporated into a variety of business applications, including R&D centers and data centers. It is possible to visualize IoRT as an area with endless potential.

5 Applications of IoT Robotics

There are a variety of domains in which IoRT can be applied. In addition to robotics, IoT has been applied in many areas as well. Using IoRT, smart homes, smart offices, smart factories, and smart cities can be created. Manufacturing equipment, search-and-rescue missions, and the military are all applications that use IoRT, including health care (especially telemedicine). Presented an application framework illustrating IoRT's potential.

5.1 Environmental and Industrial Smart Technologies

In smart homes and offices, IoT and IoRT are used to control outdoor temperatures and humidity, schedule and remind chores, and monitor patient health remotely. It is also possible to schedule and schedule meetings with the help of smart robots. With M2M communication and coordination, smart robots can be used to control and monitor manufacturing processes, as well as perform tasks like monitoring and regulating environmental conditions, lifting heavy objects and locating and mapping, they can also perform complex and safety–critical tasks that are otherwise dangerous for humans. Cleans indoor environments with a robot called Roomba. The AppBot Riley is a typical indoor/outdoor monitoring robot [37]; monitoring children's indoor activities.

5.2 Healthcare

Healthcare delivery is being transformed by the use of IoT and robots in many ways, with the goal of assisting caregivers and ensuring patient care is timely and efficient. Telemedicine, e-Health, and m-Health technologies can be enhanced by incorporating robots into their delivery to enhance in-house and remote healthcare delivery.

5.3 Military and Policing

It has been proven that IoT and robotics are useful components of the military ecosystem [38] as well as security in general. A few examples of how IoT are used are for detecting intrusion and prevention, for detecting chemicals, biological, and explosive devices, etc. [39]. Military IoT (MIoT) architecture presented at a high level. Data/information will be captured and shared in the military environment by using this architecture. Second, the collected information will be communicated between military personnel, equipment, and application systems. Experiments and operations involving robots have been carried out extensively in the military. IoT and robotics have been merged to create IoRT, which enhances robot intelligence and knowledge in the military setting.

5.4 Smart Restaurant

With the use of IoRT revolutionizing all aspects of daily life, people are changing their daily routines. There is a proposal for a robot waiting in restaurants in [40]. There is a QR code that stores information about table number and menu. Chefs and managers receive the orders together with the table number on a website. A robot delivers the orders to the customers when they are ready. Cloud-based storage keeps the information safe and automatic. A number of IoRT applications are being explored in academia, and some solutions have already been adopted by commercial companies. IoRT technology will continue to be a valuable research area as new technologies such as 5G and 6G are introduced in the upcoming years.

Intelligent environments can be created using robotics in a variety of ways, including home automation, domestic services, healthcare, transportation, logistics, and industry. SSDBs are software-defined buildings that are part of smart environments [16]. They are programmable buildings with sensors that can be used for various purposes including monitoring presence, recognizing activities and identifying people, as well as detecting emotional states of the occupants. A variety of hardware components can be used to implement the sensing IoT functions, including occupancy sensors, tracking sensors, and positioning sensors. Robots use the obtained raw data to determine their actions by applying a computation method. A distributed robotic sensing system, such as SSDB, can be used to compensate for the limited memory and sensing capabilities of robots. Robots will become more autonomous and capable as a result. In this way, robots will be able to function as actuators for a variety of purposes [28]. Using IoRT as an example, let's discuss some applications.

5.5 *Smart Home*

For a Smart Home to serve humans better, sensors and devices should be available that allow people to interact with each other and with robots appropriately. A smart home can include the following services: The monitoring of nutrition and the recommendation of meals; Assisting children with their homework; People with various diseases are monitored during their therapy, etc. Home chores can be handled by robots as part of their service.

5.6 *Smart Office*

Visitors can be greeted at a Smart Office using sensors and devices, Providing meeting support as well as controlling a comfortable microclimate and facilitating workflow. Meanwhile, robots that are programmed to interact effectively with visitors and employees can also be implemented.

5.7 *Smart Nursing House*

Patient care at Smart Nursing House is tailored to each patient's specific needs. The cost of routine tasks can be reduced for patients who need round-the-clock care through integration of IoRT services [41]. A service robot can customize work based on patients' needs by announcing events and news throughout the day. In some cases, the IoRT may require external assistance in place of maintenance staff. Intelligent environments are capable of detecting wandering elderly people, identifying behavior disturbances, and assisting them with robots. The patient's room needs to be equipped with various sensors and devices so that medications can be taken, their environment can be regulated, and the patient can be monitored.

5.8 *Smart Workshop/Factory*

Using smart workshop/factory techniques, production processes can be monitored and controlled, accounting and quality control of manufactured products can be carried out, equipment can be set up and emergency situations can be monitored and the environment can be adjusted accordingly. Additionally, different robots (as well as autonomous, collaborative, stationary and mobile) can carry out loading/unloading operations, production operations (welding, assembly, etc.), and quality control in the industrial zone, while mobile robots are also capable of controlling human-made machinery and equipment.

6 Challenges in Implementing IoRT

Among the research interests of the team are technologies that enable interoperability across heterogeneous environments, distributed computing environments, as well as energy efficiency [42] and using artificial intelligence to enhance robot autonomy [43]—Robot autonomous decision-making can be improved by using machine learning algorithms; IoRT components, such as confidentiality, authentication, authorization, and integrity, can be protected by cryptographic algorithms, preventing intrusions, identity theft, and other cyber security attacks inherited from the Internet component of the framework.

HRI (human–robot interaction) and M2M (machine-to-machine communication) are both at the point of solving the problem of connecting distributed "things. There is no set time for sharing data in IoRT or its constituent technologies, but the data must be shared at any time and in a timely manner. Edge computing and big data are becoming significant tools for managing and sharing data in complex communication situations [44]. Robots and other IoRT devices have to be cognitively autonomous as part of M2M interactions. AI and ML can help them achieve that.

Networks, the backbone of IoRT, are susceptible to cyber-attacks (as they form the backbone of networking) and must be protected from hacking. IoRT communications must be secure via point-to-point protocols to ensure robot and other "thing" communications are protected. There are several IoRT security measures, including digital certificates and signatures, but these must be adapted to work in distributed environments.

6.1 IoT Security

IoRT security measures, such as digital certificates and signatures, must be adapted for distributed environments [45]. Therefore, this paper recommends that we rethink the major security challenges, especially in robotics, in order for the Internet of Things to become a viable paradigm.

It is essential that IoT-enabled robotic health care applications are developed and implemented in a way that enforces access control and robot identification to prevent malicious behavior, which can be dangerous to humans. In addition, patients' privacy needs to be safeguarded to ensure that sensitive health data is handled effectively. A well-defined privacy and security solution is also crucial for military scenarios, due to the importance of maintaining information integrity and confidentiality. A single unauthorized user could compromise the security of entire nations if he gained access to confidential information.

6.2 Identified Challenges

Security issues, communication issues, and semantic concerns arise when smart objects and robots are interconnected as described earlier. It is imperative that a protocol stack ensures communication efficiency across bandwidth, energy, and computation within an IoT network, as well as enabling smart object and robot communication. It is undeniable that this is more challenging under extreme traffic conditions. There are two other challenges associated with the interaction of devices. To enable robots to interact with any kind of object, the former involves choosing the communication paradigm. There are actually numerous different MAC/PHY protocols out there based on a networked device's power and computing capabilities. There is now widespread WiFi availability in airport areas, allowing a wide range of devices to access the Internet, including mobile phones, tablets, laptops, and handheld devices.

7 Conclusion

With IoT devices and robots interconnected seamlessly in real-time, the Internet of Robotic Things is transforming many industries by boosting efficiency. Robots, IoT devices, and humans can now interact seamlessly in real-time in the IoRT, enhancing efficiency across multiple domains. Developing robust security protocols, optimizing algorithms, and collaborative frameworks will be necessary for IoRT's full potential to be realized. Future applications of IoRT will be found within industry 4.0, smart cities, healthcare, and beyond. Automation and intelligence are key to creating highly adaptive and intelligent systems in dynamic environments.

References

1. P.P. Ray, Internet of robotic things: concept, technologies, and challenges. IEEE Access **4**, 9489–9500 (2016). https://doi.org/10.1109/ACCESS.2017.2647747
2. P. Simoens, M. Dragone, A. Saffiotti, The internet of robotic things: a review of concept, added value and applications. Int. J. Adv. Rob. Syst. **15**(1), 1729881418759424 (2018). https://doi.org/10.1177/1729881418759424
3. S. Khalid, Internet of robotic things: a review. JASTT **2**(02), 106–118 (2021). https://doi.org/10.38094/jastt203104
4. O. Vermesan et al., Internet of robotic things intelligent connectivity and platforms. Front. Robot. AI **7**, 104 (2020). https://doi.org/10.3389/frobt.2020.00104
5. L. Vandewinckele et al., Overview of artificial intelligence-based applications in radiotherapy: recommendations for implementation and quality assurance. Radiother. Oncol.. Oncol. **153**, 55–66 (2020). https://doi.org/10.1016/j.radonc.2020.09.008

6. R.S. Batth, A. Nayyar, A. Nagpal, Internet of robotic things: driving intelligent robotics of future—concept, architecture, applications and technologies, in *2018 4th International Conference on Computing Sciences (ICCS)* (IEEE, Jalandhar, 2018), pp. 151–160. https://doi.org/10.1109/ICCS.2018.00033
7. Z. Alsulaimawi, A privacy filter framework for internet of robotic things applications, in *2020 IEEE Security and Privacy Workshops (SPW)* (IEEE, San Francisco, CA, USA, 2020), pp. 262–267. https://doi.org/10.1109/SPW50608.2020.00059
8. B. Yuan et al., Secure data transportation with software-defined networking and *k-n* secret sharing for high-confidence IoT services. IEEE Internet Things J. **7**(9), 7967–7981 (2020). https://doi.org/10.1109/JIOT.2020.2993587
9. L.A. Grieco et al., IoT-aided robotics applications: technological implications, target domains and open issues. Comput. Commun.. Commun. **54**, 32–47 (2014)
10. P. Sethi, S.R. Sarangi, Internet of things: architectures, protocols, and applications. J. Electr. Comput. Eng. **2017**, 1–25 (2017). https://doi.org/10.1155/2017/9324035
11. B. Siciliano, *Springer Handbook of Robotics*, vol. 2 (Springer, 2008), pp. 15–35, 2008.
12. G. Hu, W.P. Tay, Y. Wen, Cloud robotics: architecture, challenges and applications. IEEE Network **26**(3), 21–28 (2012)
13. S.S. Sabry, N.A. Qarabash, H.S. Obaid, The road to the internet of things: a survey, in *2019 9th Annual Information Technology, Electromechanical Engineering and Microelectronics Conference (IEMECON)* (IEEE, 2019), pp. 290–296. Accessed: 31 Oct 2024. [Online]. Available: https://ieeexplore.ieee.org/abstract/document/8876989/
14. I.U. Din et al., The internet of things: a review of enabled technologies and future challenges. IEEE Access **7**, 7606–7640 (2018)
15. N. Mohamed, J. Al-Jaroodi, I. Jawhar, A review of middleware for networked robots. Int. J. Comput. Sci. Netw. Secur. **9**(5), 139–148 (2009)
16. O. Saha, P. Dasgupta, A comprehensive survey of recent trends in cloud robotics architectures and applications. Robotics **7**(3), 47 (2018)
17. D. Giusto, A. Iera, G. Morabito, L. Atzori, *The internet of things: 20th Tyrrhenian workshop on digital communications* (Springer Science & Business Media, 2010). Accessed: 31 Oct 2024. [Online]. Available: https://books.google.com/books?hl=en&lr=&id=vUpiSRc0b7AC&oi=fnd&pg=PR5&dq=+The+Internet+of+Things:+20th+Tyrrhenian+Workshop+on+Digital+Communications&ots=JtDfGivqng&sig=tX0LMcKLxgPEFpVwZ2Y76JwM8PI
18. K.K. Patel, S.M. Patel, P. Scholar, Internet of things-IOT: definition, characteristics, architecture, enabling technologies, application & future challenges. Int. J. Eng. Sci. Comput. **6**(5) (2016). Accessed: 31 Oct 2024. [Online]. Available: http://www.opjstamnar.com/download/Worksheet/Day-110/IP-XI.pdf
19. V. Alcácer, V. Cruz-Machado, Scanning the industry 4.0: a literature review on technologies for manufacturing systems. Eng. Sci. Technol. Int. J. **22**(3), 899–919 (2019)
20. Y. Lu, Industry 4.0: a survey on technologies, applications and open research issues. J. Ind. Inf. Integr.Integr. **6**, 1–10 (2017)
21. P. Leitao, S. Karnouskos, L. Ribeiro, J. Lee, T. Strasser, A.W. Colombo, Smart agents in industrial cyber–physical systems. Proc. IEEE **104**(5), 1086–1101 (2016)
22. J. Wan et al., Software-defined industrial internet of things in the context of industry 4.0. IEEE Sens. J. **16**(20), 7373–7380 (2016)
23. P.D. Urbina Coronado, H. Ahuett-Garza, R. Morales-Menendez, P. Orta Castañón, L. D. Dávila, M.R. Flores Escalera, Connectivity of a modular electric vehicle by the use of a mobile device. Adv. Mech. Eng. **9**(7), 168781401770808 (2017). https://doi.org/10.1177/1687814017708087
24. J. Romero-Hdz, B. Saha, G. Toledo-Ramirez, D. Beltran-Bqz, Welding sequence optimization using artificial intelligence techniques: an overview. Int. J. Comput. Sci. Eng.Comput. Sci. Eng. **3**(11), 90–95 (2016)
25. K. Madhira, S. Mehta, R. Bollineni, D. Kavathia, AGWallP—automatic guided wall painting system, in *2017 Nirma University International Conference on Engineering (NUiCONE)* (IEEE, Ahmedabad, 2017), pp. 1–5. https://doi.org/10.1109/NUICONE.2017.8325614

26. H. Yan, Q. Hua, Y. Wang, W. Wei, M. Imran, Cloud robotics in smart manufacturing environments: challenges and countermeasures. Comput. Electr. Eng.. Electr. Eng. **63**, 56–65 (2017). https://doi.org/10.1016/j.compeleceng.2017.05.024
27. B.A. Shawar, E. Atwell, Chatbots: are they really useful? J. Lang. Technol. Comput. Linguist. **22**(1), 29–49 (2007)
28. H. Jawad, R. Nordin, S. Gharghan, A. Jawad, M. Ismail, Energy-efficient wireless sensor networks for precision agriculture: a review. Sensors **17**(8), 1781 (2017). https://doi.org/10.3390/s17081781
29. F.K. Shaikh, S. Zeadally, Energy harvesting in wireless sensor networks: a comprehensive review. Renew. Sustain. Energy Rev. **55**, 1041–1054 (2016). https://doi.org/10.1016/j.rser.2015.11.010
30. S. Saponara, L. Pilato, L. Fanucci, Exploiting CCTV camera system for advanced passenger services on-board trains, in *2016 IEEE International Smart Cities Conference (ISC2)* (IEEE, Trento, Italy, 2016), pp. 1–6. https://doi.org/10.1109/ISC2.2016.7580748
31. M.S. Shah, P.B. Borole, Surveillance and rescue robot using Android smartphone and the Internet, in *2016 International Conference on Communication and Signal Processing (ICCSP)* (IEEE, Melmaruvathur, Tamil Nadu, India, 2016), pp. 1526–1530. https://doi.org/10.1109/ICCSP.2016.7754413
32. R. Amin, S.H. Islam, G.P. Biswas, M.K. Khan, N. Kumar, A robust and anonymous patient monitoring system using wireless medical sensor networks. Futur. Gener. Comput. Syst.. Gener. Comput. Syst. **80**, 483–495 (2018). https://doi.org/10.1016/j.future.2016.05.032
33. R.M. Fenning, J.K. Baker, B.R. Baucom, S.A. Erath, M.A. Howland, J. Moffitt, Electrodermal variability and symptom severity in children with autism spectrum disorder. J. Autism Dev. Disord.Disord. **47**(4), 1062–1072 (2017). https://doi.org/10.1007/s10803-016-3021-0
34. M. Mazzara, I. Afanasyev, S.R. Sarangi, S. Distefano, V. Kumar, M. Ahmad, A reference architecture for smart and software-defined buildings, in *2019 IEEE International Conference on Smart Computing (SMARTCOMP)* (IEEE, Washington, DC, USA, 2019), pp. 167–172. https://doi.org/10.1109/SMARTCOMP.2019.00048
35. I. Afanasyev, et al., Towards the internet of robotic things: analysis, architecture, components and challenges, in *2019 12th International Conference on Developments in eSystems Engineering (DeSE)* (IEEE, 2019), pp. 3–8. Accessed: 31 Oct 2024. [Online]. Available: https://ieeexplore.ieee.org/abstract/document/9073162/
36. Y. Yuan, G.-Z. Cao, A. Zhu, X. Lyu, Y. Wang, Communication scheme of cloud platform for the lower limb exoskeleton rehabilitation robot, in *2020 17th International Conference on Ubiquitous Robots (UR)* (IEEE, 2020), pp. 327–332. Accessed: 31 Oct 2024. [Online]. Available: https://ieeexplore.ieee.org/abstract/document/9144989/
37. A.O. Bajeh et al., Internet of robotic things: its domain, methodologies, and applications, in *Emergence of Cyber Physical System and IoT in Smart Automation and Robotics*, ed. K.K. Singh, A. Nayyar, S. Tanwar, M. Abouhawwash. Advances in Science, Technology & Innovation (Springer International Publishing, Cham, 2021), pp. 135–146. https://doi.org/10.1007/978-3-030-66222-6_9
38. Q.P. Ha, L. Yen, C. Balaguer, Robotic autonomous systems for earthmoving in military applications. Autom. Constr.. Constr. **107**, 102934 (2019)
39. L. Yushi, J. Fei, Y. Hui, Study on application modes of military Internet of Things (MIOT), in *2012 IEEE International Conference on Computer Science and Automation Engineering (CSAE)* (IEEE, 2012), pp. 630–634. Accessed: 31 Oct 2024. [Online]. Available: https://ieeexplore.ieee.org/abstract/document/6273031/
40. T.M.N.U. Akhund, M.A.B. Siddik, M.R. Hossain, M.M. Rahman, N.T. Newaz, M. Saifuzzaman, IoT Waiter Bot: a low cost IoT based multi functioned robot for restaurants, in *2020 8th International Conference on Reliability, Infocom Technologies and Optimization (Trends and Future Directions) (ICRITO)* (IEEE, 2020), pp. 1174–1178. Accessed: 31 Oct 2024. [Online]. Available: https://ieeexplore.ieee.org/abstract/document/9197920/
41. C. Mahieu, F. Ongenae, F. De Backere, P. Bonte, F. De Turck, P. Simoens, Semantics-based platform for context-aware and personalized robot interaction in the internet of robotic things. J. Syst. Softw.Softw. **149**, 138–157 (2019). https://doi.org/10.1016/j.jss.2018.11.022

42. S. Rajendran, R.M. Lourde, Ab-initio study on FinFETs and their application in loT aided robotics, in *2016 Sixth International Symposium on Embedded Computing and System Design (ISED)* (IEEE, 2016), pp. 38–42. Accessed: 31 Oct 2024. [Online]. Available: https://ieeexplore.ieee.org/abstract/document/7977051/
43. S.H. Alsamhi, O. Ma, M.S. Ansari, Survey on artificial intelligence based techniques for emerging robotic communication. Telecommun. Syst. **72**(3), 483–503 (2019), https://doi.org/10.1007/s11235-019-00561-z
44. N.F. Thusabantu, G. Vadivu, Adoption of big data streaming techniques for simultaneous localization and mapping (SLAM) in IoT-aided robotics devices, in *Cognitive Informatics and Soft Computing*, vol. 768, ed. P.K. Mallick, V.E. Balas, A.K. Bhoi, A.F. Zobaa. Advances in Intelligent Systems and Computing. (Springer, Singapore, 2019), pp. 315–320. https://doi.org/10.1007/978-981-13-0617-4_31
45. D. Miorandi, S. Sicari, F. De Pellegrini, I. Chlamtac, Internet of things: vision, applications and research challenges. Ad Hoc Netw.Netw. **10**(7), 1497–1516 (2012)
46. L. Romeo, A. Petitti, R. Marani, A. Milella, Internet of robotic things in smart domains: applications and challenges. Sensors **20**(12), 3355 (2020)
47. F.M. Anwar, M.B. Srivastava, Precision time protocol over LR-WPAN and 6LoWPAN, in *2017 IEEE International Symposium on Precision Clock Synchronization for Measurement, Control, and Communication (ISPCS)* (IEEE, 2017, pp. 1–6. Accessed: 31 Oct 2024. [Online]. Available: https://ieeexplore.ieee.org/abstract/document/8056739/
48. G. Micoli, et al., ASAP: a decentralized slot reservation policy for dynamic 6TiSCH networks in industrial IoT, in *2019 IEEE International Conference on Communications Workshops (ICC Workshops)* (IEEE, 2019), pp. 1–6. Accessed: 31 Oct 2024. [Online]. Available: https://ieeexplore.ieee.org/abstract/document/8757078/
49. S.L. Anderson, Asimov's 'three laws of robotics' and machine metaethics. AI & Soc. **22**(4), 477–493 (2008). https://doi.org/10.1007/s00146-007-0094-5
50. S. Greene, H. Thapliyal, A. Caban-Holt, A survey of affective computing for stress detection: evaluating technologies in stress detection for better health. IEEE Consum. Electron. Mag. **5**(4), 44–56 (2016)
51. E. Guizzo, How Aldebaran robotics built its friendly humanoid robot, Pepper. IEEE Spectrum (2014)
52. J. Zhu, G. Lin, F. You, H. Liu, C. Zhou, Multiway dynamic trust chain model on virtual machine for cloud computing. China Commun. **13**(7), 83–91 (2016)

Smart Robotics and Applications Using Data Analytics—Present and Future Directions

Yazeed AlZahrani

Abstract This chapter explores the integration of smart robotics and data analytics in modern manufacturing, examining past advancements and future directions in intelligent robotic systems. With the adoption of smart manufacturing, robotic equipment has evolved from simple automation tools to sophisticated, adaptive systems powered by artificial intelligence (AI), machine learning, and data analytics. Key technologies such as industrial manipulators, collaborative robots, and unmanned aerial vehicles (UAVs) have transformed production environments, enhancing efficiency and customization. The review details how AI-driven data analytics enables autonomous decision-making in real-time, integrating cyber-physical systems in an interconnected ecosystem known as Industry 4.0. The chapter discusses existing research, methodologies, and applications in robotic learning, as well as challenges and potential research pathways. Emphasis is placed on how the Internet of Robotic Things (IoRT) will shape future manufacturing landscapes, enhancing interoperability and enabling more autonomous and precise robotic interventions across various industries.

Keywords Internet of Robotic Things (IoRT) · Internet of Things · Smart environment · Machine learning · Intelligent system · Data analytic

1 Introduction

A key element of smart manufacturing has been robotic equipment, which can be used to improve factory automation since smart manufacturing was introduced [1]. Over the past century, industrial robots have been incorporated into production lines, and they have become increasingly important as manufacturing systems have evolved. Industrial manipulators, collaborative robots, automated guided vehicles,

Y. AlZahrani (✉)
Department of Computer Engineering, College of Engineering in Wadi Addawasir, Prince Sattam Bin Abdulaziz University, Al-Kharj, Saudi Arabia
e-mail: Y.alzahrani@psau.edu.sa

R. Sharma and G. Jeon (eds.), *Data Analytics for Smart Robotics and Its Applications*, Intelligent Systems Reference Library 272,
https://doi.org/10.1007/978-3-031-87697-4_11

and unmanned aerial vehicles (UAVs) are some of the robots that have improved manufacturing systems [2]. AI, such as deep learning, GPUs, and neural processors (NPUs), will enable robots to become more intelligent. As a result of this trend, robots are being freed from old-fashioned hard-programmed scenarios and are being enabled to perform tasks that are more flexible and human-centred [3]. Robots are designed to adapt to more challenging environments by self-learning to succeed, just as machines do [4, 5]. A closer look at robot learning technology is provided here, along with a summary of background, methodologies, typical research, industrial applications, challenges, and future research directions. Using such a review, we hope to provide an overview of smart robotic manufacturing, machine learning, and robot learning, particularly recent advances in these fields. To inspire readers, manufacturing research is also emphasized alongside industrial applications and future directions.

Further, in modern digital factories, every system, machine, and product is digitally connected. Data collected from the digitization of industries covers multiple aspects of industries, making it a valuable source of information for smart industries. As Big Data infrastructures expand and develop in industries, humans will no longer be able to interpret the vast amounts of operational data generated. An emphasis is already being placed on data analytics, including the deployment of automated decision-making algorithms. Cyber-physical systems will be able to communicate with one another in the future factory, which will be made up of intelligent self-learning robots. Any time the machines need raw materials, they can request them, and they can request services as well. An economy's performance and efficiency are the main drivers of its prosperity, and better-performing and efficient industries lead to greater prosperity. By using smart technologies and maximizing machine utilization, the productivity of newly developed manufacturing industries can increase by 17–20%. Among the things manufacturers compete for to meet consumer demands are innovation, quick response times to market changes, and affordable, reliable products. As a result of this competition, smart manufacturing, digitalization of processes, and cyber-physical control have finally become a reality in manufacturing plants as well as in business centres. The use of smart devices and sensors to manufacture goods enables end customers to provide customized manufacturing instructions directly to manufacturers [2, 6]. The implementation of policies on cyber-physical systems and digital manufacturing will facilitate the advancement of manufacturing in the future. 'Industry 4.0' (I4.0) is the term used in Germany to describe the fourth industrial revolution, where interconnected machine systems interact with one another to plan production, schedule plants, customize products, manufacture flexible products, and identify and recover faults. The manufacturing industry is becoming increasingly dependent on technologies [7–9]. It is also known as "Made in China 2025," a plan and program to promote manufacturing to the next level that will be implemented by China in 2025 [10–13]. Due to the shift from hierarchical control modes to distributed manufacturing systems, a smart manufacturing system must be developed that can handle the increasing level of customization, the sudden variation in supply chains, and the smaller production lots that are needed [14]. An analysis by McKinsey Global Institute shows 60% of automation potential in the manufacturing industry, which also implies that smart manufacturing technologies can boost industries' productivity [15].

2 Related Work

Robotics emphasizes the interaction between humans and robots and how comfortable humans are in collaborating. Our discussion of the review research in Table 1 is complemented by some perspectives that differ from those listed there. It is important to note that this review places a high priority on manufacturing research, which other reviews haven't done. Second, the methodology we provide is detailed without being overly mathematical. An analysis and classification of published manufacturing research is presented. Unlike this review, we do not cover robotics, which are rarely used in manufacturing research, including humanoid robots, service robots, and medical robots. The concludes with a summary of possible research directions and challenges related to robot learning towards smart robotic manufacturing.

Using information-driven models requires multivariate models that provide in-depth insight into the system's physical behaviour. An intelligent manufacturing facility uses machines, smart sensors, and robot platforms to generate information

Table 1 Development and implementation of robotic technologies

References	Category	Key strategies and components
[19]	Controls with intelligence	Controlling impedances adaptively and applying predictive models to a variety of scenarios, interacting with the human environment
[20]	Evaluating and designing	An approach that integrates design and evaluation at every stage of development, designing and interacting with users in a user-centric manner
[21, 22]	The living space of a robot and its interaction with humans	Cyberspace can be used to enhance the perception and movement of robots by incorporating advanced sensors, artificial intelligence, and actuators
[23]	Analyzing risks	Analyzing hazards from multiple perspectives, including psychological and organizational factors, beyond physical factors
[24]	Standards compliance	A voluntary industry consensus standard should be followed; additional guidance may be obtained from supplemental technical reports
[25]	The human affective response	Monitoring human affective responses to the use of robots, improving workplaces, and minimizing adverse effects
[26]	Taking advantage of the expertise of industry leaders	Learn about building robotic systems, testing, simulation, and maintenance from industry leaders

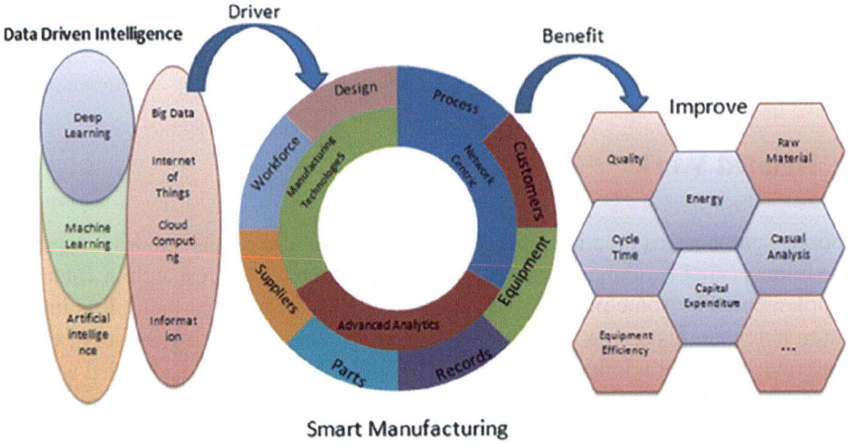

Fig. 1 Manufacturing's role in data-driven intelligence

for tracking, maintenance, and, ultimately, fundamental line management [16]. In some manufacturing units, however, this information is kept in silos. The more context records that are integrated from diverse sources of information, the better the performance and the faster production will be.

In this chapter, data modelling and analysis are discussed with an emphasis on manufacturing intelligence. In record analytics, extraordinary stages can be distinguished, such as descriptive, diagnostic, predictive, and prescriptive analytics, as demonstrated in Fig. 1.

A robot manipulator's precision performance can be summarized in two main values, according to ISO 9283, which outlines performance criteria for manipulators. Poses must be accurate and repeatable. Pose repeatability describes how well robots can return to the same position and orientation, while pose accuracy explains how precisely they move to a desired point [17]. Geometric and non-geometric parameters, as well as defects in the kinematic model, are usually responsible for pose accuracy errors, in addition to assembly and mechanical errors [17, 18]. The robot calibration process allows us to determine the parameters that affect the manipulator's accuracy so that we can modify the positioning software to make it more precise [17]. A model is developed, measurements are taken, parameters are identified, and compensation is implemented. The position of each joint is measured externally using a Laser Tracker device to determine its Cartesian pose (or end-effector pose).

3 The IoRT's Big Data Management Algorithms

It will be possible to integrate smart sensors, heterogeneous sensing devices, as well as big data algorithms and autonomous robots with cloud computing and wireless communication technologies [27, 28] in the near future. In machine learning, sensors and IoT devices are used as data sources. Turing is characterized by robot navigation, robotic algorithms on networks, smart sensors, and machine intelligence. Smart factories are built on the analysis of spatial data, the creation of virtual maps, and the integration of production systems. Several algorithms for cyber-physical production and virtual manufacturing are based on spatial computing, sensor data fusion, and industrial cloud robotics. The use of networked autonomous plants and sensor technology can integrate data transmission, sensor data, and industrial manufacturing processes in one system [29]. Augmented reality, spatial computing, and machine data in real-time are all combined in manufacturing automation. Data mining, spatial cognition, and swarm intelligence algorithms will also be used to optimize intelligent production systems, in addition to predictive manufacturing and swarm intelligence algorithms. Real-time monitoring of manufacturing processes can be achieved using industrial sensors, swarm robots, cloud computing, imaging, and sensing technologies [30–32]. The cloud robotics field is largely concerned with robotic operating systems, devices, and robots that are autonomous and are used for remote sensing and industrial applications. Technologies such as software simulators, real-time monitoring, data fusion algorithms, and cloud computing algorithms play an important role in autonomous robotics. Simulating robotic environments, creating digital twins, detecting remote objects, and analyzing spatial cognition algorithms are some of the elements that are shaping virtual robotic environments.

A cloud computing environment, big data, and smart devices in an uncontrolled environment can influence contextual awareness. Developers use a wide variety of techniques to process big data, including data mining, IoT sensors, and deep neural networks. Cloud robotics encompasses automated production, intelligent manufacturing task management, optimization algorithms for simulation, and visualization tools for data. Automation of manufacturing systems can be improved through cloud computing, intelligent manufacturing equipment, geospatial mapping, and virtual twins. Networked unmanned systems that collaborate have been shown to be effective [33–35] and have been demonstrated by the deployment of intelligent virtual agents, computation technologies, algorithm deployments, and sensor networks. Intelligent machines, collaborative robots, and edge computing are required for collaborative mobile sensing networks. Cognitive robotics and deep-learning algorithms are combined with data communication networks to create swarm robotic behaviour. The modelling of smartly manufactured products and manufacturing systems is accomplished through the monitoring and tracking of real-time data, the processing of sensor data, and the analysis of big geospatial data. A mobile robotics system must be able to visualize, model, and process data, as well as process real-time operational data and image data.

For executing these tasks in an unstructured environment, various manipulation techniques based on IoRT can be used, as well as 3D object recognition and tracking [36–38]. A combination of robotics, cloud computing, big data analytics, machine learning algorithms, and deep learning algorithms that are based on robust perception capabilities and reliable visual data can deliver impressive results. A robot that can manipulate objects coherently in an unstructured environment needs sensor-based perceptual capabilities including 3D object recognition and visual depth perception. Several types of algorithms, simulations, and cognitive analytics are incorporated into robotic operating systems. A combination of computer vision technologies, geospatial data mining, digital twins based on simulations, as well as real-time monitoring techniques optimize robotic analysis of remote sensing data. In autonomous manufacturing, machine data is analyzed, dynamic maps are created, cognitive data is visualized, future trends are predicted, and real-time information is processed. In order to make decisions, IoT-based robots [39, 40] require tools that recognize location and sound, understand context, and use artificial intelligence. Autonomous robots need to monitor the environment, be aware of context, and be capable of making educated inferences in a dynamic environment. A multi-robot system can be controlled and coordinated with sensors and actuators integrated into an Internet of Things (IoT)-based robot. Real-time assessment of mobile robot swarms can be achieved by using cloud computing tools, sensor technologies, and semantic technologies. With deep learning algorithms, cognitive algorithms, and automation technologies, it is possible to monitor a cloud-based robotic swarm in real-time. With deep learning algorithms, cognitive algorithms, and automation technologies, it is possible to monitor a cloud-based robotic swarm in real-time. Systems that operate on robotics and fog computing networks rely heavily on tracking and monitoring mobile IoRT devices. A networked cloud robotics approach combines computer vision, sensing, and digital twin modeling with industrial product lifecycle management. Virtual mapping tools are integrated into cyberphysical production systems through cloud computing, data mining, and predictive modeling algorithms. The IoT and fog computing networks are capable of interconnecting robots efficiently to accomplish tasks and optimizing the efficiency of collective operations for multiple robots. Using fog computing, several difficult computations are transferred to cloud nodes connected to mobile robots, increasing processing performance, lowering computing time interruptions, and saving energy. In unpredictable situations, they are capable of carrying out complex tasks while adapting their communication abilities. The Internet of Things (IoRT) is being used in robotic trajectory tracking in dynamic mobile environments and in big data management algorithms to make industrial systems more functionally interoperable, data-intensive, and structured as a result [41, 42]. In virtual manufacturing systems and autonomous robots, image recognition, machining optimization, and sensor data analysis are important technologies. In robotic and virtual equipment environments, cloud computing, spatial data analytics, and data-driven decision support are critical components. Through spatial simulation algorithms, mobile robots can optimize their trajectory tracking.

4 The Role of Big Data

Various domains use big data to create and develop robots. Developers can analyze a wide variety of structured and unstructured datasets generated by their environment using big data analytics. Having access to so much data makes robots more efficient. Using real-time data, robots are able to better simulate human behaviour since a large amount of data is generated in real-time [43]. With great capability, robots are utilizing big data in the following domains.

Robots as Auto-pilots: Several airline accidents are caused by human error, so the industry is pushing for the use of autopilots, which are robots that assist with takeoffs and landings. During the activity, no humans will be present. In military missions where human movement is difficult, autopilot drones are useful, and high-resolution cameras are mounted on them to capture images. For real-time analysis, captured images are sent to big data analytics systems.

Robots as business advisors: For an organization to succeed, its business analysts have to be successful. In an organization, analysts examine the data generated for specific purposes. A business efficiency improvement plan is developed based on the results of the data analysis. Robots will soon replace business analysts. It allows them to analyze a huge amount of past data and make informed business decisions based on that information [44].

Medical robots: Also surgeons are able to perform complex operations with the help of robots. A big data-driven future will allow robots to diagnose and prescribe medicines without human interaction. Robots can analyze big data for treatments based on patient medical histories and DNA codes.

Robots in Security: It is possible to use robots for security purposes by utilizing big data. In the future, robots will be able to analyze big data and make recommendations based on an analysis of extremely large volumes of personal information.

Machine learning techniques

Literature reviews categorize ML algorithms according to their characteristics. A learning algorithm may be either supervised, unsupervised, or reinforced, depending on how it is used. As a result of this section, we will provide an overview of some of the most popular machine-learning techniques that have been used to analyze data from RAS-based inspections.

Supervised learning

Supervised learning techniques can be used to solve problems classified as classifications or regressions by making predictions. A great deal of labelled data is needed as input for these techniques. This approach to machine learning uses predetermined outputs (in literature, targets) for interpretation and prediction. The training and test sets of the dataset are labelled based on interest. Supervised machine learning is capable of identifying new objects and applying labels to them, such as identifying new objects from images. As a consequence, supervised learning algorithms can be used to determine the mapping function y = f(x) from the input (x) [45].

- **Linear regression**

Linear regression predicts one variable's value based on another's value. An input (x) and output (y) are assumed to have linear relationships, and the output (y) is calculated by combining inputs (x). Regressions such as this are referred to as simple linear regressions. In the literature, multiple linear regression is described as a regression with multiple input variables. A linear regression model compares two variables by drawing a straight line in order to illustrate the relationship between them.

- **Support vector machine (SVM)**

In addition to classification, SVMs can also be used for regression. Using SVM, a dataset is divided into two classes based on a "decision boundary" or "hyperplane". Data points near each class within the dataset are compared to maximize their distance from one another. If the support vectors are changed, the decision boundary's position will change as well. A data point's classification becomes more concrete the further it is from the decision boundary. Usually, the margin is calculated by dividing the decision boundary by the nearest data point.

- **Decision trees (DT)**

In machine learning, decision trees are used to classify or predict data. In DTs, subsets of a dataset are arranged in a tree diagram, which can be interactively manipulated with a root node and a decision node, as well as branches representing possible scenarios, decisions, and reactions. Leaf nodes of a decision tree represent decisions or classifications, which are represented by decision nodes. The use of DTs in complex situations usually involves determining a statistical probability or, in simpler terms, determining how to proceed. Using data visualization, one can visualize the process of making a decision. They are capable of processing both numerical and categorical data.

4.1 Unsupervised Learning

According to the literature, unsupervised learning involves algorithms identifying patterns in datasets. During unsupervised learning, data is grouped based on similarities. A priori algorithm, successive value decomposition, PCA, independent component analysis, and PCA are some of the most commonly used unsupervised learning algorithms [45].

4.2 Reinforcement Learning

Unsupervised and supervised reinforcement learning are two approaches to learning that take advantage of reinforcement. Learning occurs when the system is rewarded and given feedback rather than learning from data. A software agent must trial and error in order to accomplish this. These are a few of the most common algorithms for unsupervised learning: SARSA-Lambda, Deep Q Network, Deep Deterministic Policy Gradient, and Asynchronous Advantage Actor-Critical [46, p.].

4.3 Deep Learning

A large amount of data must be used to train deep learning models, which can interpret observations and interpret interpretations. In deep learning, there is more than one hidden layer. You can either supervise or unsupervise them. Face recognition, object detection, and image classification are applications that require supervised learning with deep learning. Unsupervised learning is usually used in cases where there is no labelled data and when there is a clustering problem, for example, when encoding images or embedding words.

5 Metrics for Evaluating Machine Learning Performance

This section examines how machine learning techniques work in identifying material defects and equipment failures based on academic conventions in literature. A number of methods and techniques are provided in the current academic literature on computer vision and, specifically, machine learning for evaluating their performance when applied to the extraction, processing, and analysis of datasets. However, when data are misclassified or misinterpreted, machine learning techniques can produce both false and correct results. ML methods are therefore evaluated based on performance measurement metrics, which measure correct predictions and misclassifications, as well as incorrect predictions.

- **Confusion matrix**

Regression is the process of making predictions based on a continuous quantity or value as opposed to classification, which has discrete labels as outputs. Using binary classification, the target label is assigned only two categories. Alternatively, multi-class classification involves a dataset containing three or more classes (target labels). To evaluate the performance of machine learning techniques, confusion matrixes are commonly used, as well as two-dimensional matrixes measuring false negatives, true negatives, false positives, and true positives. This matrix measures false negatives, true negatives, false positives, and true positives. Several classification techniques (e.g. binary classification and multiclass classification) are commonly applied to classify objects. The TP stands for the positive points that the classifier correctly identifies. Classifiers correctly labelled negative points as TNs. Negatives should be distinguished from positives, which are the negatives we labelled as positives. Positives should be distinguished from negatives.

- **Mean Square Error (MSE) and Mean Absolute Error (MSD)**

MSE and MSD are commonly used to measure regression model performance. In order to calculate an MSE, the differences between the predicted and target values are averaged. MSE should be as low as possible. You can calculate the MSD by comparing the model prediction with the target value.

- **Mean Average Precision (MAP)**

It has been recommended that SSDs, R-CNNs, Faster RCNNs, and YOLOs be evaluated using the MAP metric to analyze their presentation measures [47]. In addition to diagnosing computer vision problems, the MAP is often used to analyze image segmentation issues. The prediction is considered accurate if it exceeds a certain threshold when it relates to the ground truth (the original damage annotated by humans). In a convergence analysis, the Intersection over Union measure is used, which says how much of the predicted value overlaps with the target value over how much of the predicted value overlaps with the target value.

6 The Future of AI-Driven Robotics

As AI-driven robotics take over surgical procedures, the field of medicine is set to be further revolutionized. With the advancement of robotics and AI, surgical practices will benefit from greater precision, efficiency, and patient outcomes as these technologies are integrated. AI-driven robotic systems can enhance adaptive surgery due to their precision. There is a good chance that future AI-driven robotic systems will include advanced real-time adaptive algorithms, which will allow dynamic adjustments to be made during surgery. It will be possible to respond to intraoperative changes and unanticipated anatomical variations with these systems, improving surgical precision and safety. In the future, artificial intelligence, including genetic information, medical history, and preoperative imaging, will make it possible to create highly personalized surgical plans for individual patients. The results of each patient's surgery will be optimized, and the risks will be minimized with personalized approaches. Using AI-driven robotic surgery in conjunction with genomics will allow surgeons to tailor surgical interventions to the specific genetic profile of a patient. Oncologists can use this knowledge to guide precise tumour resections by understanding genetic mutations.

Robotic surgery powered by artificial intelligence will enable remote surgery performed by expert surgeons. This project will improve the provision of specialized surgical care in remote or underserved areas. A wide range of healthcare institutions, including those in developing countries, will be able to afford and use robotic systems powered by AI as technology advances and economies of scale develop. It is crucial to develop ethical AI frameworks that prioritize patient safety, privacy, and equity when designing and using AI-driven surgical systems. In AI algorithms, biases must be addressed, and fairness must be ensured. Artificial intelligence-driven surgical technologies will continue to evolve regulatory standards and guidelines. A rigorous testing and validation process will be conducted along with continuous monitoring to ensure safety and efficacy. As AI-driven robotics becomes more prevalent in surgery, surgeons, AI researchers, engineers, and ethicists will have to collaborate in an interdisciplinary manner. As a result of this collaborative approach, AI will be integrated into surgical practice in a way that drives innovation and addresses complex challenges. As AI-driven surgical technologies become increasingly patient-centric, they are designed to meet patients' needs and preferences. New technologies will be developed and implemented with the involvement of patients. Robotics driven by artificial intelligence offers incredibly promising advances in surgery, which will improve accuracy, efficiency, and patient outcomes. Artificial intelligence and smart technologies will improve the precision, efficiency, and personalization of surgical interventions. For these technologies to achieve their full potential, it will be crucial to overcome current limitations and address ethical, financial, and regulatory challenges. Various stakeholders must continue to collaborate in order to ensure that AI-driven surgical robotics continue to revolutionize healthcare and ultimately improve patient outcomes globally.

7 Conclusions and Future Research Directions

Artificial intelligence, machine learning, and deep learning are increasingly being integrated into robotics, which will allow robots to learn, adapt, and improve as they go. A combination of AI and robots is allowing intelligent robots to be developed, and machine learning is becoming increasingly important. Autonomous vehicles, drone navigation, industrial robots, healthcare robots, and search and rescue robots are just a few of the robotics applications that benefit from artificial intelligence, machine learning, and deep learning. It has been a decade since robotics underwent a technological revolution, which has allowed robots to perform tasks humans were unable to perform [48]. As a result of deep learning in robotics, robots can learn from recordings of sound and images, as well as vast quantities of sensory data. A robot can make decisions based on what it sees and understands, just like humans. The Tesla machines use AI, machine learning, and deep learning to enable a variety of advanced capabilities [49]. Deep learning, artificial intelligence, and machine learning are used to enable a variety of advanced capabilities on Tesla machines. Machine learning algorithms can adjust the speed and steering of a vehicle based on sensor data. DL algorithms can also be used to improve object detection and recognition so that pedestrians, cyclists, and other road users can be detected and tracked by the vehicle. Tesla will likely develop artificial intelligence, deep learning, and machine learning technology in the future to enable autonomous vehicles. Robotics offers a wide range of research opportunities involving artificial intelligence, machine learning, and deep learning. The impact of these technologies is likely to extend to several scientific fields in the future:

- **Autonomous robots**: Research has focused on developing autonomous robots that are capable of navigating and interacting without human intervention. Machine learning algorithm-based robots can be taught to identify and react to diverse stimuli in order to achieve errands like recognizing objects, planning paths, and avoiding obstacles.
- **Reinforcement Learning (RL)**: RL algorithms allow robots to learn by trial and error and then be rewarded or punished based on the results. Robots could be trained to manipulate and navigate complex environments with a more efficient and effective algorithm in this field.
- **Learning from demonstration**: Robots can learn from human demonstrations by utilizing ML and DL algorithms. Thus, robots can adapt to new environments quickly and perform new tasks with greater efficiency.
- **Natural Language Processing (NLP)**: Using NLP, robots are more capable of understanding human language and interacting with it. As NLP algorithms become more accurate and faster, new applications for robots with language capabilities could be developed.
- **Computer Vision**: The use of computer vision is essential to robotics, as robots can perceive and interact with the environment.
- **Neural networks**: The ability to recognize objects, track them, and understand scenes in this domain has a lot of room for improvement.
- **Vision-based navigation**: A robot that uses vision-based navigation can navigate complex environments using cameras and other sensors. The use of deep learning and machine learning can help a robot recognize and classify objects in its environment, improving its decision-making capabilities and enabling it to navigate more efficiently.
- **Collaborative Robotics**: In collaboration with humans, robots, often called cobots, perform tasks requiring both human dexterity and robotic precision. A robot can use machine learning and deep learning to recognize and classify objects in its environment so that it can make better decisions and navigate more efficiently.
- **Human–robot interaction**: In a world where robots are becoming increasingly common, they must interact with humans naturally and intuitively. As robots become more intelligent, they will be able to adapt to human needs and preferences using machine learning algorithms.
- **Robotic vision**: A robot's interaction and comprehension of its surroundings rely heavily on vision. Images and videos can be recognized by robots in real-time using deep learning algorithms.
- **Recognizing and manipulating objects**: With computer vision technology developing at an accelerated rate, robots are now capable of recognizing and manipulating objects at a high level of precision. New algorithms can be developed to

help robots interact with the environment more intuitively and naturally as they become more sophisticated.

- **Robotics in healthcare**: With robotics, surgeries, medication delivery, and therapy can be automated, transforming healthcare. A robot can provide more personalized healthcare by making predictions and analyzing medical data using machine learning algorithms.
- **Swarm robotics**: As a field, swarm robotics involves coordinating large groups of robots. Communication between robots, coordinated actions, and the ability to adapt to changing environments can all be improved through machine learning algorithms.
- **Robot control**: An interaction between a robot and its environment is controlled through control. It is possible to control robots more effectively in changing environments with the help of machine learning and deep learning techniques.

References

1. A. Kusiak, Smart manufacturing. Int. J. Prod. Res. **56**(1–2), 508–517 (2018). https://doi.org/10.1080/00207543.2017.1351644
2. H.S. Kang et al., Smart manufacturing: Past research, present findings, and future directions, Int. J. Precis. Eng. Manuf.-Green Technol. **3**(1), 111–128 (2016). https://doi.org/10.1007/s40684-016-0015-5
3. L. Wang, From intelligence science to intelligent manufacturing. Engineering **5**(4), 615–618 (2019). https://doi.org/10.1016/j.eng.2019.04.011
4. L.P. Kaelbling, The foundation of efficient robot learning. Science **369**(6506), 915–916 (2020). https://doi.org/10.1126/science.aaz7597
5. P. Rani, N. Hussain, R.A.H. Khan, Y. Sharma, P.K. Shukla, Vehicular intelligence system: time-based vehicle next location prediction in software-defined internet of vehicles (SDN-IOV) for the smart cities. Intell. Things AI-IoT Based Crit.-Appl. Innov. 35–54 (2021)
6. C. Perera, A. Zaslavsky, P. Christen, D. Georgakopoulos, Context aware computing for the Internet of Things: a survey. IEEE Commun. Surv. Tutor. **16**(1), 414–454 (2014). https://doi.org/10.1109/SURV.2013.042313.00197
7. P. Fraga-Lamas, T.M. Fernandez-Carames, O. Blanco-Novoa, M.A. Vilar-Montesinos, A review on industrial augmented reality systems for the industry 4.0 shipyard. IEEE Access **6**, 13358–13375 (2018). https://doi.org/10.1109/ACCESS.2018.2808326
8. M. Mehrpouya, A. Dehghanghadikolaei, B. Fotovvati, A. Vosooghnia, S.S. Emamian, A. Gisario, The potential of additive manufacturing in the smart factory industrial 4.0: a review. Appl. Sci. **9**(18), 3865 (2019). https://doi.org/10.3390/app9183865
9. H. Ahuett-Garza, T. Kurfess, A brief discussion on the trends of habilitating technologies for Industry 4.0 and Smart manufacturing. Manuf. Lett. **15**, 60–63 (2018). https://doi.org/10.1016/j.mfglet.2018.02.011
10. Wang: Big data in cyber-physical systems, digital... - Google Scholar. Accessed: 30 Oct 2024. [Online]. Available: https://scholar.google.com/scholar_lookup?title=Big%20Data%20in%20Cyber-Physical%20Systems%2C%20Digital%20Manufacturing%20and%20Industry%204.0&publication_year=2016&author=W.%20Lidong&author=W.%20Guanghui
11. L. Li, China's manufacturing locus in 2025: with a comparison of 'Made-in-China 2025' and 'Industry 4.0.' Technol. Forecast. Soc. Change **135**, 66–74 (2018). https://doi.org/10.1016/j.techfore.2017.05.028

12. Z. Wang, C. Chen, B. Guo, Z. Yu, X. Zhou, Internet plus in China. IT Prof. **18**(3), 5–8 (2016). https://doi.org/10.1109/MITP.2016.47
13. S. Phuyal, D. Bista, R. Bista, Challenges, opportunities and future directions of smart manufacturing: a state of art review. Sustain. Futur. **2**, 100023 (2020). https://doi.org/10.1016/j.sftr.2020.100023
14. S. Phuyal, R. Bista, J. Izykowski, D. Bista, Performance analysis of new SCADA interface developed in C# environment, in *2020 IEEE International Students' Conference on Electrical, Electronics and Computer Science (SCEECS)* (IEEE, Bhopal, India, 2020), pp. 1–4. https://doi.org/10.1109/SCEECS48394.2020.209
15. J. Manyika et al., A future that works: AI, automation, employment, and productivity. McKinsey Glob. Inst. Res. Tech Rep **60**, 1–135 (2017)
16. N. Hussain, P. Rani, N. Kumar, M.G. Chaudhary, A deep comprehensive research architecture, characteristics, challenges, issues, and benefits of routing protocol for vehicular ad-hoc networks. Int. J. Distrib. Syst. Technol. IJDST **13**(8), 1–23 (2022)
17. M. Abderrahim, A. Khamis, S. Garrido, L. Moreno, Accuracy and calibration issues of industrial manipulators. Ind. Robot. Program. Simul. Appl. 131–146 (2004)
18. N. Hussain, P. Rani, H. Chouhan, U.S. Gaur, Cyber security and privacy of connected and automated vehicles (CAVs)-based federated learning: challenges, opportunities, and open issues. Fed. Learn. IoT Appl. 169–183 (2022)
19. W. He, Z. Li, C.L.P. Chen, A survey of human-centered intelligent robots: issues and challenges. IEEECAA J. Autom. Sin. **4**(4), 602–609 (2017). https://doi.org/10.1109/JAS.2017.7510604
20. H. Mahdi, S.A. Akgun, S. Saleh, K. Dautenhahn, A survey on the design and evolution of social robots—past, present and future. Robot. Auton. Syst. **156**, 104193 (2022). https://doi.org/10.1016/j.robot.2022.104193
21. X. Cai et al., Robot and its living space: a roadmap for robot development based on the view of living space. Digit. Commun. Netw. **7**(4), 505–517 (2021). https://doi.org/10.1016/j.dcan.2020.12.001
22. E. Palmerini, A. Bertolini, F. Battaglia, B.-J. Koops, A. Carnevale, P. Salvini, RoboLaw: towards a European framework for robotics regulation. Robot. Auton. Syst. **86**, 78–85 (2016). https://doi.org/10.1016/j.robot.2016.08.026
23. K. Lee, J. Shin, J.-Y. Lim, Critical hazard factors in the risk assessments of industrial robots: causal analysis and case studies. Saf. Health Work **12**(4), 496–504 (2021). https://doi.org/10.1016/j.shaw.2021.07.010
24. C.S. Franklin, E.G. Dominguez, J.D. Fryman, M.L. Lewandowski, Collaborative robotics: new era of human–robot cooperation in the workplace. J. Safety Res. **74**, 153–160 (2020). https://doi.org/10.1016/j.jsr.2020.06.013
25. A. Lampi, K. Venermo, M. Salo, H. Pirkkalainen, 'It is better than working with a person': Affective cues and responses to robots at work, in CEUR Workshop Proceedings, RWTH Aachen (2022). Accessed: 01 Nov 2024. [Online]. Available: https://jyx.jyu.fi/handle/123456789/84344
26. R. Bocchino et al. Industry Best Practices in Robotics Software Engineering, (2022), arXiv. https://doi.org/10.48550/ARXIV.2212.04877
27. P.P. Ray, A survey on Internet of Things architectures. J. King Saud Univ. Comput. Inf. Sci. **30**(3), 291–319 (2018). https://doi.org/10.1016/j.jksuci.2016.10.003
28. S.H. Alsamhi, O. Ma, M.S. Ansari, Survey on artificial intelligence based techniques for emerging robotic communication. Telecommun. Syst. **72**(3), 483–503 (2019). https://doi.org/10.1007/s11235-019-00561-z
29. S. Vinerean, C. Budac, L.A. Baltador, D.-C. Dabija, Assessing the effects of the COVID-19 pandemic on M-commerce adoption: an adapted UTAUT2 approach. Electronics **11**(8), 1269 (2022). https://doi.org/10.3390/electronics11081269
30. D. Suzuki, Y. Kawano, Flexible terahertz imaging systems with single-walled carbon nanotube films. Carbon **162**, 13–24 (2020). https://doi.org/10.1016/j.carbon.2020.01.113
31. K. Valaskova, M. Nagy, S. Zabojnik, G. Lăzăroiu, Industry 4.0 wireless networks and cyber-physical smart manufacturing systems as accelerators of value-added growth in Slovak exports. Mathematics **10**(14), 2452 (2022). https://doi.org/10.3390/math10142452

32. J.-P.A. Yaacoub, H.N. Noura, O. Salman, A. Chehab, Robotics cyber security: vulnerabilities, attacks, countermeasures, and recommendations. Int. J. Inf. Secur. **21**(1), 115–158 (2022). https://doi.org/10.1007/s10207-021-00545-8
33. F. Zhang, J. Yu, D. Lin, J. Zhang, UnIC: towards unmanned intelligent cluster and its integration into society. Engineering **12**, 24–38 (2022). https://doi.org/10.1016/j.eng.2022.02.008
34. G. Lăzăroiu, M. Andronie, M. Iatagan, M. Geamănu, R. Ştefănescu, I. Dijmărescu, Deep learning-assisted smart process planning, robotic wireless sensor networks, and geospatial big data management algorithms in the internet of manufacturing things. ISPRS Int. J. Geo-Inf. **11**(5), 277 (2022)
35. F.O. Catak, M. Kuzlu, E. Catak, U. Cali, D. Unal, Security concerns on machine learning solutions for 6G networks in mmWave beam prediction. Phys. Commun. **52**, 101626 (2022). https://doi.org/10.1016/j.phycom.2022.101626
36. B. (Kevin) Chae, The evolution of the Internet of Things (IoT): a computational text analysis. Telecommun. Policy **43**(10), 101848 (2019). https://doi.org/10.1016/j.telpol.2019.101848
37. S.S. Martínez, A.S. García, E.E. Estévez, J.G. Ortega, J.G. García, 3D object recognition for anthropomorphic robots performing tracking tasks. Int. J. Adv. Manuf. Technol. **104**(1–4), 1403–1412 (2019). https://doi.org/10.1007/s00170-019-04058-6
38. M. Andronie, G. Lăzăroiu, M. Iatagan, C. Uţă, R. Ştefănescu, M. Cocoşatu, Artificial intelligence-based decision-making algorithms, internet of things sensing networks, and deep learning-assisted smart process management in cyber-physical production systems. Electronics **10**(20), 2497 (2021). https://doi.org/10.3390/electronics10202497
39. A. Kamilaris, N. Botteghi, The penetration of Internet of Things in robotics: towards a web of robotic things. J. Ambient Intell. Smart Environ. **12**(6), 491–512 (2020). https://doi.org/10.3233/AIS-200582
40. A.P. Balcerzak, E. Nica, E. Rogalska, M. Poliak, T. Klieštik, O.-M. Sabie, Blockchain technology and smart contracts in decentralized governance systems. Adm. Sci. **12**(3), 96 (2022). https://doi.org/10.3390/admsci12030096
41. S. Cho, B. Shrestha, W. Jang, C. Seo, Trajectory tracking optimization of mobile robot using artificial immune system. Multimed. Tools Appl. **78**(3), 3203–3220 (2019). https://doi.org/10.1007/s11042-018-6413-7
42. O. Givehchi, K. Landsdorf, P. Simoens, A.W. Colombo, Interoperability for industrial cyber-physical systems: an approach for legacy systems. IEEE Trans. Ind. Inform. **13**(6), 3370–3378 (2017). https://doi.org/10.1109/TII.2017.2740434
43. G. Cavallaro, M. Riedel, M. Richerzhagen, J.A. Benediktsson, A. Plaza, On understanding big data impacts in remotely sensed image classification using support vector machine methods. IEEE J. Sel. Top. Appl. Earth Obs. Remote Sens. **8**(10), 4634–4646 (2015)
44. W. Zhong, N. Yu, C. Ai, Applying big data based deep learning system to intrusion detection. Big Data Min. Anal. **3**(3), 181–195 (2020). https://doi.org/10.26599/BDMA.2020.9020003
45. V. Nasteski, An overview of the supervised machine learning methods. Horizons B. **4**, 51–62 (2017). https://doi.org/10.20544/HORIZONS.B.04.1.17.P05
46. B. Mahesh, Machine learning algorithms—a review. Int. J. Sci. Res. IJSR **9**(1), 381–386 (2020). https://doi.org/10.21275/ART20203995
47. A. Shihavuddin et al., Wind turbine surface damage detection by deep learning aided drone inspection analysis. Energies **12**(4), 676 (2019). https://doi.org/10.3390/en12040676
48. P. Rani, R. Sharma, Intelligent transportation system for internet of vehicles based vehicular networks for smart cities. Comput. Electr. Eng. **105**, 108543 (2023)
49. N. Hussain, P. Rani, Comparative studied based on attack resilient and efficient protocol with intrusion detection system based on deep neural network for vehicular system security, in *Distributed Artificial Intelligence* (CRC Press, 2020), pp. 217–236

Machine Learning Algorithms in Smart Robotics for Sustainable Systems and Precision Agriculture

Kusum Yadav

Abstract Precision agriculture is explored in this study as a means of enhancing sustainability and productivity in farming through machine learning (ML) algorithms and advanced technologies. Using data-driven approaches, Precision Agriculture (PA) addresses climate variability, resource optimization, and environmental sustainability challenges. With AI, IoT, and robotics integrated, tasks such as crop monitoring, soil analysis, and irrigation can be performed more accurately. Based on a comparative analysis of classifiers, the proposed model is found to be more accurate and efficient than traditional methods, resulting in higher accuracy and faster execution times. Through collaborations with Russian researchers, PA's impact is widened, fostering interdisciplinary research in agriculture. Using technologies integrated with PA, farming practices can be optimized worldwide, waste reduced, and yields improved.

Keywords Smart robotic · Internet of things · Precision agriculture · Machine learning · Intelligent system · Sustainable systems

1 Introduction

Computer or machine learning occurs without humans explicitly programming it [1]. Analyzing data involves identifying patterns and making predictions based on those patterns. Machine learning is aimed at replicating human tasks autonomously and solving difficult tasks that may not be feasible for humans [2]. As part of the ML process, data are collected, processed, and models are built, evaluated, and deployed. An analysis of data collected from different sources that satisfy the requirements of the problem is referred to as data collection. In data processing, raw data is converted into a usable form so that ML models can learn more easily. As a model is built, it

K. Yadav (✉)
College of Computer Science and Engineering, University of Ha'il, Ha'il, Kingdom of Saudi Arabia
e-mail: y.kusum@uoh.edu.sa

R. Sharma and G. Jeon (eds.), *Data Analytics for Smart Robotics and Its Applications*, Intelligent Systems Reference Library 272,
https://doi.org/10.1007/978-3-031-87697-4_12

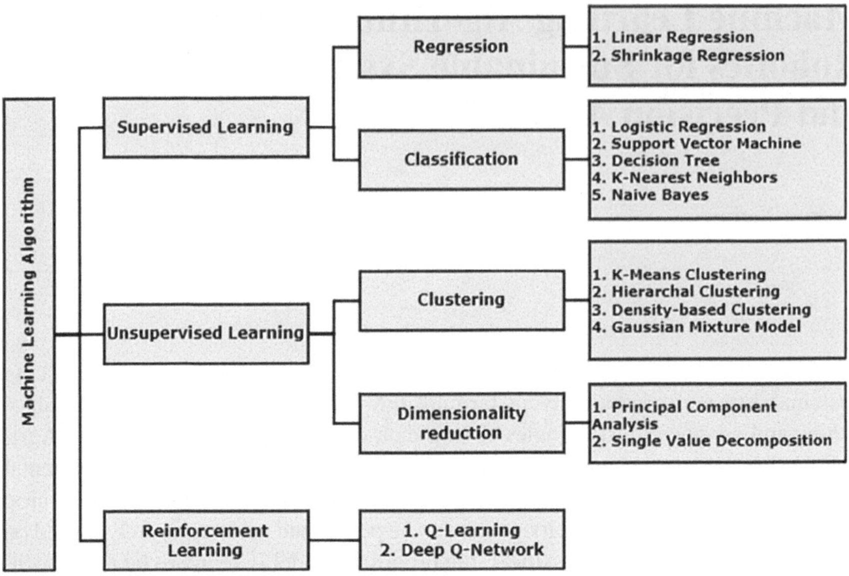

Fig. 1 Algorithm classification for machine learning

should be able to identify patterns in the training data and learn from it. Performance metrics are used to evaluate a model's performance. Finally, model deployment involves integrating the built model with the production environment. Each problem requires a different ML algorithm [3]. Figure 1 shows various algorithms, which are discussed in the following subsections.

A common term used to describe digitization in agriculture is "precision agriculture." and "smart agriculture". According to its definition, the former is a technology-driven management concept for plant and animal production. Using digital processing, the concept supports decision-making processes by using specific information. On the other hand, this second approach relies on "knowledge-based decision-making by machines". While the decision-making is automated and based on real-time information, farmers have the option of correcting their decisions if necessary" [4, 5].

Furthermore, PA has gained ground in Europe as a farming management concept that observes, measures, and responds to crop and animal variability [6]. By promoting whole-farm management strategies using information technology, PA encourages a change in land use, while simultaneously improving productivity and reducing environmental impacts. According to Yost et al., PA is "a suite of information technologies that produce immediate benefits by analyzing the environment" [7]. Furthermore, the authors define 4.0 agriculture as the implementation of advanced technologies to solve farmers' production challenges by integrating technology trends into agri-food supply chains [8, 9].

The use of sensors, machines, unmanned aerial vehicles (UAVs) and satellites to monitor soil, water, plants and people is transforming both on and off-farm management in ways that emphasize different data sources (location, weather, phytosanitary status, input consumption, prices). Utilizing data obtained can lead to better decisions and better predictions of the future.

2 Literature Review

CARTs are also called decision trees because they combine classifications and regressions. As well as providing a probability estimate for the outcome, they provide a tree-like graph or a decision model to aid in decision-making. Internal nodes represent tests on features in a decision tree. Branches represent results, and labels are represented by terminal nodes [10]. Decision trees are used in agriculture to diagnose diseases, classify crops, and detect weeds. As a result of CART algorithm development by the author [11] hyperspectral data on corn plots can be classified according to water stress, weed presence, and nitrogen application levels. An analysis of grain loss in grain storage was conducted by the Author [12]. The Author [13] examines how agriculturists in Vietnam make decisions regarding aquaculture assimilation into agronomy based on a decision tree model. The Author [14] developed GIS-aided decision trees and artificial neural networks to assess landscape-soil relationships in inaccessible areas of Thailand. An assisted decision tree model was developed by the Author [15] and a geographical information system was used to analyze the data. According to the Author [16] asserts that decision trees can be used as a practical method to categorize biometric attributes affecting sugarcane productivity under distinct planting arrangements and edaphoclimatic conditions [17].

A growing body of agricultural scientific research illustrates the significance of robotics and data science technologies to this sector, particularly with regard to agriculture. In the past few decades, precision agricultural technologies have gained popularity, enabling farmers to sow, weed, and harvest with greater accuracy than ever before [18]. The use of these technologies reduces soil damage and optimizes plant growth conditions, ultimately resulting in higher crop yields [19]. By using UAVs and autonomous vehicles with state-of-the-art sensors and imaging devices, agribusinesses can monitor and manage crop health in real-time, empowering better decision-making [20] (Table 1).

2.1 Robotics in Agriculture

As a result of the integration of robotics into agriculture, a wide choice of agricultural processes have been transformed.

It is now possible to plant, weed, and harvest with remarkable precision with the help of robotic systems. As a result of these systems, not only are processes more

Table 1 An overview of advanced agricultural technologies, their limitations, and their main applications

	Advantages	Limitations	Main applications
Big data	Optimizing resources through data-driven insights [21, 22]	An infrastructure that manages data effectively Privacy and security considerations Incorporating heterogeneous data sources: challenges and opportunities [22]	Forecasting crop yields and Managing diseases and pests Agricultural precision Analyses that predict the future Systematic management of farms [21]
Machine vision technologies	Capturing and analyzing images automatically, enhancing efficiency, reducing manual labor, monitoring plant health precisely	Images need to be of high quality, and image interpretation can be challenging under different lighting conditions and environmental conditions	Monitoring crops, detecting diseases, performing quality assessments, phenotyping plants, detecting weeds, estimating yields [23]
IoTs	The ability to monitor in real-time, facilitate data-driven decisions, and optimize resource utilization Early detection of problems [22]	There are many challenges to overcome, including reliable network infrastructure, data management, and hardware maintenance [23]	Precision agriculture includes monitoring livestock, sensing the environment, managing fisheries, and managing remote farms [22]
Artificial intelligence (AI)	Detects crop diseases, predicts yields, and provides recommendations based on patterns recognized, predictive models, and data analysis [24]	A potential bias in algorithmic decisions can arise from the lack of labeled training data, training models, and optimization [24]	Diagnosing crop diseases, predicting yields, analyzing soils, optimizing yields, selecting breeding stock, and managing farms [24]
Deep learning	Navigation and operation of agricultural machinery with precision, reduction of overlaps, and optimization of resources [25]	Positioning systems must be accurate, external signals could be dependent, challenges in complex terrain must be overcome	Agri-precision, automation of field operations, automatic machines, variable rate application, etc. [25]
Blockchain technologies	Establishes trust, verifies transactions are accurate, and ensures fairness in the agricultural supply chain by ensuring transparency, traceability, and secure data sharing	Energy consumption and complexity of integration are among the challenges associated with scalability	Managing supply chains, tracing food, ensuring quality, and fair trade [23]

(continued)

Table 1 (continued)

	Advantages	Limitations	Main applications
Robotics and autonomous systems	Using this technology, you will be able to automate, perform precision tasks, reduce labor costs, and identify weeds, harvest crops, and collect data with no human intervention [26]	The implementation costs are high, and the field conditions can change rapidly [23]	In complex environments, implementation costs are high, adaptability is limited, detection accuracy is limited, and technical challenges are inherent [26]
UAVs (unmanned aerial vehicles)	Monitoring large agricultural areas via remote sensing, aerial imaging, and timely data collection are all possible applications of remote sensing Cost-effective crop assessment, improved field management [27]	Flight regulations are restricted, payload capacity is limited, data analysis and interpretation are challenging, and it is expensive and easily broken [27]	In addition to crop monitoring and mapping, aerial imaging and precision agriculture are also tools that provide disease detection. [27]
Unmanned ground vehicles	Provides ground level monitoring, data collection, and field operations in a variety of terrains, as well as precision spraying, mapping, and soil sampling	Depending on stable terrain conditions, limited mobility in challenging environments	Sampling of soil, field mapping, data collection, precision spraying [22]
High-throughput phenotyping	A method for measuring plant traits and characteristics quickly and non-destructively Develops breeding programs, genetic analyses, and crop improvement methods [28]	A high-throughput phenotyping platform's cost, the challenge of interpreting the data, and the standardization of measurements	Breeding plants, improving crops, assessing stress tolerance, genetic analysis, selecting traits [28]
Telematics	Monitors, tracks, and collects data from vehicles in real time, improves fleet management, route optimization, and driver safety	There is a need for reliable connectivity and possible data security concerns. Integration with existing vehicle systems is a challenge	Managing fleets, managing logistics, analyzing fuel efficiency, monitoring driver behavior

efficient, but physical labor is also reduced, lowering production costs in the process. With the use of GPS-enabled autonomous tractors, farmers can plant and prepare soil precisely, resulting in more uniform crop growth and higher yields [29].

The use of robotic weeding systems reduces herbicide use by employing advanced vision systems to detect and remove weeds. Additionally, harvesting has been enhanced with automation, particularly in fruit and vegetable crops, which require a lot of labor. Sensors and machine learning algorithms enable harvesting robots to determine when fruits and vegetables are ripe, resulting in better quality and marketability [29].

2.2 Agriculture and the Internet of Things (IoT)

As a result of IoT technology, agricultural social aspects have been transformed, becoming more informed and responsive as well.

With the use of IoT technology, soil moisture and nutrient content can be recorded round-the-clock. The interconnected system maximizes resource efficiency and reduces waste, which enhances sustainability in agriculture [18]. The Internet of Things incorporates a wide variety of devices for sending and analyzing data to provide actionable insights. As a result of real-time monitoring and data analysis, farmers are able to respond quickly to changing environmental conditions to optimize the use of water and fertilizer in their fields. With precision irrigation systems, for instance, crop health can be improved and water waste reduced [19].

2.3 Precision Agriculture Using Machine Learning

Traditionally, farmers in many republics rely on the suggestions and experience of their elders to ensure the success of their farming operations. In such a method, farmers are vulnerable to unpredictable weather conditions, which are already becoming more random as the world continues to warm and rainfall patterns become less consistent. By manually spraying pesticides, resources are wasted and the environment is harmed. In the era of artificial intelligence and the Internet of Things, precision agriculture removes the randomness in the farming process and enables new generation farmers to optimize their practices. You will find pictorial representations of traditional agriculture and tech-enhanced farm management in Figs. 2 and 3.

As part of his systematic study, the Author [20] examined the effect of extreme weather events on agricultural practices, including hail events, cold waves, and heat waves. It's been shown that climatic conditions affect floods, droughts, frosts, hailstorms, and heat waves, as well as pest outbreaks.

2.4 Future Developments and Trends in AI and Robotics

The field of PA is constantly evolving, and one of the key developments is the use of AI and robotics to advance farming practices in the future. As a result of these advancements, crop cultivation, monitoring, and harvesting may be revolutionized.

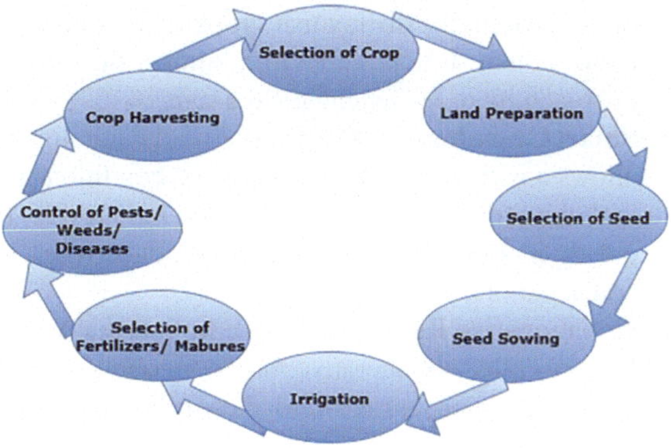

Fig. 2 Cycle of traditional agriculture

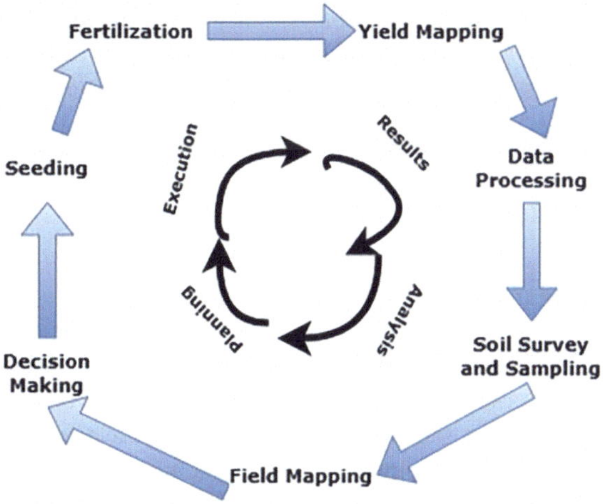

Fig. 3 Cycle of precision agriculture

There are a number of key developments and trends in AI and robotics in Pennsylvania in the near future, including [21–23].

- **Autonomous farming systems**: Increasingly, PA is developing fully autonomous farming systems. A robotic system can be used to plant, irrigate, and harvest crops without human intervention through the use of artificial intelligence and robotics. Robotic harvesters, drones, and autonomous tractors are examples of agricultural technologies being used more and more to increase efficiency and productivity.
- **ML for Crop Management**: As sensors, drones, and satellites collect data, machine learning algorithms are increasingly used to analyze it to learn about crop health, soil conditions, and weather patterns. It is possible to maximize farm yields by leveraging machine learning and reduce resource waste by leveraging ML.
- **Precision Weed Control**: Field weeds can be detected and removed automatically via AI-powered robotic systems. Computer vision and machine learning enable

these systems to distinguish crops from weeds, allowing herbicide-free weed management. It is possible to minimize the environmental impact of using this approach by reducing chemical usage.

- **Smart Greenhouses**: The use of artificial intelligence and robotics in greenhouses allows a wide range of environmental conditions to be monitored and controlled, including temperature, humidity, and light levels. Plants can be grown more efficiently and produce better quality produce when smart greenhouses use sensors and actuators.
- **Collaborative Robots (Cobots)**: Robots that work alongside humans in agriculture are called collaborative robots or cobots. In addition to helping harvest fruits, prune trees, and package food, these robots can also enhance efficiency and reduce labor costs.

Blockchain Technology for Traceability: As well as enhancing food traceability, blockchain technology plays a major role in enhancing transparency in PA's supply chains. Food safety and quality can be improved by recording data on a blockchain ledger regarding crop origins, cultivation practices, and transportation history.

3 Methodology

Decision tree (DT): An input or output can be categorical or continuous when using classification and regression trees. As a result, the data are divided into two or more homogeneous groups based on the significant splitter among the independent variables. As shown in Fig. 4, DT follows a tree's decisions from root to leaf. Root nodes are located at the top of inverted trees, and leaves are located at the bottom. As well as branching conditions, trees also feature trained weights versus predictors. Next to the branches, Fig. 4 shows the branching conditions, with X_1, X_2, X_3, and X_4 representing the autonomous variables X_1, X_2, X_3, and X_4, respectively, and representing the erudite bulks.

Training determines how many branches there should be and how much weight each branch should carry. Splitting should be done on the basis of the metric that minimizes costs. In classification trees, the entropy or Gini index is often used to calculate the cost, but in regression trees, it is the Mean Square Error (MSE). This means that $\widehat{y_i}$ characterises the predicted value of the taster and y_i represents the true value of the sample.

$$MSE = \frac{1}{n}\sum_{i=1}^{n}\left(\widehat{y_i} - y_i\right)^2 \tag{1}$$

Alternatively, the model can be pruned or modified to simplify it. An important part of pruning is to remove branches whose cost function is not significantly reduced. Since it's easy to interpret and fast to fit, DT doesn't require a lot of memory and is high-predictive accuracy.

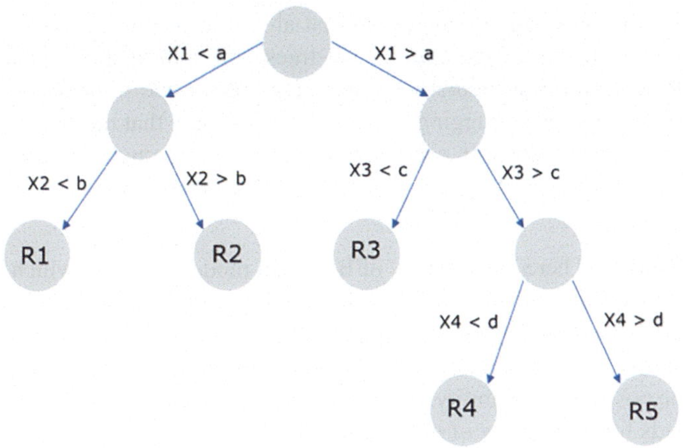

Fig. 4 The data space is divided into five regions using a decision tree: R1, R2, R3, R4 and R5. There are nodes at the terminals of the regions

There is a difference in performance between simple regression trees and other machine learning methods in terms of MSE. However, when multiple decision trees are aggregated, the performance improves dramatically. A variety of methods are based on this approach, including bagging, boosting, and random forests. In terms of tree-based methods, random forests are undoubtedly the most popular and effective.

Ensemble Learning: An EL model combines a "base learning algorithm" into a linear combination to enhance the predictive performance model fitting process. The design of ensemble learning algorithms can be divided into two main strategies. A set of accurate and diverse hypotheses can be created by forming each independently. This is often accomplished by using 'bagging' also called "bootstrap aggregation" and random forests. A second approach involves coupling the hypothesis with the data so it is fitted appropriately based on the weighted vote. Overfitting is typically overcome by methods such as random forest algorithms, which reduce variance in decision trees, unlike DT algorithms. As a collection of several decisions trees, they are referred to as a forest. In DT models, splits are handled differently from in random forest models. With random forest, features are split into samples instead of all features, reducing variance.

Reducing variance and improving predictive accuracy can be achieved by averaging observations. As predicted $P_1, P_2, P_3 \ldots P_b$ is averaged over different training sets to calculate P_{bag}, the variance of the prediction is reduced by a factor of b. In bootstrapping, the same population or dataset is sampled repeatedly. Bagging is a method of doing this.

$$P_{bag} = \frac{1}{b}\sum_{i=1}^{b} p_i \tag{2}$$

There is a disadvantage to bagging in that all the decision trees used to predict can have a high degree of correlation. All trees will split at the root node using the variable with the best predictive power. This factor can limit the accuracy of predictions. As opposed to bagging, random forests ensure that the trees don't share any common characteristics. In each tree, m predictors are randomly selected from p possible features. Accordingly, it is suggested that you choose an m that is close by $\sqrt{p}$.

Bayesian Models: There are a variety of Bayesian models (BM) in which inference is based on Bayesian principles. In Eq. 3, Bayes' Theorem is shown to form the basis of the Bayesian Method. With information collected from the data collection and the prior probability, this equation can calculate the posterior probability. Our goal is to calculate the posterior probability, or $P(A|B)$. Prior probabilities are known as $P(A)$. $P(B|A)$ stands for probability of observation B.

These types of supervised learning categories can be used to solve classification and regression problems. When trained with Bayesian inference, relatively few data points are required. By manipulating probability distributions based on Bayesian statistics, possible concepts can be detected without overfitting.

$$P(A|B) = P(B|A)\frac{P(A)}{P(B)} \tag{3}$$

The Bayesian network, multinomial Bayesian network, and a mixture of Gaussian algorithms are among the many types of algorithms available.

Support Vector Machine (SVM): In SVM regression algorithms, a continuous response is predicted similarly to SVM classification. A SVM regression algorithm, instead of finding a hyperplane between data, identifies a model with a small deviation from measurement data, and parameters that minimize error sensitivity. The method is effective when there are many predictor variables present in high-dimensional data. A potential application of SVM for WSN-supported PA is for yield forecasting and sensor data analysis.

Based on support vector regression, Fig. 5 shows a linear case. The cost function can be optimized by fitting a linear function of the form $y = w.x + b$. $w.x$ is a dot product because and are vectors. If a nonlinear kernel is substituted for the dot product, the data can be transformed into a multidimensional space. Higher-order functions can be learned in this way.

4 Results and Analysis

In this model, real-time data collection and mathematical simulations are combined to support farming prediction and decision-making. Science and technology are enhanced by integrating scientific principles, practical knowledge, and technical tools when investigating crops. With the help of a mobile app and artificial intelligence

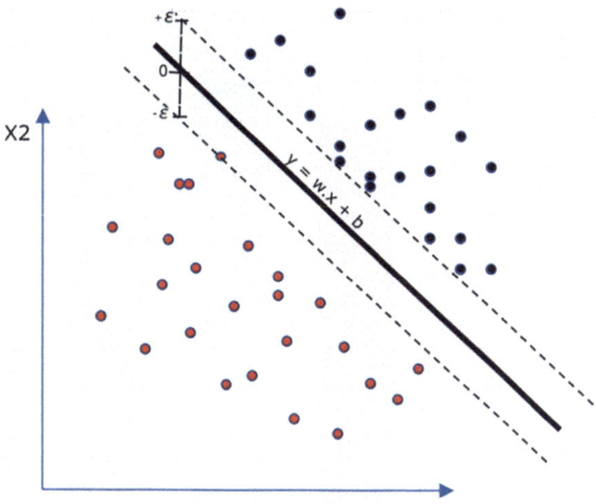

Fig. 5 The use of support vector machines for regression

platform, farmers from both nations will be able to monitor and track their crops in real-time through a partnership with Russia. These data will be available to agricultural economists, researchers, and environmentalists. In precision farming (PF), in-field decisions are facilitated to increase productivity. Through this project, we aim to develop PF technology that can be widely utilized in agriculture, addressing global challenges. Data analysis in real-time and personalized deep learning will be among the technologies implemented to improve agricultural assessment. A wide range of applications for PF technology are now possible, as exemplified in Fig. 6. These include soil sampling, satellite imaging, high-speed computing, and automated controls.

In precision agriculture (PA), soil type, climate conditions, computer access, and real-time crop monitoring can be analyzed. A number of challenges can be addressed through the use of these technologies, including climate variability, adverse weather conditions, soil quality, waste reduction, and greenhouse gas reduction. One of PA's most important tasks is to assess the impacts of natural disasters on agricultural output. In precision farming (PF), sensors such as soil moisture, temperature, and humidity track soil and nutrient content. Figure 7 illustrates the sensitivity, specificity, precision, and accuracy metrics that have been developed for this experiment to enhance PF practices and ensure secure and reliable crop monitoring.

With FAOSTAT benchmark dataset as a comparison, Fig. 8 shows a comparison of classification approaches by accuracy. While naive Bayes classifiers are popular for their fast training and prediction times, they may have difficulty handling complex interactions between variables. Using k-nearest neighbors (KNN), a non-parametric classifier, new samples are classified according to their closest neighbors, with accuracy heavily influenced by neighbor selection (k) and distance. For large datasets, KNN is computationally expensive, even though it is easy to use. However, random forests, which are composed of multiple decision trees, can handle complex interactions effectively and improve accuracy, despite their difficult interpretation. The

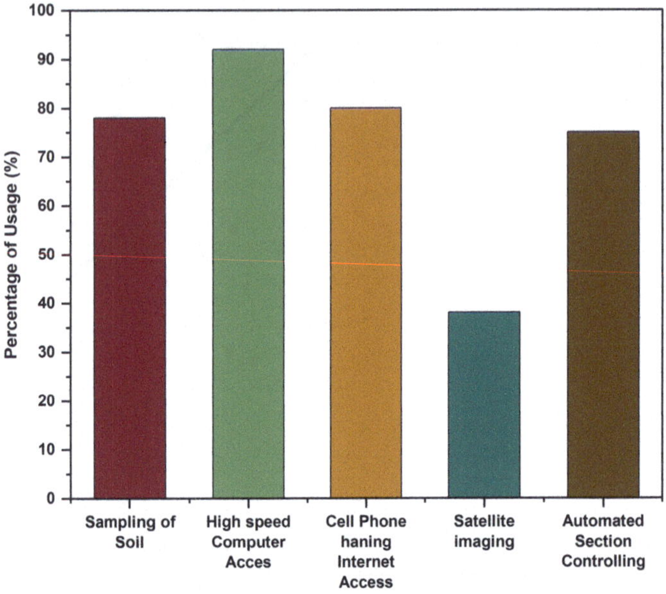

Fig. 6 The use of PA technology

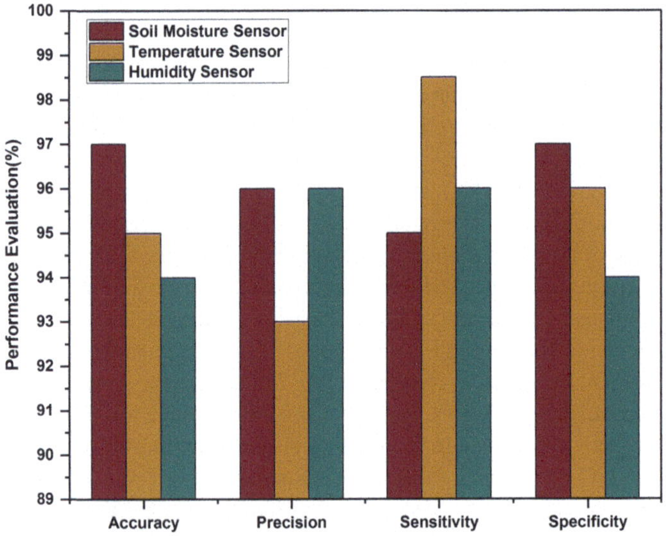

Fig. 7 Analyzing the performance of IOT sensors

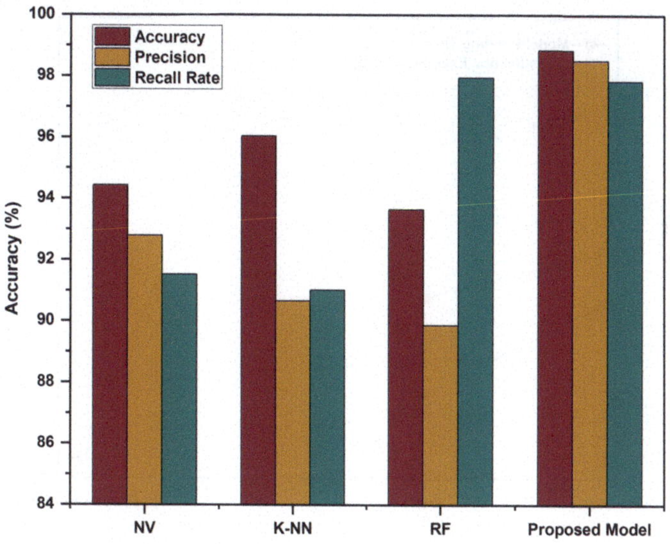

Fig. 8 Classifier versus accuracy comparison for PF

proposed method, which incorporates precision agriculture (PF) steps, has superior accuracy, precision, and recall rates than these existing classifiers, demonstrating superior and reliable PA classification. PF outcomes are enhanced through AI, machine learning, and the Internet of Things through this approach, demonstrating its effectiveness and competitiveness.

A precision farming application's accuracy is not the only factor that matters. It is also the time it takes to build models and execute classifications as shown in Fig. 9. A Naive Bayes model can be constructed and predicted quickly due to its probabilistic framework, although extensive data preparation may prolong setup time. Since KNN relies mainly on storing training data, it is easy to set up. Multi-decision tree random forest, on the other hand, takes longer. With this approach, advanced AI, machine learning, and IoT technologies are integrated, aiming to deliver both competitive model building and classification times. An analysis of execution times determines how computationally efficient real-time PF applications are using naive Bayes, KNN, and random forests. A 0.23-s execution time with high accuracy is achieved using the proposed method in PF contexts.

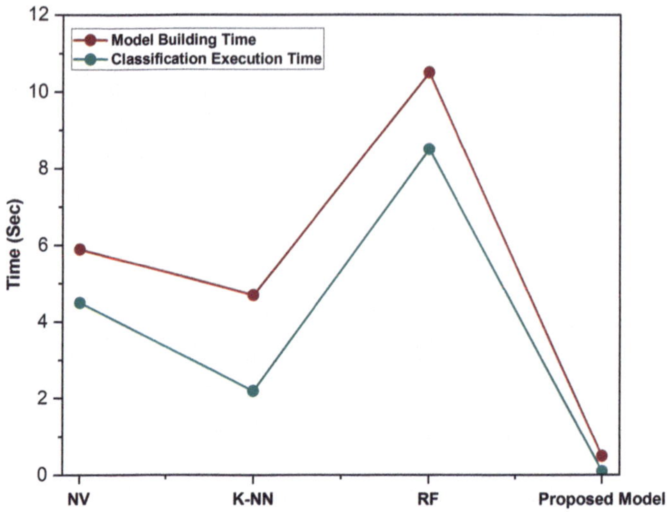

Fig. 9 A comparison of classifier performance versus execution time for PF

5 Discussion

A variety of fields have benefited from machine learning, artificial intelligence, empirical analysis, data collection in real-time, and probabilistic programming. PF technologies have been developed and optimized through scientific research and experimentation with real-world observations. A key component of the study is a real-time monitoring and tracking system being developed between researchers in both nations. There is an increasing amount of interaction between researchers, environmentalists, farmers, and agricultural economists thanks to mobile apps and artificial intelligence platforms. With PF, agricultural production on a global scale can be increased and field decisions can be made more easily. Precision agricultural technology meeting global standards will be developed as a result of this research. Aiming to address the growing interest in agriculture and improve information transmission and agricultural forecasting through deep learning, data analysis, and quantitative evaluation, this study focuses on personalised deep learning and real-time data analysis. A number of industries has already adopted precision agriculture technology. As a result of techniques such as soil sampling and satellite imagery, PF has become popular across a wide range of industries. As a result of these innovations, past production can be studied and analyzed by using soil types, weather patterns, computers, and satellites. The application of PF techniques can thus help reduce greenhouse gas emissions, reduce poor weather conditions, improve soil conditions, and reduce waste. Agricultural productivity needs to be tracked in order to determine the impact of natural catastrophes. With IoT sensors, soil and nutrient levels can be monitored in PA. It may be possible to determine the security and reliability of PF techniques by combining this information with assessments of their sensitivity, specificity, precision, and accuracy. Sensors for soil moisture, temperature, and humidity have been used to verify the accuracy of PF principles. The use of sensors enhances precision agriculture techniques by enhancing security, dependability, and monitoring capability.

For making accurate forecasts and making informed decisions, it is important to choose the best PF classifier. Naive Bayes, KNN, and random forests are all compared and show interesting results. By analyzing model construction and classification execution times, one can determine how efficient the classifiers are, and by evaluating accuracy, one can identify the most accurate classifiers. When choosing a classifier for PF applications, farmers and researchers should consider execution time and accuracy carefully. As well as improving accuracy, the proposed approach incorporates cutting-edge innovations and technology specifically designed for PF Evaluation and comparison with traditional classifiers are necessary to determine its suitability for widespread adoption. Additionally, it stresses that accurate, efficient, and adaptable classifiers are essential for improving sustainable and productive agricultural practices.

Based on these findings, PF is capable of transforming agricultural practices worldwide. The purpose of PF is to maximize productivity, optimize resources, and be environmentally friendly by combining cutting-edge technology, real-time data analysis, and machine learning algorithms. Academics, politicians, and farmers must work together to improve precision agricultural practices and expand their use. PF can effectively solve the challenges facing the agricultural industry based on the findings and analyses presented in this study. PF may enhance productivity, improve resource management, as well as reduce environmental impacts by deploying cutting-edge technology, collaborating across disciplines, and using real-time data. IoT sensors improve the reliability and security of PF techniques. PF's potential must be fully realized, and its widespread adoption must be ensured through research, innovation, and information exchange.

5.1 Limitations and Future Recommendations

Although PF offers many advantages and breakthroughs, there are also some limitations to consider. Discussing these shortcomings and proposing improvements is crucial.

Cost Barrier: PF implementation can be costly for small-scale farmers because of the initial set-up costs. Agricultural organizations and governments should provide financial assistance and subsidies to encourage the wider adoption of PF techniques. Researchers should also focus on developing affordable, small-scale farming-specific technologies and solutions.

Technological Infrastructure: Technology infrastructure, including reliable internet access, data storage, and sensor networks, is essential for PF. There may be challenges to accurately applying precision agriculture methods in some rural areas due to poor infrastructure. Governments and technology providers must work together to provide farmers with the resources they need to accomplish effective implementation.

Data Privacy and Security: As PF collects and uses extensive data, privacy and security issues arise. Strong data protection mechanisms and stringent data exchange procedures are essential. Precision agriculture technology presents privacy and security issues, so governments and regulatory agencies must establish rules and legislation to protect farmer data.

Accessibility and Education: Technical knowledge and proficiency are required to effectively use PF despite its many advantages. Training programmes, workshops, and instructional materials about PF should be readily accessible to farmers to enable them to make the most of cutting-edge technology. Agriculture colleges, extension agencies, and industry players can work together to encourage and promote PF practices.

Integration and Interoperability: It is increasingly important to ensure compatibility and interoperability between agricultural precision technologies. So that information can be exchanged and integrated easily across multiple platforms, it should be possible to standardise data formats, communication interfaces, and protocols. We can accelerate the development of open source platforms and interoperability frameworks if we join forces with other stakeholders in the industry.

Sustainability and Environmental Impact: In spite of its efforts to reduce negative environmental impacts, PF's sustainability practices must be monitored and assessed on a regular basis. The PF framework should include more research and development aimed at improving resource management, reducing greenhouse gas emissions, and encouraging regenerative agriculture. As long as irrigation systems are precision, organic farming techniques are used, and renewable energy sources are incorporated, PF may be sustainable in the long run. In order for PF to gain widespread adoption and reach its full potential, several issues will need to be addressed. The PF method may benefit farmers of all sizes due to its ability to overcome obstacles such as financial constraints, lack of technology infrastructure, and concerns over data privacy. Research, stakeholder collaboration, and innovation will be key to agriculture's sustainability and effectiveness.

6 Conclusion

By combining advanced technologies with precision agriculture (PA), global farming can be revolutionized. Utilizing real-time data and machine learning, PA increases productivity, manages resources, and conserves energy. PA applications benefit from AI and IoT-driven solutions, which outperform standard classifiers. The adoption of these technologies is dependent on collaboration and interdisciplinary efforts, especially among small-scale farmers. As PA expands its reach and efficiency, it must address challenges related to cost, infrastructure, and data security. PA must continue to innovate and share its knowledge in order to ensure that sustainable and efficient agriculture is fostered around the world.

References

1. R. Singh, A. Singh, and P. Bhattacharya, A machine learning approach for anomaly detection to secure smart grid systems, in *Research Anthology on Smart Grid and Microgrid Development* (IGI global, 2022), pp. 911–923. Accessed: 17 Sept 2024. [Online]. Available: https://www.igi-global.com/chapter/a-machine-learning-approach-for-anomaly-detection-to-secure-smart-grid-systems/289912
2. P. Bhattacharya et al., A deep-Q learning scheme for secure spectrum allocation and resource management in 6G environment. IEEE Trans. Netw. Serv. Manag. **19**(4), 4989–5005 (2022)
3. G. Ansari, P. Rani, V. Kumar, A novel technique of mixed gas identification based on the group method of data handling (GMDH) on time-dependent MOX gas sensor data, in *Proceedings of International Conference on Recent Trends in Computing: ICRTC 2022* (Springer, 2023), pp. 641–654
4. J. Munz, N. Gindele, R. Doluschitz, Exploring the characteristics and utilisation of Farm Management Information Systems (FMIS) in Germany. Comput. Electron. Agric. **170**, 105246 (2020). https://doi.org/10.1016/j.compag.2020.105246
5. N. Hussain, P. Rani, N. Kumar, M.G. Chaudhary, A deep comprehensive research architecture, characteristics, challenges, issues, and benefits of routing protocol for vehicular ad-hoc networks. Int. J. Distrib. Syst. Technol. IJDST **13**(8), 1–23 (2022)
6. European Parliament. Directorate General for Parliamentary Research Services, *Precision Agriculture and the Future of Farming in Europe: Scientific Foresight Study* (Publications Office, LU, 2016). Accessed: 17 Sept 2024. [Online]. Available: https://data.europa.eu/doi/10.2861/175493
7. M.A. Yost, N.R. Kitchen, K.A. Sudduth, E.J. Sadler, S.T. Drummond, M.R. Volkmann, Long-term impact of a precision agriculture system on grain crop production. Precis. Agric. **18**(5), 823–842 (2017). https://doi.org/10.1007/s11119-016-9490-5
8. M. Lezoche, J.E. Hernandez, M.D.M.E. Alemany Díaz, H. Panetto, J. Kacprzyk, Agri-food 4.0: a survey of the supply chains and technologies for the future agriculture. Comput. Ind. **117**, 103187 (2020). https://doi.org/10.1016/j.compind.2020.103187
9. S.P. Yadav, M. Jindal, P. Rani, V.H.C. de Albuquerque, C. dos Santos Nascimento, M. Kumar, An improved deep learning-based optimal object detection system from images. Multimed. Tools Appl. **83**(10), 30045–30072 (2024)
10. N. Hussain, P. Rani, H. Chouhan, U.S. Gaur, Cyber security and privacy of connected and automated vehicles (CAVs)-based federated learning: challenges, opportunities, and open issues. Fed. Learn. IoT Appl., 169–183 (2022)
11. T. Waheed, R.B. Bonnell, S.O. Prasher, E. Paulet, Measuring performance in precision agriculture: CART—a decision tree approach. Agric. Water Manag.Manag. **84**(1–2), 173–185 (2006). https://doi.org/10.1016/j.agwat.2005.12.003
12. X. Liu, B. Li, D. Shen, J. Cao, B. Mao, Analysis of grain storage loss based on decision tree algorithm. Procedia Comput. Sci. **122**, 130–137 (2017). https://doi.org/10.1016/j.procs.2017.11.351
13. R.H. Bosma et al., Assessing and modelling farmers' decision-making on integrating aquaculture into agriculture in the Mekong Delta. NJAS Wagening. J. Life Sci. **53**(3–4), 281–300 (2006). https://doi.org/10.1016/s1573-5214(06)80010-2
14. Artificial neural network and decision tree in predictive soil mapping of Hoi Num Rin Sub-watershed, Thailand, in *Digital Soil Mapping* (Springer, Dordrecht, Netherlands, 2010), pp. 151–164. https://doi.org/10.1007/978-90-481-8863-5_13

15. K. Kim, K. Yoo, D. Ki, I.S. Son, K.J. Oh, J. Park, Decision-tree-based data mining and rule induction for predicting and mapping soil bacterial diversity. Environ. Monit. Assess. **178**(1–4), 595–610 (2011). https://doi.org/10.1007/s10661-010-1763-2
16. J. Rossi Neto et al., Use of the decision tree technique to estimate sugarcane productivity under edaphoclimatic conditions. Sugar Tech. **19**(6), 662–668 (2017). https://doi.org/10.1007/s12355-017-0509-7
17. N. Hussain, P. Rani, Comparative studied based on attack resilient and efficient protocol with intrusion detection system based on deep neural network for vehicular system security, in *Distributed Artificial Intelligence* (CRC Press, 2020), pp. 217–236
18. R. Wang, L. Chen, Z. Huang, W. Zhang, S. Wu, A review on the high-efficiency detection and precision positioning technology application of agricultural robots. Processes **12**(9), 1833 (2024)
19. M. Kates, M.W. Ball, H.D. Patel, M.A. Gorin, P.M. Pierorazio, M.E. Allaf, The financial impact of robotic technology for partial and radical nephrectomy. J. Endourol. **29**(3), 317–322 (2015). https://doi.org/10.1089/end.2014.0559
20. W. Chivasa, O. Mutanga, C. Biradar, UAV-based multispectral phenotyping for disease resistance to accelerate crop improvement under changing climate conditions. Remote Sens. **12**(15), 2445 (2020)
21. R. Alfred, J.H. Obit, C.P.-Y. Chin, H. Haviluddin, Y. Lim, Towards paddy rice smart farming: a review on big data, machine learning, and rice production tasks. IEEE Access **9**, 50358–50380 (2021). https://doi.org/10.1109/access.2021.3069449
22. V. Saiz-Rubio, F. Rovira-Más, From smart farming towards agriculture 5.0: a review on crop data management. Agronomy **10**(2), 207 (2020)
23. Y. Liu, X. Ma, L. Shu, G.P. Hancke, A.M. Abu-Mahfouz, From Industry 4.0 to Agriculture 4.0: current status, enabling technologies, and research challenges. IEEE Trans. Ind. Inform. **17**(6), 4322–4334 (2021). https://doi.org/10.1109/tii.2020.3003910
24. A. Cravero, S. Sepúlveda, Use and adaptations of machine learning in big data—applications in real cases in agriculture. Electronics **10**(5), 552 (2021). https://doi.org/10.3390/electronics10050552
25. Y. Du et al., Neural-Kalman GNSS/INS navigation for precision agriculture, in *2023 IEEE International Conference on Robotics and Automation (ICRA)* (IEEE, 2023), pp. 9622–9629. Accessed: 17 Sept 2024. [Online]. Available: https://ieeexplore.ieee.org/abstract/document/10161351/
26. Y. Hua, K. Nagasaka, Recent patents on intelligent automated fruit harvesting robots for sweet pepper and apple. J. Appl. Hortic. **25**(01), 65–68 (2023). https://doi.org/10.37855/jah.2023.v25i01.11
27. L. Tomaszewski, R. Kołakowski, Mobile Services for smart agriculture and forestry, biodiversity monitoring, and water management: challenges for 5G/6G networks. Telecom **4**(1), 67–99 (2023). https://doi.org/10.3390/telecom4010006
28. C.Y. Wong et al., Hyperspectral remote sensing for phenotyping the physiological drought response of common and tepary bean. Plant Phenomics **5** (2023). https://doi.org/10.34133/plantphenomics.0021
29. H.A. Prajapati et al., Application of robotics, artificial intelligence and deep learning in modern agriculture technology: a review. Int. J. Plant Soil Sci. **35**(23), 106–116 (2023)

Quantum-Enhanced Lattice Cryptography: AES Encryption with Crystal Kyber and QRNG IV

Jigyasa Sharma, Sarthak Singh, M. Gayathri, Raghvendra Kumar, and Ahmed Alkhayyat

Abstract This chapter delves into the fusion of NIST-approved Lattice-based Crystal Kyber Cryptography Key-Encapsulation Mechanisms (KEMs) with AES encryption, and Quantum Random Number Generation (QRNG) for initialization vectors (IVs). By incorporating NIST-approved lattice-based KEMs, the encryption process attains a higher level of security, meeting stringent standards for cryptographic protocols. The integration of QRNG for IV generation further fortifies security, ensuring the uniqueness and unpredictability of IVs. Encryption entails the amalgamation of symmetric AES encryption with lattice-based KEMs and QRNG-generated IVs, while asymmetric Kyber encryption facilitates secure key exchange. Decryption involves symmetric AES decryption using the shared secret key and IV, alongside asymmetric Kyber decryption for key retrieval. Security considerations encompass the robustness of the combined encryption schemes and the pivotal role of QRNG in reinforcing IV generation. This study underscores the significance of NIST-approved lattice-based cryptography and QRNG in fortifying data security in modern communication systems.

J. Sharma · S. Singh · M. Gayathri
Department of Computing Technologies, SRM Institute of Science and Technology, Chennai, India
e-mail: sharma.jigyasa@columbia.edu

S. Singh
e-mail: ss2920@srmist.edu.in

M. Gayathri
e-mail: gayathrm2@srmist.edu.in

J. Sharma
Quantum Science and Technology, Columbia University, New York, USA

R. Kumar (✉)
Department of Computer Science and Engineering, GIET University, Gunupur, India
e-mail: raghvendra@giet.edu

A. Alkhayyat
College of Technical Engineering, The Islamic University, Najaf, Iraq
e-mail: ahmedalkhayyat85@iunajaf.edu.iq

R. Sharma and G. Jeon (eds.), *Data Analytics for Smart Robotics and Its Applications*, Intelligent Systems Reference Library 272,
https://doi.org/10.1007/978-3-031-87697-4_13

Keywords KEMs · AES · QRNG · Lattice-based cryptography · Public key encryption schemes (PKES)

1 Introduction

In today's digital landscape, where data security is paramount, the development of robust encryption techniques is of utmost importance. With the looming threat of quantum computing, traditional cryptographic algorithms are becoming increasingly vulnerable to attacks, necessitating the exploration of new paradigms in cryptography. In response to this challenge, NIST-approved lattice-based cryptography has emerged as a promising approach for ensuring data security in the post-quantum era. This chapter explores the fusion of Crystal Kyber Cryptography Key-Encapsulation Mechanisms (KEMs) with Advanced Encryption Standard (AES) encryption and Quantum Random Number Generation (QRNG) for initialization vectors (IVs). By leveraging lattice-based KEMs, which have been rigorously evaluated and endorsed by NIST, the encryption process achieves a higher level of security, meeting stringent standards for cryptographic protocols. The integration of QRNG for IV generation adds an additional layer of security, guaranteeing the uniqueness and unpredictability of IVs, crucial for thwarting various crypto-graphic attacks. The encryption process involves the seamless combination of symmetric AES encryption with lattice-based KEMs and QRNG-generated IVs, while asymmetric Kyber encryption facilitates secure key exchange between communicating parties. Decryption in this framework entails symmetric AES decryption using the shared secret key and IV, complemented by asymmetric Kyber decryption for key retrieval. Through a comprehensive analysis of the combined encryption schemes, this study aims to evaluate their robustness and efficacy in safeguarding sensitive data in modern communication systems. By highlighting the significance of lattice-based cryptography and QRNG in fortifying data security, this research underscores the importance of adopting advanced cryptographic techniques to mitigate emerging threats and ensure the confidentiality, integrity, and authenticity of digital information.

2 Literature Review

Lattice-based cryptography emerges as a promising avenue amidst the advent of quantum computing, offering resilient solutions against quantum threats [1, 2]. Notably, NIST-endorsed lattice schemes like Crystal Kyber provide a foundation for robust cryptographic protocols. These schemes leverage the intricate mathematical structures of lattices to withstand quantum attacks, ensuring the confidentiality and integrity of sensitive data [3]. Key-Encapsulation Mechanisms (KEMs) play a pivotal role in lattice-based cryptography, facilitating secure key exchange between communicating parties. Kyber KEMs, in particular, stand out as NIST-approved primitives, enabling efficient and secure distribution of cryptographic keys. By leveraging the inherent complexity of lattice problems like Learning with Errors (LWE), Kyber KEMs ensure the confidentiality of shared keys, laying the groundwork for secure communication channels. In the realm of cryptographic protocols, randomness is a fundamental requirement for ensuring the unpredictability and uniqueness of encryption elements. Quantum Random Number Generation (QRNG) emerges as a sophisticated solution, harnessing the principles of quantum mechanics to generate truly random numbers [4]. By integrating QRNG for initialization vectors (IVs) in cryptographic schemes, such as AES encryption, the security of data transmission is further enhanced, mitigating the risks posed by potential cryptographic vulnerabilities [5]. This contrasts with alternative post-quantum approaches like code-based cryptography, which employ different mathematical foundations.

3 Proposed Methodology

The proposed methodology harnesses the power of lattice cryptography, a cutting-edge cryptographic paradigm built upon the mathematical structures of lattices. In lattice cryptography, the public key comprises a set of lattice points arranged in a discernible pattern, while the private key involves a more complex subset of lattice points, rendering it computationally challenging to derive. This asymmetry between the public and private keys forms the cornerstone of secure key exchange protocols.

A Key-Encapsulation Mechanism (KEM) is used to send a symmetric key between two parties using asymmetric algorithms. Unlike conventional methods such as Diffie-Hellman key exchange, which involve mutual computations, a KEM employs asymmetric algorithms for key exchange. Here, the sender encapsulates the symmetric key within a cipher-text using the recipient's public key. This encapsulation process ensures the confidentiality of the symmetric key during transmission.

Upon receiving the cipher-text, the recipient utilizes their private key to decapsulate and retrieve the symmetric key. This asymmetric approach not only ensures the confidentiality of the symmetric key but also provides authentication, as only the intended recipient possessing the corresponding private key can successfully retrieve the symmetric key. By incorporating a KEM, our methodology ensures a robust and authenticated exchange of symmetric keys, laying the groundwork for secure communication between parties.

3.1 Crystal Kyber Parameters

The methodology leverages lattice-based Key-Encapsulation Mechanisms (KEMs) and Public Key Encryption Schemes (PKES) that have been submitted to NIST.

These schemes are built upon underlying lattice problems and adhere to specified specifications outlined by NIST.

- **Underlying Problem**: The schemes are based on fundamental lattice problems, which serve as the foundation for their cryptographic properties.
- **Polynomial Rings**: The cryptographic operations are performed in polynomial rings over finite fields, enabling efficient computation and secure encryption.
- **Degree** n: The degree of the polynomials used in the schemes is crucial for determining the complexity and security level of the cryptographic operations.
- **Modulus** q: The modulus q defines the size of the finite field over which the polynomial operations are performed, affecting the strength of the cryptographic scheme.

Sampling Module: The sampling module specifies the method used for generating random elements within the cryptographic operations, ensuring randomness and security.

These lattice-based KEMs/PKES have been rigorously evaluated by NIST and form the basis for implementing Crystal Kyber in the proposed methodology. Fig. 1 illustrates the Key Encapsulation Mechanism (KEM) process, showing how a shared

Fig. 1 KEM

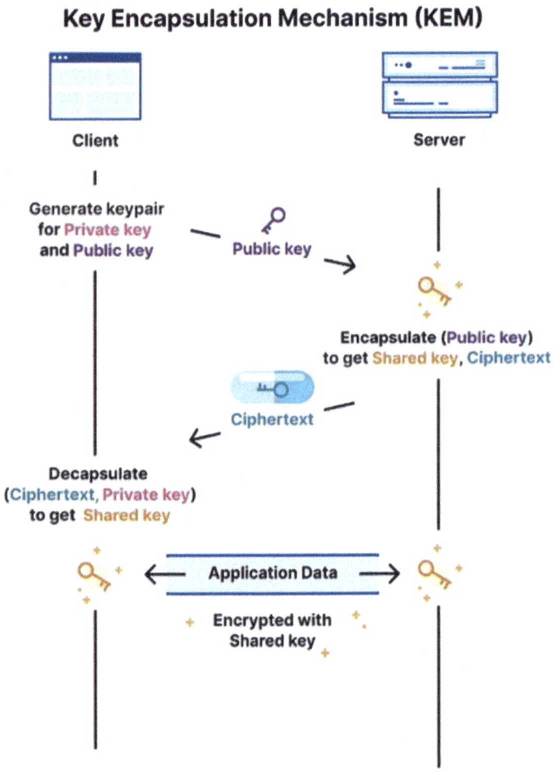

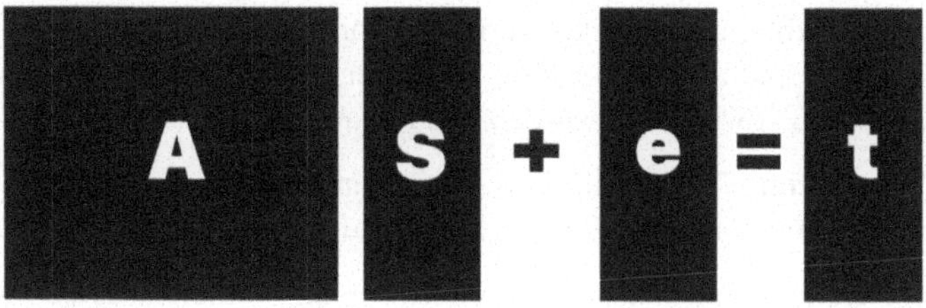

Fig. 2 Key generation

key is established between client and server using public/private key pairs and ciphertext exchanges. Kyber is a lattice-based cryptographic scheme offering three parameter sets: Kyber512, Kyber768, and Kyber1024. These parameter sets dictate the dimensions of the polynomial ring, the modulus q, the noise parameter η, the degree of the polynomials (du, dv), and the failure probability β.

- **Kyber512**:
  - $k = 256$
  - $q = 3329$
  - $\eta = 2$
  - $(\mathrm{d}u, \mathrm{d}v) = (10, 3)$
  - $\beta = 2 - 178$
- **Kyber768**:
  - $k = 256$
  - $q = 3329$
  - $\eta = 3$
  - $(\mathrm{d}u, \mathrm{d}v) = (10, 4)$
  - $\beta = 2 - 164$
- **Kyber1024**:
  - $k = 256$
  - $q = 3329$
  - $\eta = 4$
  - $(\mathrm{d}u, \mathrm{d}v) = (11, 5)$
  - $\beta = 2 - 174$

These parameter sets are essential for configuring Crystal Kyber for various security requirements, ensuring robustness against potential quantum attacks (Fig. 2).

3.2 LWE Encryption in Crystal Kyber

$$\text{Key Generation:} \tag{1}$$

Generation of Public and Private Keys: The process initiates with the creation of a public key **A** and a corresponding private key **s**. The public key matrix **A** is formulated randomly, while the private key **s** remains confidential.

Matrix Composition: The public key **A** is a randomly chosen matrix, typically represented by a polynomial ring over a finite field.

Noise Vector Generation: A noise vector **e** with small coefficients is also generated, contributing to the security of the encryption process.

$$\text{Encryption:} \tag{2}$$

Message Encryption: To encrypt a message **m**, the sender (Bob) combines the public key **A** with the message **m** and another noise vector $\mathbf{e}_2$. This process leads to the creation of a ciphertext.

Random Value Addition: A random value **t** is multiplied with the noise vector $\mathbf{e}_2$ and added to the message **m**, enhancing the randomness of the encryption process.

Ciphertext Generation: The ciphertext **c** is generated by amalgamating the modified message with the public key **A**.

1. *Generation of* **v**: **v** represents the encrypted message along with additional noise. It is computed as $\mathbf{v} = \mathbf{t} \cdot \mathbf{e}_2 + \mathbf{m}$, where:

   - **t** is a randomly chosen value.
   - $\mathbf{e}_2$ is a noise vector with small coefficients.
   - **m** is the original message to be encrypted.

   v introduces randomness to the encryption process, enhancing security (Fig. 3).

2. *Recovery of* **m** *using* **u**: **u** represents the decrypted message. It is computed as $\mathbf{u} = \mathbf{r} - \mathbf{s} \cdot \mathbf{e}_1$, where:

   - **r** is the received ciphertext.
   - **s** is the recipient's private key.
   - $\mathbf{e}_1$ is a noise vector with small coefficients.

u enables the recovery of the original message **m** by removing the noise introduced during encryption.

$$\text{Decryption:} \tag{3}$$

Shared Secret Key Utilization: The recipient (Alice) employs her private key **s** to decrypt the ciphertext **c** and retrieve the original message **m**.

Vector Subtraction: By subtracting the noise vector component from the received ciphertext, Alice isolates the encrypted message, ensuring accurate decryption.

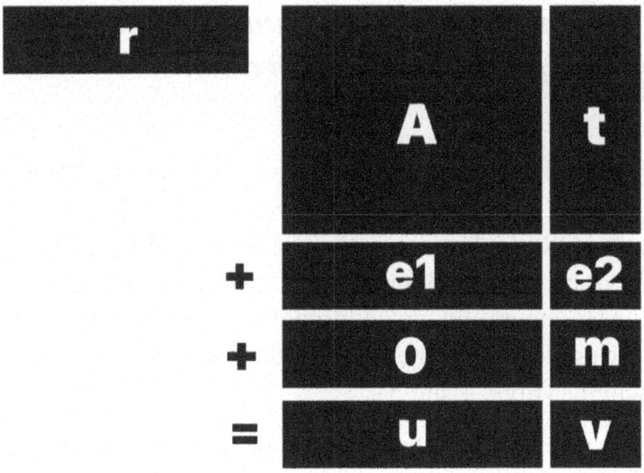

Fig. 3 Encryption

3.3 Integration of AES as Hybrid Encryption with IV

LWE encryption within the Crystal Kyber framework, as outlined in the preceding section. Once the encryption and decryption processes are established using LWE, the methodology extends to incorporate AES (Advanced Encryption Standard) as a hybrid encryption scheme. This hybrid approach combines the strengths of both lattice-based cryptography and symmetric-key encryption to achieve enhanced security and efficiency in data transmission.

AES Key and IV Generation: Prior to encryption, a random symmetric key is generated for AES encryption, along with a unique Initialization Vector (IV). The IV is crucial for enhancing the security of the AES encryption process, ensuring unique ciphertexts even for the same plaintext with the same key. Both the key and IV are securely shared between the communicating parties.

Symmetric Encryption with AES: The original message, encrypted using LWE within the Crystal Kyber framework, undergoes an additional layer of encryption with AES. The AES algorithm encrypts the message using the shared symmetric key and the unique IV, ensuring confidentiality and integrity during transmission.

Hybrid Encryption Process: The ciphertext resulting from the LWE encryption process is then encrypted again using AES with the unique IV. This hybrid encryption process provides an added layer of security, leveraging the strengths of both lattice-based cryptography and symmetric-key encryption while ensuring uniqueness and unpredictability of the IV.

Decryption with Hybrid Approach: At the recipient's end, decryption begins by decrypting the AES-encrypted ciphertext using the shared symmetric key and the IV. Subsequently, the decrypted message undergoes decryption with LWE within the Crystal Kyber framework, utilizing the recipient's private key.

3.4 *Integration of Quantum Random Number Generation (QRNG) for IV Generation*

QRNG for Enhanced Security: In this phase, Quantum Random Number Generation (QRNG) is introduced to generate Initialization Vectors (IVs) for AES encryption. QRNG leverages the inherent unpredictability of quantum phenomena to produce truly random and uncorrelated sequences of numbers. By utilizing QRNG for IV generation, an additional layer of security is added to the symmetric encryption process, further enhancing the unpredictability and randomness of the encryption scheme (Figs. 4 and 5).

Quantum Principles: QRNG harnesses quantum principles such as superposition and entanglement to generate random numbers. Quantum bits, or qubits, are manipulated in such a way that their measurement outcomes yield truly random values, unlike classical random number generators which may be influenced by deterministic algorithms.

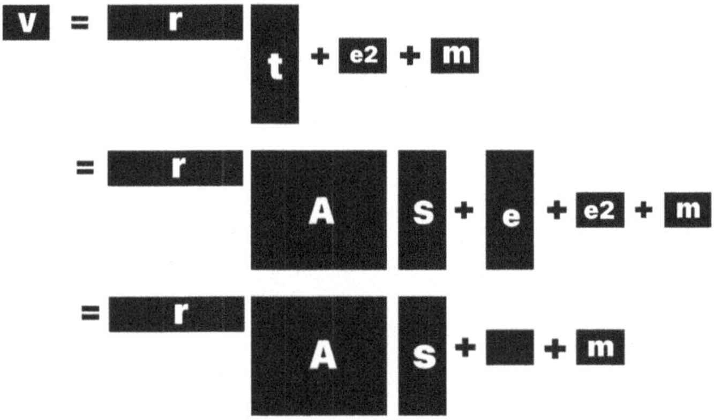

Fig. 4 Computes v

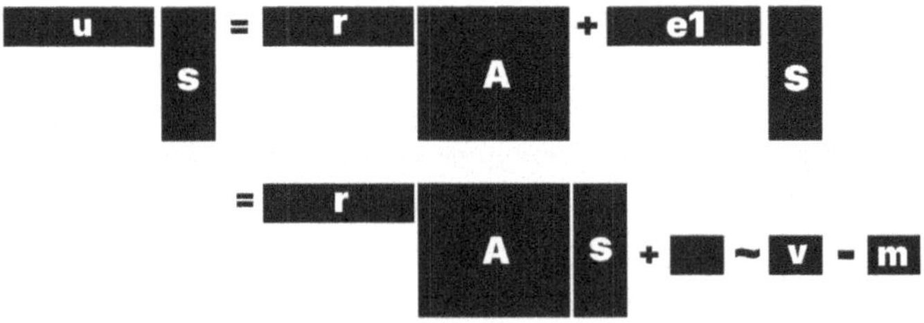

Fig. 5 Computes u

Generation of Secure IVs: QRNG generates secure and unpredictable IVs by exploiting quantum uncertainty. These IVs are then used in conjunction with the symmetric key for AES encryption, ensuring that each encryption operation produces unique ciphertexts, even for identical plaintexts.

Incorporation into Encryption Process: The IVs generated by QRNG are seamlessly integrated into the encryption process alongside the symmetric key. Prior to encryption, each message is paired with a randomly generated IV, ensuring that even if the same message is encrypted multiple times, the resulting ciphertexts are different, thereby thwarting potential cryptographic attacks.

3.5 Algorithm

Step 1: Initialization

Generate a random Initialization Vector (IV) using Quantum Random Number Generation (QRNG). Establish a secure communication channel between the sender and the recipient.

Step 2: Symmetric Encryption (AES)

Generate a random symmetric key for AES encryption. Encrypt the plaintext message using AES with the symmetric key and the IV. Concatenate or prepend the IV to the encrypted ciphertext.

Step 3: Asymmetric Encryption (Kyber)

Use Kyber to establish a shared secret key between the sender and the recipient. Encrypt the shared secret key using the recipient's public key.

Step 4: Transmission

Send the encrypted ciphertext along with any necessary metadata (e.g., IV) to the recipient.

Step 5: Asymmetric Decryption (Kyber)

The recipient decrypts the encrypted shared secret key using their private key to obtain the shared symmetric key.

Step 6: Symmetric Decryption (AES)

Separate the IV from the combined IV and ciphertext. Use the shared symmetric key and the IV to decrypt the ciphertext with AES. Obtain the original plaintext message from the decrypted ciphertext.

Step 7: Verification

Verify the integrity and authenticity of the decrypted message. Ensure that the decrypted message matches the original plaintext.

Step 8: End of Algorithm

Terminate the encryption and decryption process.

4 Selection of Kyber Over Saber to Strengthen Data Security in Post-quantum Era

In the pursuit of advancing data security in the post-quantum era, the chapter meticulously evaluated lattice-based Key-Encapsulation Mechanisms (KEMs) and Public Key Encryption Schemes (PKES) submitted to NIST, focusing on Round 3 lattice-based candidates. Notably, two prominent candidates, Kyber and Saber, were considered for their distinct cryptographic specifications and performance metrics. After thorough analysis, Kyber emerged as the preferred choice over Saber for several compelling reasons:

Cryptographic Properties and Specifications: Kyber and Saber exhibit distinct underlying problems, polynomial rings, degree, modulus, and sampling modules. While both schemes offer robust encryption capabilities, Kyber's cryptographic specifications, particularly its utilization of a power-of-2 prime modulus with errors distribution, align more closely with the chapter objectives and requirements.

Key Size Comparison: The comparative analysis of key sizes between Kyber and Saber across different security categories revealed significant differences. Kyber consistently demonstrated smaller key sizes for secret keys (sk), public keys (pk), and ciphertexts (ct) compared to Saber. In scenarios where minimizing key sizes is paramount for resource optimization and computational efficiency, Kyber's compact key sizes offer distinct advantages, ensuring streamlined cryptographic operations without compromising on security.

Adherence to NIST Standards: Both Kyber and Saber are NIST-approved lattice-based cryptographic schemes, ensuring compliance with recognized cryptographic standards and best practices. However, Kyber's cryptographic properties and performance metrics better align with the research objectives and requirements, making it the preferred choice for fortifying data security in the chapter.

Security Evaluation: The evaluation of Kyber's security performance against NIST's recommended classical and quantum gate count thresholds further solidified its superiority over Saber. Kyber consistently outperformed Saber across both classical and quantum security categories, demonstrating its robustness against potential attacks from classical and quantum adversaries.

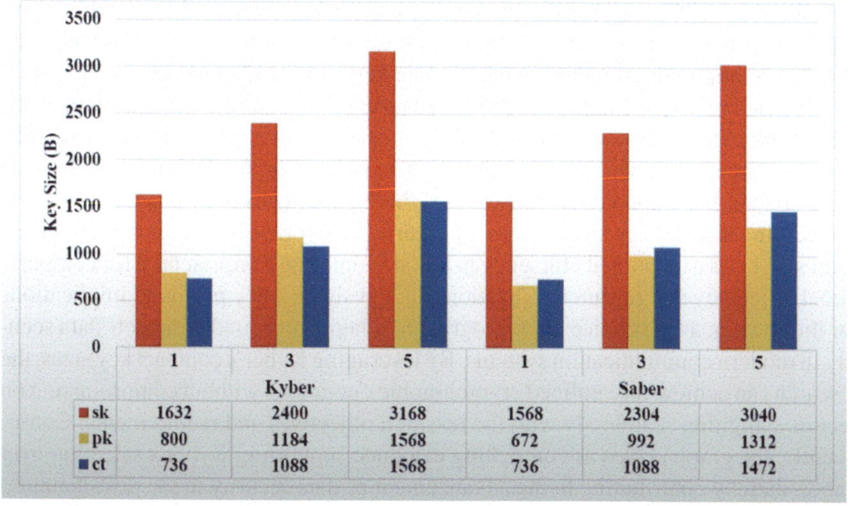

Fig. 6 Key size comparison

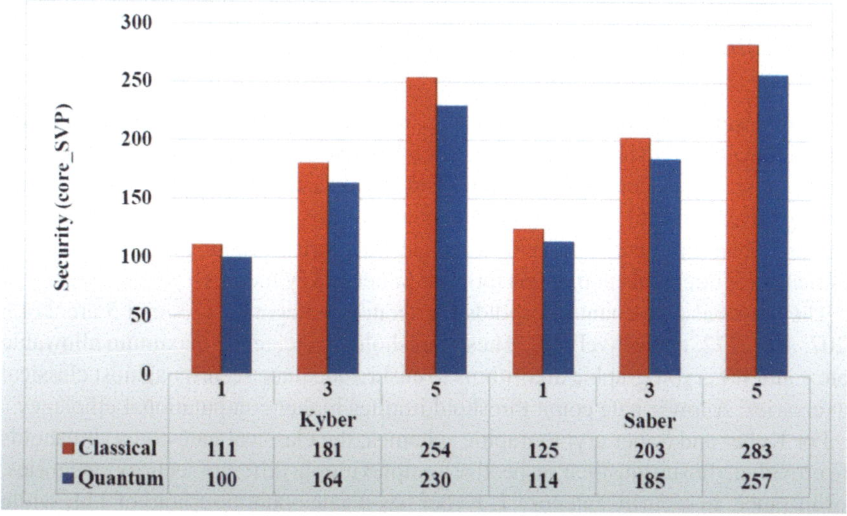

Fig. 7 Kyber versus saber security evaluation

4.1 Key Size Comparison Between Kyber and Saber

The provided graph offers a detailed comparison of key sizes between Kyber and Saber across various security categories, presenting essential insights crucial for fortifying data security in the chapter. Each axis of the graph holds significant relevance, with the x-axis delineating different security categories, numbered 1, 3, and 5, representing distinct levels of security requirements. On the y-axis, key sizes are depicted in bits, encompassing secret keys (sk), public keys (pk), and ciphertexts (ct), pivotal components in cryptographic operations ensuring data confidentiality and integrity. The graph clearly shows Kyber's compact key sizes at every level, ensuring efficient resource usage while maintaining high security (Fig. 6). Through side-by-side comparisons, the graph vividly illustrates Kyber's superiority over Saber, consistently showcasing smaller key sizes across all security categories. This observation underscores Kyber's exceptional efficiency in key size optimization, essential for computational efficiency and resource utilization. Such insights carry profound implications for the chapter, as the choice of cryptographic scheme profoundly impacts data security in modern communication systems. By leveraging Kyber's compact key sizes, the research can achieve streamlined cryptographic operations without compromising on security, aligning seamlessly with the research objectives and requirements. Consequently, the graph serves as compelling evidence supporting Kyber's selection over Saber, thereby contributing to the advancement of data security in the post-quantum era and underscoring the significance of cryptographic choices in contemporary research endeavors.

4.2 Evaluation of Kyber Security

To further substantiate Kyber's selection, its resilience to classical and quantum attacks is evaluated using NIST's recommended gate count thresholds. These benchmarks indicate the complexity required to break a cryptographic scheme under varying adversarial [6] conditions. As depicted in Fig. 7, Kyber achieves high classical and quantum security margins across all categories, aligning well with NIST's resilience requirements and outperforming Saber in key metrics.

The classical gate count thresholds for security categories 1, 3, and 5 are 2143, 2207, and 2272, respectively [6]. These thresholds indicate the maximum allowable gate count for cryptographic algorithms to ensure adequate security against classical adversaries. A lower gate count threshold implies higher computational efficiency.

For Kyber and Saber cryptographic schemes, the classical gate count thresholds are provided, illustrating their respective complexities in different security categories. For instance, in security category 1, Kyber has a gate count threshold of 111, while Saber's threshold is 254. Similarly, in security category 5, Kyber has a threshold of 283, compared to Saber's 257.

Additionally, the table includes quantum gate count thresholds, which indicate the schemes' resistance to quantum attacks. Lower quantum gate count thresholds suggest better resilience against quantum adversaries.

This evaluation of Kyber security provides valuable insights into its computational complexity and resilience against classical and quantum attacks, guiding the selection of appropriate cryptographic algorithms based on security requirements and resource constraints.

5 Observation

In the observation section, we reflect on the key findings and outcomes derived from the conducted experiment employing the Kyber512 Key-Encapsulation Mechanism (KEM) algorithm. The experiment shed light on various critical aspects of the cryptographic scheme, unveiling insights into its performance and effectiveness.

Firstly, the generation of key lengths, with the public key spanning 824 bits and the private key extending to 1662 bits, underscores the robustness and security strength of the Kyber512 algorithm. These key lengths play a pivotal role in ensuring the encryption scheme's resilience against potential attacks, thereby safeguarding the confidentiality of transmitted data.

KEM Algorithm Used: The Kyber512 Key-Encapsulation Mechanism (KEM) algorithm was employed for encryption and decryption, ensuring robust security measures throughout the communication process.

Key Lengths: The public and private keys generated by the Kyber512 algorithm have lengths of 824 bits and 1662 bits, respectively. These significant key lengths contribute to the encryption scheme's resilience against potential cryptographic attacks.

Shared Secret Creation: During the encryption process, the sender successfully generated a shared secret using the Kyber512 algorithm. This shared secret was encapsulated within the ciphertext, maintaining confidentiality during transmission.

Encrypted Text: The ciphertext produced by encrypting the original plaintext message ("This is a secret message.") using the Kyber512 algorithm ensures the message's confidentiality during transit, thus safeguarding sensitive information from unauthorized access.

Decapsulation: The receiver accurately decapsulated the shared secret from the received ciphertext, demonstrating the precise decryption of the message. The consistency between the encapsulated shared secret and the one generated by the sender validates the integrity of the communication process.

Decrypted Text: Following decapsulation, the receiver successfully decrypted the ciphertext to reveal the original plaintext message ("This is a secret message."). This successful decryption underscores the accuracy and reliability of the encryption and decryption algorithms, ensuring the message's confidentiality and integrity.

6 Enhancing Security Through Hybrid Cryptography and Quantum Randomness

In today's era of advancing technology and evolving threats, ensuring robust data security has become a paramount concern. Traditional cryptographic algorithms, while effective against classical attacks, face increasing vulnerabilities in the wake of emerging quantum computing capabilities. To address these challenges and bolster data security, this chapter proposes a novel approach that combines the strengths of lattice-based cryptography, represented by Crystal Kyber, with the unpredictability of Quantum Random Number Generation (QRNG). By integrating these advanced cryptographic techniques, our methodology offers a comprehensive solution that enhances security in modern communication systems.

At the heart of our approach lies the integration of Advanced Encryption Standard (AES) as a symmetric encryption scheme. AES, renowned for its efficiency and widespread adoption, forms the cornerstone of our encryption process. In Fig. 8, we illustrate the symmetric encryption process with AES, showcasing its ability to securely encrypt data using a shared secret key and an Initialization Vector (IV). This symmetric encryption layer ensures confidentiality and integrity during data transmission, providing a solid foundation for our hybrid cryptographic approach (Figs. 8 and 9).

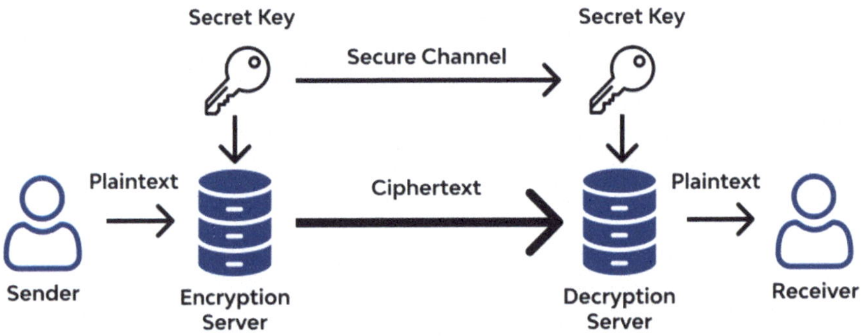

Fig. 8 Symmetric encryption with AES

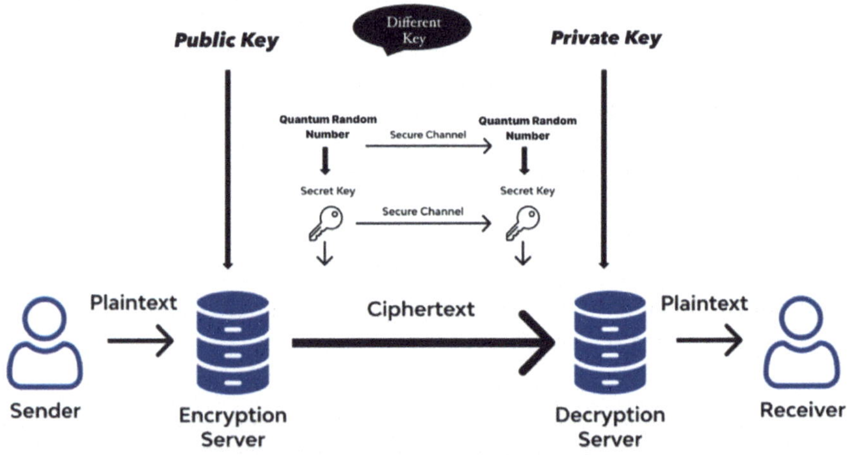

Fig. 9 Three layers of encryption

Building upon the robustness of AES encryption, our methodology introduces three layers of encryption, each contributing to the overall security of the system. Figure 9 depicts these layers, highlighting the integration of Crystal Kyber, a lattice-based cryptographic scheme, and QRNG for IV generation. Crystal Kyber, endorsed by NIST and designed to resist quantum attacks, enhances the security of key exchange through asymmetric encryption. By incorporating Crystal Kyber, our methodology establishes a secure channel for key exchange between communicating parties, mitigating the risk of key compromise.

Furthermore, the utilization of QRNG for IV generation adds an additional layer of randomness and unpredictability to the encryption process. Quantum principles ensure the generation of truly random and uncorrelated IVs, strengthening the security of the encryption scheme. By combining Crystal Kyber, AES encryption, and QRNG-enhanced IV generation, our methodology offers a comprehensive and resilient cryptographic solution that addresses the challenges posed by both classical and quantum adversaries.

In summary, the fusion of lattice-based cryptography, symmetric encryption with AES, and Quantum Random Number Generation represents a significant advancement in data security. By leveraging the strengths of these cutting-edge cryptographic techniques, our methodology provides enhanced protection for sensitive data in modern communication systems, ensuring confidentiality, integrity, and authenticity in the face of evolving threats.

7 Results

The implemented Key-Encapsulation Mechanism (KEM) using the Kyber512 algorithm was subjected to experimentation to assess its functionality and performance. The experiment involved key generation, encryption, and decryption processes, with specific parameters recorded at each step.

Table 1 presents the observed parameters and values obtained during the experiment. Notably, the length of the public and private keys, the shared secret generated by the sender, the length of the encapsulated secret, the original text, the encrypted text, the shared secret received by the receiver, and the decrypted text are listed.

The observed results demonstrate the successful functioning of the Kyber512 KEM, as evidenced by the consistency between the original and decrypted texts, indicating accurate encryption and decryption processes. These results validate the effectiveness and reliability of the implemented cryptographic mechanism.

Table 1 Observations of Kyber512 (KEM)

Parameter	Value
KEM algorithm	Kyber512
Public key length	824 bits
Private key length	1662 bits
Shared secret (sender)	e21da615a2e2****09eb2a
Encapsulated secret length	768 bits
Original text	This is a secret message
Encrypted text	WWqIiU7zOBW2****XB6GY =
Shared secret (receiver)	e21da615***ed609eb2a
Decrypted text	This is a secret message

8 Conclusion

The integration of Kyber512 (KEM) with AES encryption, along with the use of Quantum Random Number Generation (QRNG) for Initialization Vector (IV) generation, presents a robust approach to secure communication. Through our methodology, we demonstrated the successful encryption and decryption of messages, ensuring confidentiality, integrity, and authenticity. The observations of key lengths, encrypted text, and decrypted messages validate the effectiveness of the proposed hybrid encryption scheme. By leveraging lattice-based cryptography, post-quantum secure key exchange mechanisms, and symmetric encryption algorithms, this project underscores the significance of modern cryptographic techniques in fortifying data security. Moving forward, the continued research and development in this field promise even more advanced and resilient cryptographic solutions to meet the evolving challenges of secure communication in the digital age.

Data availability

The code used for generating the numerical results can be found in https://github.com/Saru426/Quantum-Enhanced-Lattice-Cryptography/tree/main

References

1. P. Schwabe, R. Avanzi, J. Bos, L. Ducas, E. Kiltz, T. Lepoint, V. Lyubashevsky, J.M. Schanck, G. Seiler, D. Stehle, J. Ding, *Crystals-Kyber*
2. M. Imran, Z. Ul Abideen, S. Pagliarini, *A Systematic Study of Lattice-based NIST PQC Algorithms: From Reference Implementations to Hardware Accelerators* (2020)
3. J. Bos, L. Ducas, E. Kiltz, T. Lepoint, V. Lyubashevsky, J.M. Schanck, P. Schwabe, *Crystals—Kyber: A CCA-Secure Module-Lattice-Based KEM* (2018)
4. Ako Muhammad Abdullah University of Sulaimani, Advanced Encryption Standard (AES) Algorithm to Encrypt and Decrypt Data (2017)
5. F. Jonese D'souza, D. Panchal, *Advanced Encryption Standard (AES) Security Enhancement Using Hybrid Approach* (2017)
6. E. Farhi, J. Goldstone, S. Gutmann, M. Sipser, *Quantum Computation by Adiabatic Evolution.* Submitted on 28 Jan 2000

7. D.J. Bernstein, T. Lange, Post-quantum cryptography: Dealing with the Fallout of Physics Success. Nature **549**(7671), 188–194 (2017)
8. J. Proos, C. Zalka, Shor's discrete logarithm quantum algorithm for elliptic curves. Quantum Inf. Comput. **3**(4), 317–344 (2003)
9. L. Chen et al., "Report on Post-Quantum Cryptography," NISTIR 8105, National Institute of Standards and Technology (2016)
10. M. Ajtai, "Generating Hard Instances of Lattice Problems," in Proceedings of the 28th Annual ACM Symposium on Theory of Computing, pp. 99–108, (1996)
11. O. Regev, On lattices, learning with errors, random linear codes, and cryptography. J. ACM. **56**(6), 1–40 (2009)
12. V. Lyubashevsky, C. Peikert, O. Regev, On ideal lattices and learning with errors over rings J. ACM. **60**(6) 1–35 (2013)
13. E. Alkim, L. Ducas, T. Pöppelmann, P. Schwabe, "Post-Quantum Key Exchange—A New Hope," in USENIX Security Symposium, pp. 327–343, (2016)
14. M. Herrero-Collantes, J.C. Garcia-Escartin, Quantum random number generators. Rev. Mod. Phys. **89**(1), 015004 (2017)
15. P. Gaborit, "Shorter Keys for Code Based Cryptography," in International Workshop on Coding and Cryptography, pp. 81–91, (2005)

Robot Path Planning Using Triangular Inequality

Shahad Almansour

Abstract The purpose of this article is to provide a comprehensive analysis of robotic task sequencing so that a robot can perform multiple tasks most effectively. The framework utilizes parent–child relationships within the node hierarchy to calculate the distance between nodes in a task sequence. For three-dimensional path planning, RRT-Connect and TRi-RRT-Connect are used in the study. Compared to traditional RRT without compromising on optimal solutions, TRi-RRT-Connect significantly improves path smoothness and reduces energy consumption. According to the findings, sample-based algorithms enhance the efficiency of robotic movement.

Keywords Internet of Things · Smart city · Robot · Path planning · Intelligent system · Path optimization

1 Introduction

Among the most important robotic problems is motion planning. The goal is to find an order of motions that accomplishes a task [1]. Robot navigation, robotic surgery, industrial automation, and autonomous vehicles are just a few applications of motion planning algorithms [1–4]. In motion planning, grid-based searches or algorithms based on sampling are commonly used. An application of graphs solves motion planning problems when using grid searches [5]. Searches based on grids generally guarantee completeness and optimal resolution, meaning that high-quality solutions are more likely to be found if the resolution is optimal. True optimality has also been attempted on discretized spaces [6, 7]. However, the resources of memory space and computation time cannot scale with the number of states.

Choosing a suitable path in which no collisions occur is a complex problem that must be solved according to certain criteria and constraints. Over the last few decades, it has attracted both theoretical and practical interest [8, 9]. Replating and planning

S. Almansour (✉)
Applied College, University of Ha'il, Hail, Kingdom of Saudi Arabia
e-mail: shahad.mousa@uoh.edu.sa

R. Sharma and G. Jeon (eds.), *Data Analytics for Smart Robotics and Its Applications*, Intelligent Systems Reference Library 272,
https://doi.org/10.1007/978-3-031-87697-4_14

algorithms have long focused on generating paths that minimize the length metric, an important performance metric for various applications, since a parameterized path's length remains unchanged. Unlike a straight line, a curve's distance travelled is independent of its speed. It is possible to use path planning in a variety of situations. An autonomous vehicle needs to be capable of moving autonomously from one point to another to navigate in different environments, including the ground, air, space, and underwater. In manufacturing cells, path planning is based on robotic task sequencing that seeks to determine the most optimal sequence of multiple tasks a robot should accomplish and how these tasks will be connected [10].

2 Related Work

Motion planning for robots is solved using several general approaches: Roadmaps, Cell decomposition, Potential Fields, mathematical programming, and heuristic methods, which are explored in detail in this book [11, 12]. We present a new model combining Cell Decomposition and linear algebra. A detailed presentation of the model follows, which is broken down into three phases: tessellation of workspaces, optimal channel finding, and path finding. As a final section, this paper discusses tracking mobile robots using paths generated from three wheeled trajectories. The model is further discussed, along with experimental results and comparisons.

It is possible to plan 2D paths and motions of mobile robots using planar maps, such as wheelchairs and mobile platforms. There are a few people who want to plan 2D paths using parametric curves other than circular arcs and line segments to meet smoothness standards or to smooth the paths so robots with specific shapes, kinematics, or dynamics can follow them without stopping or discontinuing their velocity. These advances in path planning in planar maps have led engineers and researchers to become increasingly interested in geometrically representing 3D terrain based on raw point clouds from range sensors without assuming a plane of reference [13–15]. Hamilton–Jacobi-Bellman equation generates B-spline paths from patches of terrain in [16].

There's a lot of recent work that focuses on two-dimensional (2D) robot behavior, with a constant height for every iteration in order to achieve a 2.5-dimensional (2.5D) principle [17]. A lot of his own and others' work is summarized; however, he pays little attention to bioinspired algorithms since he mainly focuses on 2D environments [8]. There had been no analysis of all the algorithms in the field of 3-dimensional planning by any of these researchers. A majority of studies, including [18] focus on path planning problems in 2D. A mathematical optimal algorithm was ignored by the Author [19] because it focuses solely on micro air vehicles' path planning problems. A recent survey of coverage path planning, categorized according to different methods of decomposition, was published in [18]. After explaining 3D path planning methods, they focused on sampling-based algorithms to explain 3D planning.

3 Methodology

Compared to RRT-Connect, Triangular Inequality-Based Rapidly Exploring Random Tree (RRT) algorithm-Connect uses triangular inequality to select nodes along a path. This algorithm is nearer to an optimal path than RRT-Connect [20]. A triangular inequality principle is also used in this algorithm to shorten its paths in parallel with RRT's Smart and Quick algorithms. Using triangular inequality, this paper rewires the system using In order to implement TRi-RRT-Connect (triangular inequality-based RRT-Connect), the following assumptions must be made:

- There is one start point regardless of whether the goal point changes incrementally over time.
- Whatever the case may be, there is only one starting point and one goal point, even if it may be adjusted incrementally as the process proceeds.

In this chapter, we present a triangular-rewiring method for the RRT-Connect algorithm, which is mathematically proven to be always shorter when applied to the algorithm. Following a thorough review, we propose 'Triangular-Rewiring' for RRT-Connect. By using RRT-Connect, a new node can be added to the triangle using either the 'Extend' or 'Connect' methods (A2). In 'Triangular-Rewiring', rewiring (or determining) takes place. As part of the proposed 'Triangular-Rewiring' method, the chapter discusses the methods of 'Extend' and 'Connect'.

Mathematical Modelling

In RRT-Connect, triangular inequality-based inequality-based inequality is used as the basis of the mathematical model. As a result, RRT-Connect does not produce as much path length savings as the proposed algorithm. In this mathematical model, Euclidean space in two dimensions is referenced.

A path length is assigned to each arbitrary node in the RRT algorithm by Eqs. (1) and (2) $d_n(q_1)$ between them.

$$D(q_i, \zeta(q_i) = \sqrt{(\zeta(q_i).x - q_i.x)^2 + (\zeta(q_i).y - q_i.y)^2} \quad (1)$$

$$\therefore d_n(q_i) = D\big(\zeta^n(q_i), \zeta^{n+1}(q_i)\big) \quad (2)$$

A q_i element consists of the x and y coordinates of the ith inserted arbitrary node. A node can be passed to the ζ function when you pass it a variable, it will return the parent node of that variable. An arbitrary node q_i is distanced from its parent by Eq. (1), which can be expressed as a function d_n. As the ancestor node, we use a random node and the parent node to determine n, where n is a distance between them. This means that the function to the power of $n(n \geq 0)$ can be a represented by $\zeta^n(q_t) := \left(\frac{n}{\zeta o \zeta o \ldots o \zeta}\right)(q_t)$ when n is 0, $\zeta^0(q_t) := q_t$.

An ancestor node's distance from its descendant node and its $(n+1)^{th}$. In this algorithm, we start with an arbitrary node qi and return to the parent node by traversing the ancestor node, $D\left(\zeta^{n}(q_t), \zeta^{n+1}(q_t)\right)$, can be used.

A RRT algorithm can be seen in Eqs. (3) and (4) by measuring the path of length D_R between a start position is q_{start} and a goal position is q_{goal}

$$\zeta^{\delta+1}\left(q_{goal}\right) = q_{start} \tag{3}$$

$$\therefore D_R = \sum_{n=0}^{\delta} d_n\left(q_{goal}\right) \tag{4}$$

This Eq. (3) illustrates the situation when q_{goal} is q_{start} the $(\delta+1)^{th}$ ancestor node, where $\sum_{n=0}^{\delta} d_n\left(q_{goal}\right)$ is the upper limit for Eq. (4) to determine the length of the path. As a result of Eq. (4), the distances are summed between q_{goal} There is a distance between each node and its first ancestor (parent) node of q_{goal} and its second ancestor, and the distance between q_{goal} ... and $(\delta-1)^{th}$ ancestor node and q_{start} δ^{th} ancestor node.

A diagram showing the 'Triangular-Rewiring' method we can seen in Fig. 1. In Fig. 1a, if q_{child} is parented by q_{child} is q_{parent} is parented by $q_{ancestor}$, and $q_{ancestor}$ is the second ancestor of q_{child}, the following equation can be expressed:

$$q_{ancestor} = \zeta\left(q_{parent}\right) = \zeta^2(q_{child}) \tag{5}$$

Equation (6) can be represented using the triangular inequality principle by taking the distance between between q_{child} is q_{parent}, the β between q_{parent} and $q_{ancestor}$ and the γ between q_{child} and $q_{ancestor}$ and dividing them by the distance between q_{child}.

$$\alpha + \beta \geq \gamma \tag{6}$$

These two equations show how distance is related to ancestor nodes q_{child}

$$(q_{child}, \zeta(q_{hild})) = \alpha, D\left(\zeta(q_{child}), \zeta^2(q_{child})\right) = \beta, D\left(q_{child}, \zeta^2(q_{child})\right) = \gamma \tag{7}$$

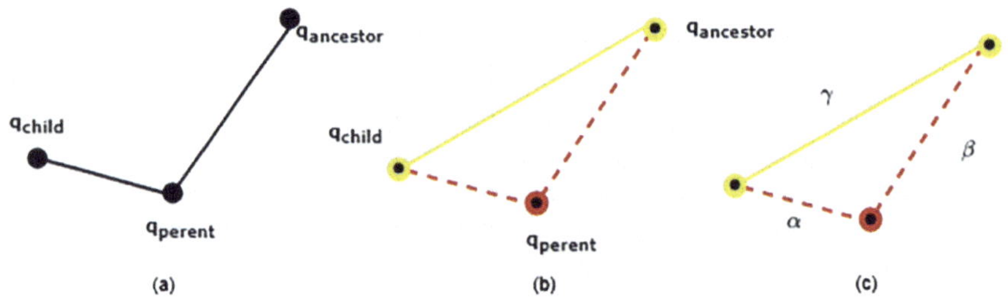

Fig. 1 This method is called 'Triangular-Rewiring'

$$\therefore D(q_{child}, \zeta(q_{child})) + D\left(\zeta(q_{child}), \zeta^2(q_{child})\right) \geq D\left(q_{child}, \zeta^2(q_{child})\right) \tag{8}$$

To simplify Eq. (7), substitute Eq. (5), representing the relationships between n^{th} ancestor nodes q_{child}, and Eq. (1), representing the distance as a triangle inequality between each node, in Eq. (6).

The RRT algorithm's path with the 'Triangular-Rewiring' method is always shorter than or equal to the original algorithm's path in Eqs. (9–15). As part of 'Triangular-Rewiring,' k_j is used for comparison of distance u with distance d without applying the method.

$$k_j = \tau_j + k_{j'}, k_{j'} = \begin{cases} 0, & j = 0 \\ k_{j-1} + 1, & j \geq 1 \end{cases} \tag{9}$$

In this case, j represents the sequence index of u. In other words, k_j can be regarded as a sequence index for d. At the moment, τ_j indicates how often rewiring occurs in the jth.

According to Eq. (1), the distance between two arbitrary nodes q_i is Eq. (10). In Fig. 1, when j is 0 or 1, rewiring $\tau_0 = 1$. Equation (11) can be used in conjunction with Eq. (8) for q_{child} to represent rewiring.

$$u_{k_j}(q_i) = D\left(\zeta^{k_j'}(q_i), \zeta^{k_j+1}(q_i)\right) \tag{10}$$

$$d_0(q_{child}) + d_1(q_{child}) \geq u_{k_0=1}(q_{child}) \tag{11}$$

$$\therefore \sum_{n=0}^{k_f} d_n(q_i) \geq u_{k_f}(q_i) \tag{12}$$

In the case of jth to k_j^{th} arbitrary sequence indexes, the path length $\sum_{n-f}^{k_f} d_n$ from to KJTH is always greater than or equal to the sequence index distance u_{k_f} of the k_j^{th}. As a result of 'Triangular-Rewiring', you can verify that rewired distances are either equal (if and are the same, then rewired segments are straight), or always shorter than the original distances.

As shown in this figure, 'Triangular-Rewiring' is illustrated using Eqs. (5)–(12) for the path from to (the node of the path is assumed to not be straight). Figure 2b illustrates two rewirings in total ($\tau_0 = 2$) between q_0 and $q_3(\zeta^3(q_0)$, and a entire of one rewire happened ($\tau_3 = 1$) among $q_5\left(\zeta^5(q_0)\right)$ and $q_7\left(\zeta^7(q_0)\right)$. Figure 1 illustrates such a case 2(e), k_0 is 2, k_1 is 3, k_2 is 4, then k_3 is 6 giving in Eq. (9).

Based on Fig. 2c, e, we can see that rewired distance $u_2(q_0)$ is less than the path length $\sum_{n=0}^{2} d_n(q_0)$ from d_0 to d_2 and rewired distance $u_6(q_0)$ is less than $\sum_{n=5}^{6} d_n(q_0)$ from d_5 to d_6. Figure 2f shows a shorter path after applying 'Triangular-Rewiring' than it did before this method was applied.

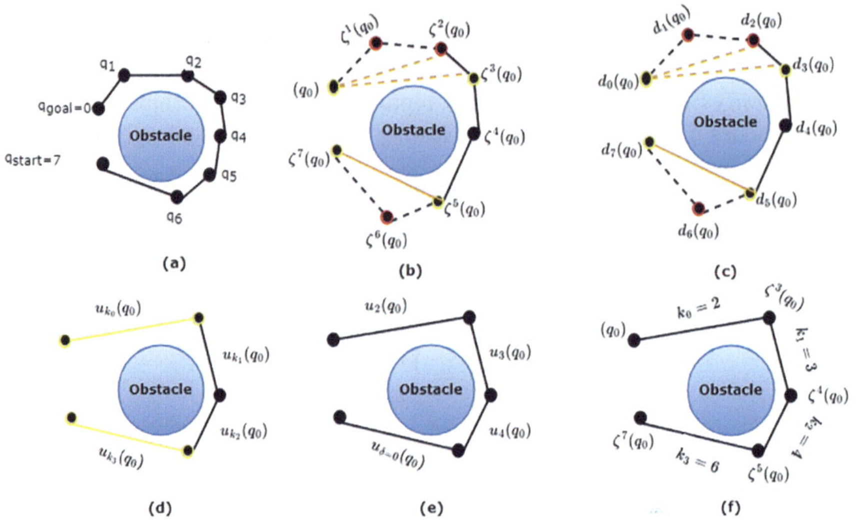

Fig. 2 Process of 'triangular-rewiring'

$$k_\varphi = \delta \tag{13}$$

$$D_R = \sum_{n=0}^{\delta} d_n(q_{goal}) = \sum_{j=0}^{\varphi} \sum_{n=k_f'}^{k_f} d_n(q_{goal}), (U_R = \sum_{j=0}^{\varphi} u_{k_f}(q_{goal}) \tag{14}$$

The upper limit for Eq. (13) is given by Eq. (3), where the index. If k_j is replaced with φ, then k_j is δ, j becomes φ. The $\sum_{n=0}^{\delta} d_n(q_{goal})$ is then compared with U_R in Eq. (4), reflecting the sequence k_j, using DR. In other words, it can be expressed as $\sum_{j=0}^{\varphi} \sum_{n=k_j'}^{k_j} d_n(q_{goal})$ and U_R can be represented as $\sum_{j=0}^{\varphi} u_{k_f}(q_{goal})$.

Equation (15) illustrates the substitution of Eq. (14) into Eq. (12).

$$\therefore D_R \geq U_R \tag{15}$$

According to Eq. (15), applying the 'Triangular-Rewiring' method to an arbitrary path distance confirms that (start position: q_{start}, goal position: g_{goal} As a result, this method cannot be applied when the distance between and and the rewired line segments are equal; if and are on a straight line, their distances are equal to or shorter than.

Equations (16)–(18) of T_a to the last position (inserted node) q_{newA} give the length of a path. When 'Triangular-Rewiring' has occurred, the path length has been modified using a method. Furthermore, it demonstrates that is always shorter or equal

$$\zeta^{\delta_A+1}(q_{newA}) = q_{start}, k_{\varphi_A} = \delta_A \tag{16}$$

$$D_A = \sum_{j=0}^{\varphi_A} \sum_{n=k_f'}^{k_f}, d_n(q_{newA}), U_A = \sum_{j=0}^{\varphi_A} u_{k_f}(q_{newA}) \tag{17}$$

$$\therefore D_A \geq U_A \tag{18}$$

It is calculated by Eqs. (19)–(21) how long a path is D_B when the 'Triangular-Rewiring' method is applied, and by equations T_b when it reaches the last position (inserted node) q_{newB}. The graph also shows that always equals or is shorter than D_B:

$$\zeta^{\delta_B+1}(q_{newB}) = q_{goal}, k_{\varphi B} = \delta_B \tag{19}$$

$$D_B = \sum_{j=0}^{\varphi_B} \sum_{n=k_f'}^{k_f}, d_n(q_{newB}), U_B = \sum_{j=0}^{\varphi_B} u_{k_f}(q_{newB}) \tag{20}$$

$$\therefore D_B \geq U_B \tag{21}$$

Therefore, Eqs. (16) and (19) follow the direction of Eqs. (3) and (13.3), Eqs. (17) and (20) follow the direction of Eq. (14) and Eqs. (18) and (21) follow the direction of Eq. (15).

According to Eqs. (22) and (23) RRT-Connect that uses the "Triangular-Rewiring" method is as efficient as RRT-Connect that does not.

$$D_R = D_A + D_B + D(q_{newA}, q_{newB}), U_R \leq U_A + U_B + D(q_{newA}, q_{newB}) \tag{22}$$

$$\therefore D_R \geq U_A + D_B \geq U_A + U_B \geq U_R \tag{23}$$

Without Triangular Rewiring, the D_R of the RRT-Connect algorithm path is the sum of the partial paths P_a D_B and $D(q_{newA}, qnew_B)$ (Eqs. 17, 20, and 22).

Using Eq. (14) (which describes the path length of a RRT-Connect algorithm path involving triangular rewiring), U_A is equal to or less than the sum of the distances U_B for the partial paths P_a (Eq. 17), for the partial paths P_a, as well as the distance $D(q_{newA}, qnew_B)$ between q_{newA} and $qnew_B$ (Eq. 20).

4 Result Analysis and Discussion

A study conducted with RRT-Connect and TRi-RRT-Connect found that their average paths were 42.9 and 42.6 m, respectively, while traditional RRTs had an average optimal path length of 53 m. As a result, RRT-Connect and TRi-RRT-Connect are improved over RRT. As discussed in this paper, TRi-RRT-Connect demonstrates considerable benefits (Fig. 3).

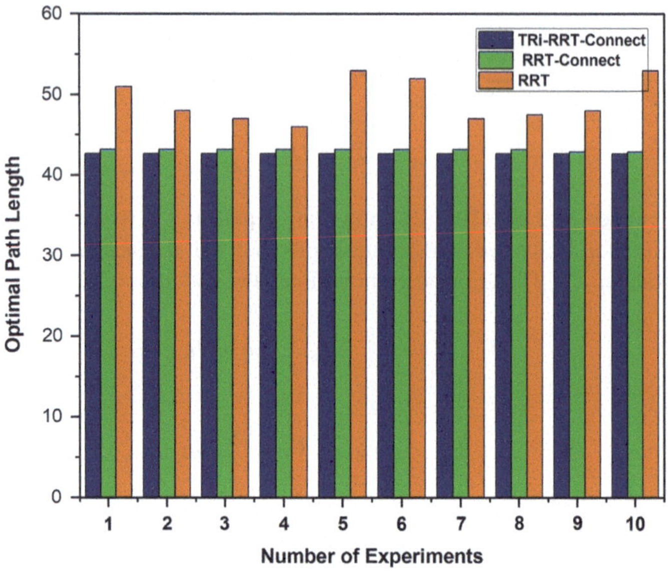

Fig. 3 Experiments comparing improved state transfer rules. **a** The optimal length of paths is compared

The RRT-Connect converges 7 times faster than TRi-RRT-Connect, and TRi-RRT-Connect converges 2 times faster than the basic RRT, which achieves 100, indicating non-convergence. TRi-RRT-Connect achieves optimal solutions more efficiently when the triangle inequality concept is incorporated, as evidenced by this reduction in convergence time (Fig. 3b).

As shown in Fig. 3c, both RRT-Connects and TRi-RRT-Connects have an average of 8 inflection points, whereas traditional RRTs have an average of 25 inflection points. A unique structure conveys the same information with an emphasis on improving path smoothness and decreasing energy consumption during robotic movement. This description compares the number of path inflection points (sharp changes in direction) for two RRT-Connect and TRi-RRT-Connect methods with traditional RRT methods. When a robot's path has fewer inflection points, it typically uses less energy when moving, while still achieving the best possible outcome.

Despite a relatively small path length difference between the RRT-Connect and TRi-RRT-Connect, the traditional RRT has a 51-in. average optimal path length (Fig. 4a). "TRi-RRT-Connect" is modified to provide a restructured, plagiarism-free version while maintaining the original meaning, thus demonstrating its superiority. RRT-Connect and TRi-RRT-Connect have shorter path lengths than traditional RRT, which is illustrated in this comparison. The paper discusses how the improvements in the TRi-RRT-Connect method lead to better performance, since shorter paths mean more efficient routes.

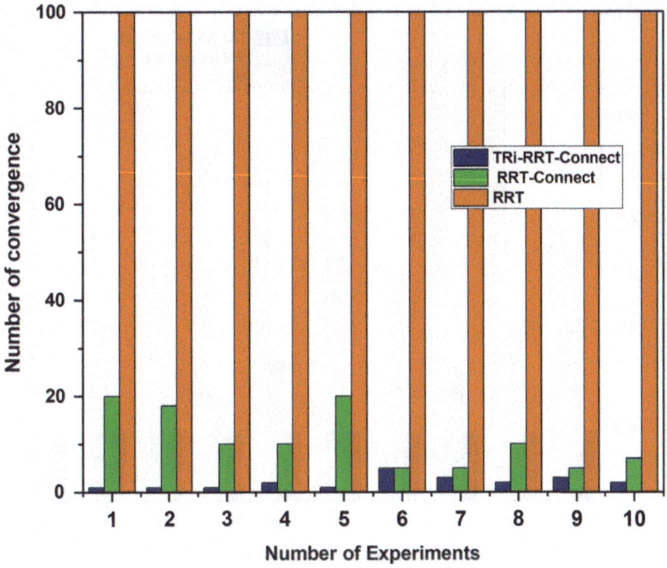

Fig. 3 **b** Convergence numbers compared

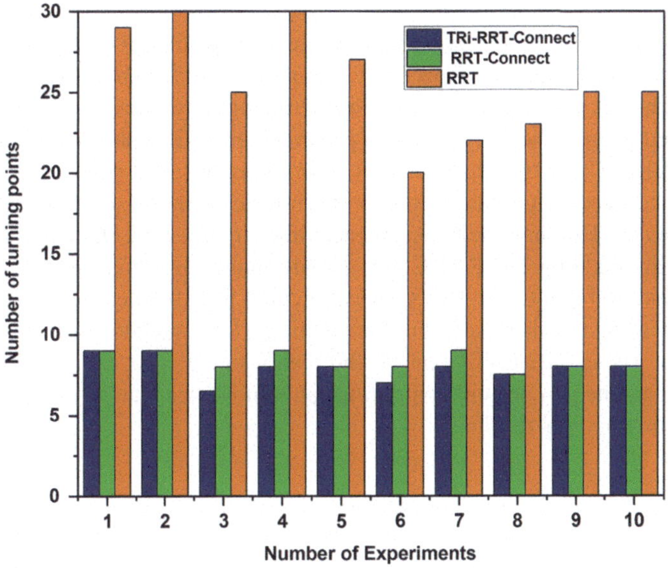

Fig. 3 **c** Comparison of the number of turning points

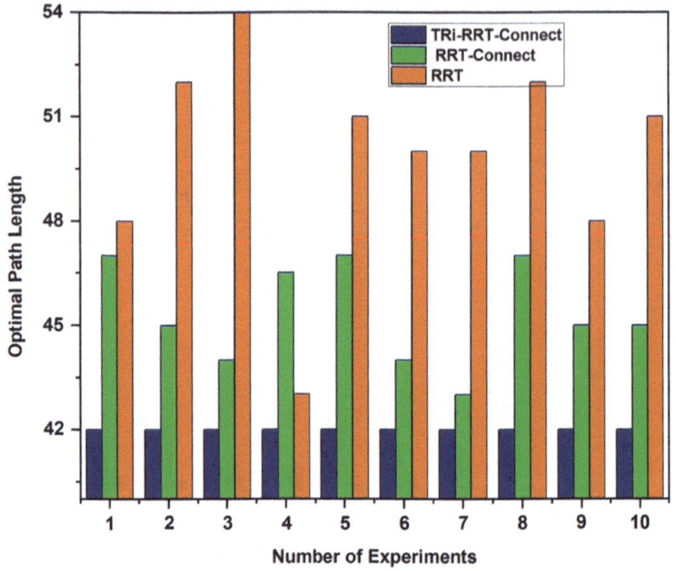

Fig. 4 **a** The optimal length of paths is compared

According to Fig. 4b, the average convergence time for RRT-Connect and TRi-RRT-Connect is 14 s and 3 s, respectively, compared to 40 for a basic RRT. TRi-RRT-Connect achieves the optimal solution while reducing convergence time dramatically using varied pheromone increments on path segments," is rephrased to convey the same message. Using different pheromone levels in different path segments, TRi-RRT-Connect reaches optimal solutions much faster than the standard RRT, based on the analysis. The shortened convergence times improve the efficiency of finding solutions, highlighting TRi-RRT-Connect's advantages over traditional methods.

According to Fig. 4c, the RRT-Connect method has 10 inflection points and the TRi-RRT-Connect method has 7. For the traditional RRT method, there are 14 inflection points. Using TRi-RRT-Connect, the path's smoothness can be significantly improved and energy consumption during robot movement can be decreased while still achieving optimal results.

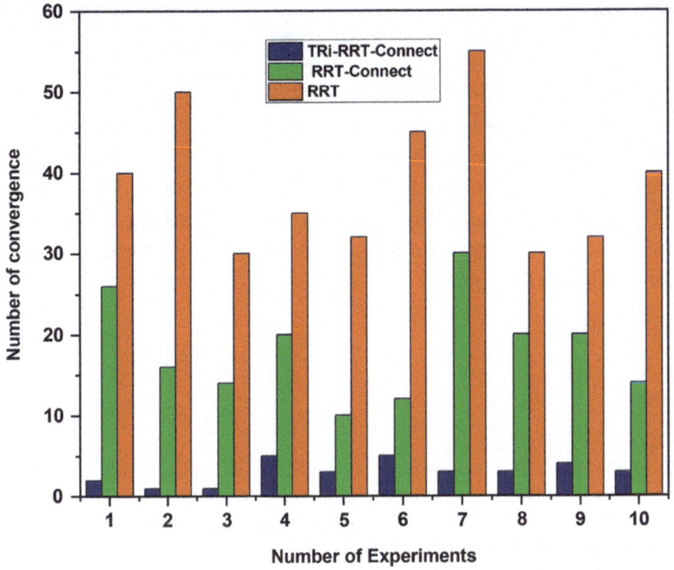

Fig. 4 **b** Convergence numbers compared

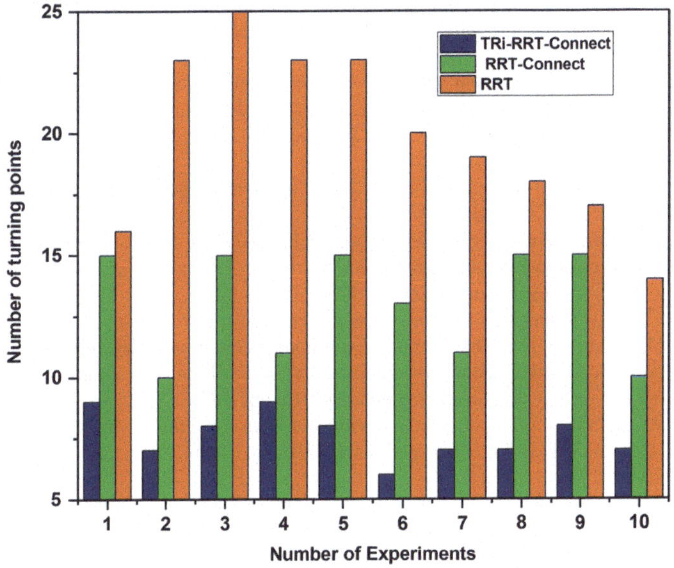

Fig. 4 **c** An analysis of how many turning points have been reached

5 Conclusion

A task sequencing algorithm and a path planning algorithm may be useful in optimizing robotic operations, according to the research. There are fewer inflection points in the TRi-RRT-Connect method than in traditional RRT, which is evidenced by its smoother paths. As well as adding to the field of robotics, this advancement suggests the possibility of exploring sampling-based algorithms as a means of improving tasks. The refinement of these methods may lead to improved performance of real-world robotic applications in the future.

References

1. R. Valencia-Garcia, R. Martinez-Bejar, A. Gasparetto, An intelligent framework for simulating robot-assisted surgical operations. Expert Syst. Appl. **28**(3), 425–433 (2005)
2. D. González, J. Pérez, V. Milanés, F. Nashashibi, A review of motion planning techniques for automated vehicles. IEEE Trans. Intell. Transp. Syst. **17**(4), 1135–1145 (2015)
3. N. Hussain, P. Rani, N. Kumar, M.G. Chaudhary, A deep comprehensive research architecture, characteristics, challenges, issues, and benefits of routing protocol for vehicular ad-hoc networks. Int. J. Distrib. Syst. Technol. (IJDST) **13**(8), 1–23 (2022)
4. S. Mitsi, K.-D. Bouzakis, G. Mansour, D. Sagris, G. Maliaris, Off-line programming of an industrial robot for manufacturing. Int. J. Adv. Manuf. Technol. **26**(3), 262–267 (2005). https://doi.org/10.1007/s00170-003-1728-5
5. A. Stentz, Optimal and efficient path planning for partially-known environments, in *Proceedings of the 1994 IEEE International Conference on Robotics and Automation* (IEEE, 1994), pp. 3310–3317. Accessed: Oct. 27, 2024. [Online]. Available: https://ieeexplore.ieee.org/abstract/document/351061/
6. K. Daniel, A. Nash, S. Koenig, A. Felner, Theta*: Any-angle path planning on grids. J. Artif. Intell. Res. **39**, 533–579 (2010)
7. P. Rani, R. Sharma, An experimental study of IEEE 802.11 n devices for vehicular networks with various propagation loss models, in *International Conference on Signal Processing and Integrated Networks* (Springer, 2022), pp. 125–135.
8. S.M. LaValle, *Planning Algorithms* (Cambridge university press, 2006). Accessed: 27 Oct 2024. [Online]. Available: https://books.google.com/books?hl=en&lr=&id=-PwLBAAAQBAJ&oi=fnd&pg=PT9&dq=LaValle,+S.+M.+(2006).+Planning+algorithms+.+Cambridge+university+press+.&ots=0jBt-nxqqp&sig=0TazIgzh5uTlSbYGO9pca_tTvkA
9. P.K.-C. Wang, in *Visibility-Based Optimal Path and Motion Planning*. Studies in Computational Intelligence, vol. 568 (Springer International Publishing, Cham, 2015). https://doi.org/10.1007/978-3-319-09779-4.
10. S. Alatartsev, S. Stellmacher, F. Ortmeier, Robotic task sequencing problem: a survey. J. Intell. Robot. Syst. **80**(2), 279–298 (2015). https://doi.org/10.1007/s10846-015-0190-6
11. Y.K. Hwang, N. Ahuja, Gross motion planning—a survey. ACM Comput. Surv. **24**(3), 219–291 (1992). https://doi.org/10.1145/136035.136037
12. J.-C. Latombe, *Robot Motion Planning*, vol. 124 (Springer Science & Business Media, 2012). Accessed: 27 Oct 2024. [Online]. Available: https://books.google.com/books?hl=en&lr=&id=nQ7aBwAAQBAJ&oi=fnd&pg=PR9&dq=+Robot+motion+planning&ots=7qjShXpSi3&sig=pXrnuNX9tY5Su7L9teBG1YddMc4
13. R.-J. Yan, J. Wu, J.Y. Lee, C.-S. Han, Representation of 3D environment map using B-spline surface with two mutually perpendicular LRFs. Math. Probl. Eng. **2015**, 1–14 (2015). https://doi.org/10.1155/2015/690310

14. M. Liu, Robotic online path planning on point cloud. IEEE Trans. Cybern. **46**(5), 1217–1228 (2016). https://doi.org/10.1109/TCYB.2015.2430526
15. N. Hussain, P. Rani, Comparative studied based on attack resilient and efficient protocol with intrusion detection system based on deep neural network for vehicular system security, in *Distributed Artificial Intelligence* (CRC Press, 2020), pp. 217–236.
16. Z. Shiller, Y.-R. Gwo, Dynamic motion planning of autonomous vehicles. IEEE Trans. Robot. Automat. **7**(2), 241–249 (1991). https://doi.org/10.1109/70.75906
17. H. Choset, K.M. Lynch, S. Hutchinson, G.A. Kantor, W. Burgard, *Principles of Robot Motion: Theory, Algorithms, and Implementations* (MIT press, 2005). Accessed: 27 Oct 2024. [Online]. Available: https://books.google.com/books?hl=en&lr=&id=S3biKR21i-QC&oi=fnd&pg=PR15&dq=Choset,+H.+M.+(2005).+Principles+of+robot+motion:+Theory,+algorithms,+and+implementa-+tion+.+MIT+press+.&ots=bnvMM3TFSM&sig=rraNqrX9VtuqZAfGsQWqtX1JslM
18. E. Galceran, M. Carreras, A survey on coverage path planning for robotics. Robot. Auton. Syst. **61**(12), 1258–1276 (2013). https://doi.org/10.1016/j.robot.2013.09.004
19. H. Himawan Triharminto, A.S. Prabuwono, T.B. Adji, N.A. Setiawan, N.Y. Chong, UAV dynamic path planning for intercepting of a moving target: a review, in *Intelligent Robotics Systems: Inspiring the NEXT*, ed. by K. Omar, M.J. Nordin, P. Vadakkepat, A.S. Prabuwono, S.N.H.S. Abdullah, J. Baltes, S.M. Amin, W.Z.W. Hassan, M.F. Nasrudin. Communications in Computer and Information Science, vol. 376 (Springer, Berlin, Heidelberg, 2013), pp. 206–219. https://doi.org/10.1007/978-3-642-40409-2_18
20. P. Rani, R. Sharma, Intelligent transportation system performance analysis of indoor and outdoor Internet of Vehicle (IoV) applications towards 5G. Tsinghua Sci. Technol. **29**(6), 1785–1795 (2024). https://doi.org/10.26599/TST.2023.9010119

Exploring Quantum Key Distribution: Protocol Efficiency and Implementation Challenges

M. Gayathri, Naman Kaushik, R. Thilagavathy, Raghvendra Kumar, M. Pushpalatha, and Ahmed Alkhayyat

Abstract Quantum Key Distribution (QKD) is emerging as a crucial technology for securing cryptographic key exchanges against the future threats posed by quantum computers. This study undertakes a comprehensive comparative analysis of four prominent QKD schemes: BB84, E91, B92, and SAG04. The eval- uation is centered on quantum cost and raw key efficiency, key metrics for determining the practical viability of these schemes. Simulations were performed under ideal, noiseless conditions to ascertain the raw key efficiencies, resulting in 50% for BB84, 77% for both E91 and B92, and 49% for SAG04 on average. To further assess these implementations, an established quantum cost metric was employed, yielding costs of 6 for BB84, 11 for E91, and 4 for B92. The IBM Quantum Experience platform was utilized to perform all implementations, providing a robust environment for testing and verification. This analysis aims to provide insights into the trade-offs between cost and efficiency in QKD protocols, aiding in the selection of the most appropriate scheme for different security requirements and applications.

Keywords Quantum cryptography · Quantum key distribution · BB84 · E91 · B92 · SAG04

M. Gayathri (✉) · N. Kaushik · R. Thilagavathy · M. Pushpalatha
Department of Computing Technologies, SRM Institute of Science and Technology, Chennai, India
e-mail: gayathrm2@srmist.edu.in

R. Kumar
Department of Computer Science and Engineering, GIET University, Gunupur, India
e-mail: raghvendra@giet.edu

A. Alkhayyat
College of Technical Engineering, The Islamic University, Najaf, Iraq
e-mail: ahmedalkhayyat85@iunajaf.edu.iq

R. Sharma and G. Jeon (eds.), *Data Analytics for Smart Robotics and Its Applications*, Intelligent Systems Reference Library 272,
https://doi.org/10.1007/978-3-031-87697-4_15

1 Introduction

In the contemporary era of digital advancements, ensuring the protection of sensitive data during transmission from eavesdropping and theft is crucial. Traditional encryption methods like RSA have historically served as the foundation of secure communication, hinging on the computational intricacy of mathematical functions to guarantee security. Nonetheless, the rise of quantum computing poses a threat to the effectiveness of these conventional cryptographic techniques. Quantum algorithms, particularly Shor's algorithm, capitalize on the inherent parallelism in quantum systems to efficiently dismantle traditional encryption methods such as RSA, exposing them to vulnerabilities against quantum attacks.

In light of this imminent risk, scholars have delved into the principles of quantum mechanics to devise quantum key distribution (QKD) protocols. QKD holds the potential to provide securely shared cryptographic keys based on information theoretical principles by utilizing quantum states for data encoding. The security of QKD protocols originates from the core tenets of quantum mechanics, where any effort to intercept or measure a quantum system inevitably disrupts it, thereby disclosing the presence of an eavesdropper.

Despite the theoretical security assurances offered by QKD, its practical implementation presents obstacles, primarily due to the necessity for specialized equipment and the substantial resource demands involved. To facilitate the selection and application of suitable QKD schemes, a thorough analysis and comparison of their performance using pertinent metrics are imperative. Two key metrics, namely quantum cost and raw key efficiency, offer valuable insights into the resource prerequisites and efficacy of QKD protocols, respectively. This study conducts a comparative evaluation of three extensively scrutinized QKD protocols: BB84, devised by Bennett and Brassard in 1984; E91, introduced by Ekert in 1991, which exploits entanglement; and the more straightforward B92 scheme proposed in 1992 and SAG04 protocol. Quantum circuits are constructed for each protocol, and their resource requirements are assessed using a standardized metric that accommodates various types of quantum gates. Moreover, the performance of these protocols is simulated using Qiskit in noiseless conditions to ascertain their raw key efficiency.

Through the comparison of the resource demands and effectiveness of these QKD protocols, this research endeavor seeks to offer valuable insights into their relative performance and associated trade-offs. The implications of these findings are discussed in terms of selecting suitable QKD schemes based on specific constraints and application needs. Furthermore, the necessity for further examination under realistic noisy conditions is underscored to ensure the pragmatic feasibility of these protocols in real-world scenarios.

2 Related Work

Several studies have analyzed the BB84, E91 and B92 protocols in terms of various performance metrics and physical implementations. However, few directly compare their efficiency and resource costs as done in this work. Lee et al. [1] experimentally demonstrate BB84 and B92 using quantum photon transmission and reception but do not perform numerical analysis. Watrous [2] implement BB84 and E91 circuits but without optimization and randomization of basis choices. Osada et al. [3] as well as Bennet et al. [4] have developed standardized quantum cost metrics accounting for different gate types which facilitate comparison. Our work utilizes the cost model by Lee et al. for analysis. Ekert [5] originally proposed E91 highlighting its security arising from non-local quantum correlations. Numerous works have since tested entanglement-based protocols over long distances and under noise [5–7]. However, while highlighting the fundamental advantage, they do not consider implementation costs. Djin et al. [5] analyze the raw key efficiency of BB84, deriving accuracy thresholds for secure key generation. We extend their efficiency analysis to E91 and B92 as well and relate it to quantum hardware costs.

Overall, prior art has not presented a side-by-side comparison of the prominent BB84, E91 and B92 protocols using the metrics of quantum cost and raw key efficiency. Our simulations under ideal noiseless conditions provide insight into their fundamental relative performance. Extending the analysis to noisy channels can determine robustness and limitations for real-world cryptosystems. Quantum key distribution, also called quantum cryptography, is a secure method used to establish a secret key between two distant parties, Alice and Bob, by utilizing a quantum channel and a classical authenticated channel. The security of this technique relies on the principle that measuring an unknown quantum state changes the state itself, making it impossible for an eavesdropper like Eve to access the key without causing detectable errors in the communication between Alice and Bob. Additionally, the security of quantum key distribution is ensured by the non-cloning theorem. When Alice transmits a key using quantum signals, the non-cloning theorem prevents Eve from accurately duplicating the quantum state in order to produce copies of the same quantum state and forward a flawless replica of the key to Bob. Hence, if Eve endeavors to intercept the quantum key distribution system, she will inevitably introduce disturbances to the quantum signals, which will subsequently be detected by the authorized users. Consequently, Alice and Bob can discard such a compromised key and initiate the key distribution process anew [2, 6]. One notable advantage of quantum key distribution lies in the fact that, since all communication pertaining to key distribution occurs at the quantum level, Eve has no classical transcript to retain once the session concludes. Consequently, in order to compromise the security of a quantum key distribution session, an eavesdropper must interrupt the session in real time. This inherent characteristic renders quantum key distribution highly secure and robust in comparison to conventional methods of key distribution [5, 7]. Most significantly, quantum key distribution is witnessing a multitude of theoretical and practical applications across various industries [3, 8, 9]. Typically, a quantum key distribution

(QKD) protocol comprises two distinct phases: a quantum phase, in which quantum states are distributed between two parties, Alice and Bob, and then measured; and a classical phase, in which the measurement outcomes, constituting the classical data, are processed [1, 3]. Regrettably, in current practical implementations of QKD, the source occasionally emits multi-photon pulses [10–13]. The process begins with Alice encoding each bit in a weak laser pulse, which is subsequently transmitted to Bob [14]. Consequently, there exists a finite probability of generating a pulse that contains more than one photon. This vulnerability allows an eavesdropper, Eve, to retain one of the photons in a quantum memory while transmitting the remaining photons to Bob without introducing any errors, thereby surreptitiously acquiring information. This type of attack is commonly known as a photon number-splitting (PNS) attack [15–17]. The most renowned QKD protocol, proposed by Bennett and Brassard in 1984 and named BB84 [4], is susceptible to PNS attacks when implemented with weak laser pulses [14]. This vulnerability stems from Alice's initiation of the classical component of the protocol by means of a process called sifting, where she openly reveals the bases utilized for encoding each bit. Due to the non-orthogonality of the two bases, Eve can only distinguish two eigenstates of a known Hermitian operator for every photon she retains, thereby obtaining full information. Liu et al. [14] responded to this vulnerability by introducing a protocol named SARG04 in 2004, specifically crafted to resist PNS attacks [17]. While SARG04 and BB84 have similarities in quantum processing, the SARG04 protocol differs in its classical sifting procedure. A comparison has been drawn in the provided example concerning the Average key efficiency among the BB84, B92, and E91 protocols. It is noted that the B92 protocol demonstrates a lesser number of states compared to BB84 and E91 protocols [18, 19]. This vulnerability emerges as a result of Alice's initiation of the classical segment of the protocol via a process called sifting, where she openly reveals the bases utilized to encode each bit. Because the two bases are not orthogonal, Eve can only distinguish two eigenstates of a known Hermitian operator for each photon she keeps, thereby achieving complete access. While the abstracts analyzed do not explicitly mention the average raw key efficiency specific to the BB84 protocol, the B92 protocol is noted for having an idealized maximum efficiency of 25% over the quantum channel, a characteristic it shares with the E91 protocol. The Quantum key distribution protocols BB84, B92, and SARG04 introduce a technique for creating a secure key using both single photon and two-photon components, a feature lacking in BB84 which relies solely on single photons [17, 20]. SARG04 notably outperforms the BB84 protocol in terms of zero-error resilience against Photon Number Splitting (PNS) attacks. Furthermore, to overcome the difficulties in realising genuine single-photon sources and detectors, a subset of Quantum Key Distribution (QKD) protocols called continuous-variable (CV) QKD was created. Data encoded in the quadratures of the quantized electromagnetic field, including compressed or coherent states, fall under this category. Distributed-phase- reference (DPR) protocols, which emphasise the need of coherence in subsequent pulses for guaranteeing security, were created to address practical demands. Photon arrival timings or the phase difference between successive weak coherent pulses can be used by DPR protocols to encode information. Two such protocols that are successful against PNS assaults are the coherent-one-way (COW) protocol and the differential-phase-shift (DPS) protocol [21, 22]. An alternative practical approach to address issues related to multi-photon sources is the decoy state method, aimed at preventing Eve from adjusting her attack strategies based on Alice's conditions. However, during the post-processing stage, Alice and Bob can assess their parameters based on that specific information. Due to various practical challenges, innovative protocols like measurement-device-independent (MDI) quantum key distribution (QKD) have been created [23]. The MDI-QKD protocol has the capacity to eradicate all detector side-channels from the measurement unit, an aspect that is commonly acknowledged as being among the most susceptible elements of QKD systems. Interestingly, a modification of the MDI-QKD, referred to as the twin-field QKD, was formulated, demonstrating the capability to expand quadratically alongside channel transmittance, marking another noteworthy advancement in the actualization of long-distance quantum communications. This protocol has been extensively explored both theoretically and experimentally, effectively demonstrating its exceptional advantages [24]. The progress in the field of QKD is also evident through the establishment of intercity QKD networks. Within the field of quantum information science, one of the fastest-growing subfields is quantum cryptography. Significant progress has been made in the subject in terms of both theoretical protocols and real-world applications. The present state of quantum cryptography and its recent advancements are examined in this thorough review, which covers a number of important topics including quantum key distribution (QKD), device-independent protocols, satellite-based systems, high-rate protocols, and quantum networks [5].

3 Key Concepts

3.1 No-Cloning Theorem

This fundamental principle of quantum mechanics states that an arbitrary unknown quantum state cannot be duplicated exactly. QKD protocols often exploit this theorem to detect eavesdropping attempts, as any attempt to clone quantum states would introduce errors.

Quantum Superposition

Qubits, the quantum bits used in QKD, can exist in multiple states simultaneously, thanks to superposition. This property is harnessed to encode information in different bases during key exchange.

3.2 Quantum Cost

It is used to evaluate the resource needs of quantum key distribution (QKD) protocols is called quantum cost. It considers a variety of variables, including the quantity and kind of quantum gates used in the protocol's execution. This measure facilitates the comparison of various QKD protocols and provides insight into their resource-usage efficiency, which is essential for the scalability and real-world application of quantum communication systems.

3.3 Raw Key Efficiency

One of the most important metrics for assessing a quantum key distribution (QKD) protocol's efficacy is raw key efficiency. By dividing the total number of sent bits by the proportion of secured shared key bits, it determines how well the protocol creates secure key bits. It is crucial for evaluating the security and practicality of a QKD system. Increased raw key efficiency suggests a more efficient technique for quantum communication-based secure key generation.

3.4 Quantum Entanglement

Entangled particles exhibit correlated properties, regardless of the distance between them. QKD protocols, such as the E91 scheme, use entangled particles to distribute secure keys, relying on the correlation between measurements on entangled qubits.

3.5 Bell States

Bell states are specific entangled quantum states that play a crucial role in QKD, particularly in protocols like E91. These states are maximally entangled and serve as a foundation for establishing secure key pairs.

4 Methodology

4.1 Quantum Key Distribution

Quantum Key Distribution (QKD) is a quantum communication method that utilizes the principles of quantum mechanics to secure the distribution of cryptographic keys. This revolutionary method allows two parties to generate a secret key that can be used for secure communication, with the guarantee that any eavesdropping would be detected.

The primary advantage of QKD over traditional key distribution methods is its ability to detect eavesdroppers. If an eavesdropper tries to measure the quantum states being transmitted, this measurement will disturb the quantum states due to the Heisenberg Uncertainty Principle and can be detected by the communicating parties.

QKD protocols such as BB84, proposed by Bennett and Brassard in 1984, and E91, proposed by Ekert in 1991, are among the most well-known and widely studied. The BB84 protocol uses two sets of orthogonal quantum states, while the E91 protocol uses entangled pairs of quantum states, taking advantage of the correlations predicted by quantum mechanics to create a secret key.

4.2 BB84 Protocol

The BB84 protocol [4] leverages the no-cloning theorem of quantum mechanics to enable secure key exchange. Alice generates a random bit and basis to encode as one of four possible qubit states which Bob measures in a random basis. If the basis matches, the shared bit is secured. The principal concept behind this protocol revolves around the encoding of each individual bit from a specifically chosen random string into the polarization basis of a lone photon, a phenomenon that can be effectively represented using qubits. It is crucial to note that the polarization state of the photon is inherently delicate and cannot be accurately measured without causing the destruction of said photon, thereby indicating the fragile nature of the information it contains. This particular characteristic plays a pivotal role in thwarting potential eavesdropping attempts, as the inability to access the information acts as a deterrent for adversaries seeking unauthorized access. The rationale behind the security of the BB84 protocol lies in the intricate balance of probabilities, which essentially hinders the extraction of information by unauthorized entities.

To gain a comprehensive understanding of the BB84 protocol, it is imperative to delve into specific key points that shed light on its operational dynamics. One recent study showcased the successful implementation of the BB84 protocol on the IBM QX platform utilizing three distinct bases, marking a significant milestone in quantum communication research. The first step in the protocol involves Alice meticulously selecting a random assortment of 4n data bits and amalgamating them into a coherent bit-string. Subsequently, she proceeds to encode her data bits using four distinct states, namely—0i,—1i,—+ i, and—i, before transmitting them to Bob for further processing. Upon receiving the qubits, Bob acknowledges their reception and proceeds to randomly measure them in either the X or Z basis. Once both parties disclose the basis used for measurement, a predetermined number of 2n bits are earmarked for the key generation process.

As an additional security measure, n bits are designated for monitoring any potential eavesdropping activities, serving as a crucial checkpoint to detect any external interference. It is important to highlight that Bob is also aware of these designated n bits, thereby ensuring a collaborative approach towards maintaining the integrity of the communication channel. In the event of discrepancies during the comparison process, where the selected bits exceed the permissible threshold, the protocol is promptly terminated to prevent any compromise of the shared key. Nonetheless, the remaining n bits play a pivotal role in creating a super-set for m bits, which are subsequently utilized in the information reconciliation and privacy amplification phases, ultimately fortifying the security of the key exchange mechanism.

The circuit (Fig. 1) uses a single qubit with Hadamard gates (H) for superposition and basis changes. Randomization is via H gates and measurements. Conditional gates based on the random bases apply appropriate H operations in the sender and receiver to recreate BB84. The cost analysis counts the gate primitives based on the metric in [1] (Fig. 2).

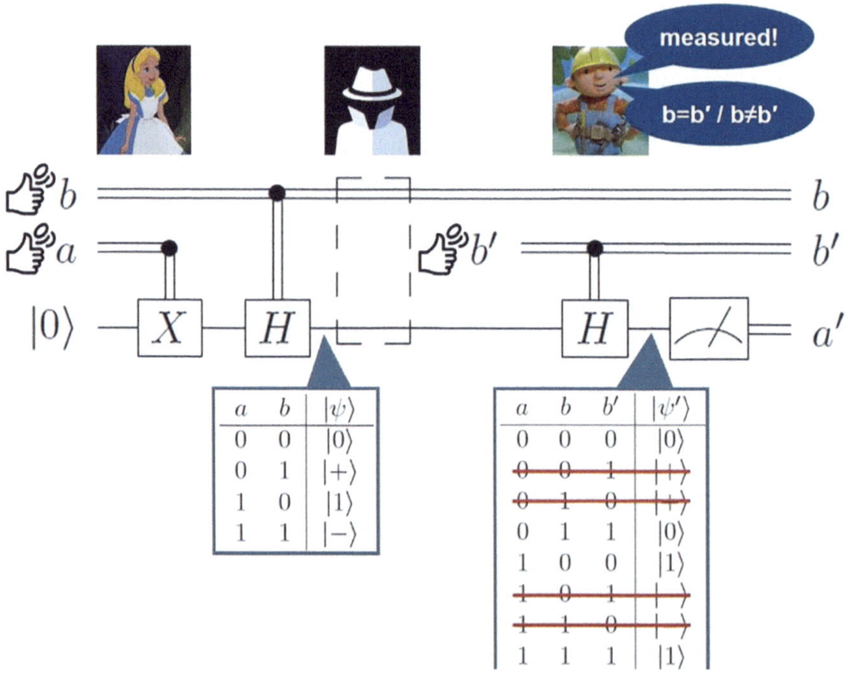

Fig. 1 Figure 1: BB84

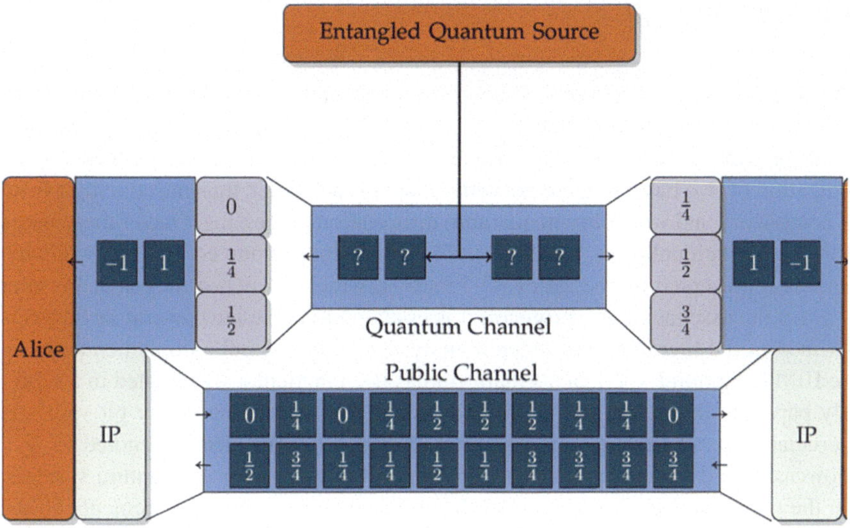

Fig. 2 Figure 2: E91

4.3 E91 Protocol

The E91 scheme [5] exploits quantum entanglement and Bell states for cryptography. Alice and Bob each have one qubit from an entangled pair. Measuring one qubit collapses the superposition of the pair, securing the correlated key when bases match.

The implementation generates a Bell state using Hadamard and CNOT gates. Basis choice bits control receiver side H gates. Quantum cost is assessed as total gates, excluding classical randomization (Fig. 3).

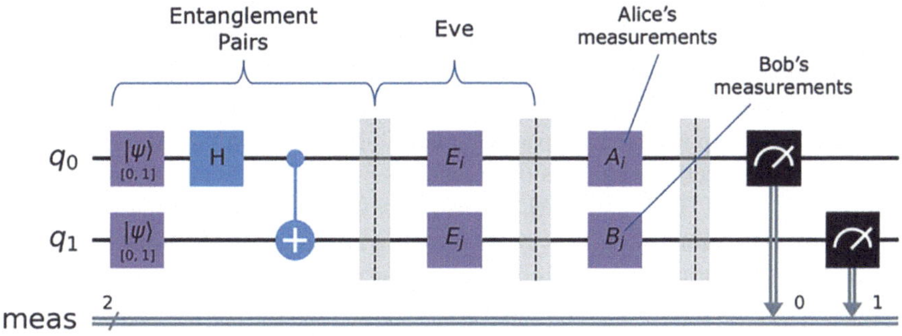

Fig. 3 E91 Qiskit quantum circuit

4.4 B92 Protocol

The B92 protocol [5] operates on non-orthogonal states, with Alice sending one of two and Bob distinguishing them in one of two bases. Only a subset of Bob's measurements provide a secure key. The primary concept behind this protocol involves the utilization of two bases that are not orthogonal to each other, thus introducing a level of complexity and versatility in quantum information processing. One of these bases is the computational basis, a fundamental element in quantum computing, while the other is represented by the states—+ i and—i, which are non-orthogonal. In order to grasp the essence of this protocol, it is essential to delve into several key aspects outlined in the literature. An indepth analysis of the protocol's implementation on the IBM Quantum Experience's platform has been meticulously detailed in a scholarly paper. In the execution of this protocol, Alice's encoding of her bit values is a crucial step, where 0 is represented as 0i and 1 is equivalently denoted as + i, showcasing a distinct approach to information representation in quantum systems. On the recipient end, Bob is granted the authority to perform measurements on the qubits in a randomized manner, utilizing either the computational basis with states 0i and 1i, or the non-orthogonal basis with states + i and i, thereby showcasing the flexibility inherent in quantum protocols. Bob's subsequent action involves the construction of a cryptographic key solely from those bit positions where the qubits are in the states 1i or i, underlining the significance of specific quantum states in key generation processes. Furthermore, to establish secure communication channels, Bob employs a public medium to disclose the precise positions of the qubits that he has retained, a strategic move aimed at facilitating the key generation process through mutual coordination and information sharing. This illustration represents a test arrangement for free-space quantum communication spanning a distance of around 50 m. It comprises setups belonging to Alice and Bob, interconnected via telescopes through a free-space channel. Alice's configuration involves a Sagnac based entangled photon source (PPKTP crystal) that produces pairs of entangled photons. Diverse optical elements such as SPAD detectors, polarizing beam splitters (PBS), quarter-wave plates (QWP), half-wave plates (HWP), and a collimator for the preparation and manipulation of photon states. The entangled photon pairs are divided, with one photon (V2, H2) directed through a dual-PBS setup and a HWP@45° towards Bob's setup via the free-space channel. Bob's setup comprises:—A refractor telescope that collects the photons from Alice.—Similar optical elements like collimators, PBS, QWP, HWP, and SPAD detectors for the measurement and examination of the received photon states. This experimental arrangement is formulated to showcase and explore various quantum communication protocols, including quantum key distribution (QKD), utilizing entangled photon pairs transmitted through free space across a moderate distance of approximately 50 m. The circuit (Fig. 4) sets the sender state using the input and an H gate for superposition. The receiver attempts identification via an H gate and measures, with the result indicating if the bit is secured. Only quantum operations are included for cost calculations (Fig. 5).

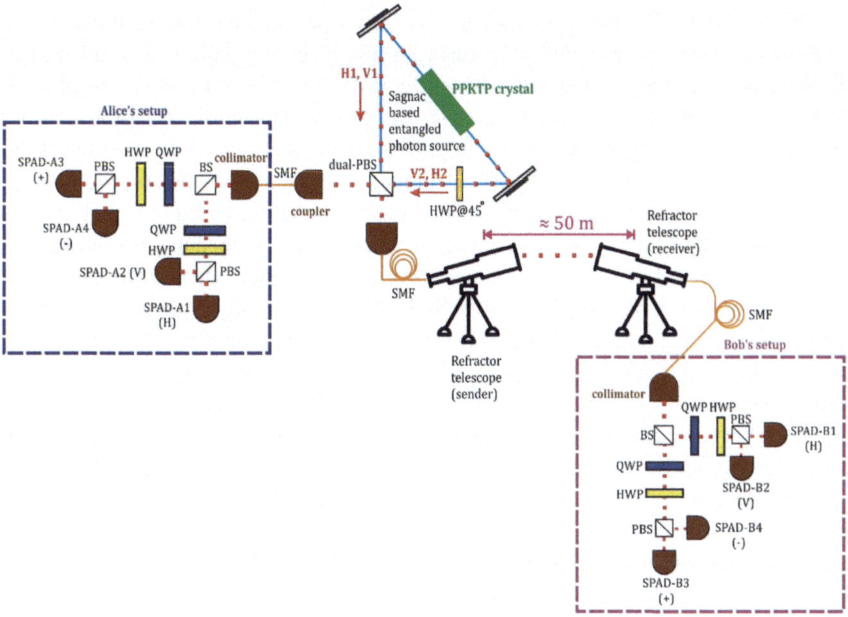

Fig. 4 This illustration represents a test arrangement for free-space quantum communication spanning a distance of around 50 m using BBM92 protocol

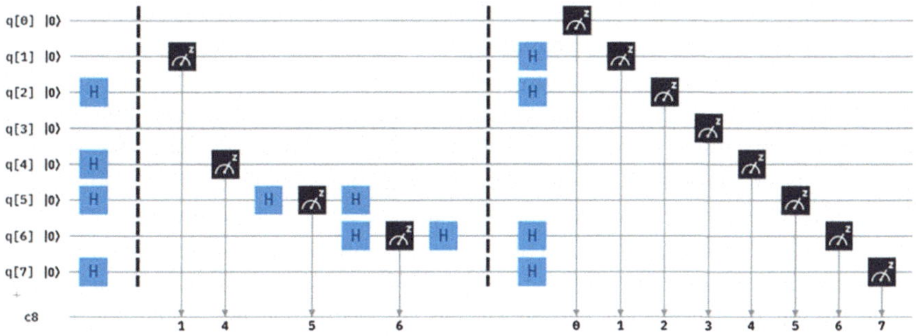

Fig. 5 BB92 Qiskit quantum circuit

4.5 *SAG04 Protocol*

The SARG04 protocol was developed by researchers employing the four states of BB84 with alternative information encoding to provide a more robust protocol for attenuated laser pulses instead of single photon sources. In 2004, Scarani et al. proposed the SARG04 protocol. The SARG04 protocol was developed with the primary objective of enhancing the robustness of the BB84 protocol, a cornerstone in quantum key distribution schemes. While both protocols utilize the same four states as in BB84, SARG04 incorporates a distinct information encoding mechanism. In the BB84 protocol, a laser pulse featuring lower-level attenuation is employed to encode a bit in a photon. However, should a pulse contain more than one photon, a potential vulnerability arises as an eavesdropper could execute a photon number splitting attack by transmitting a single photon to the receiver after splitting off the extra photon. This manipulation poses a threat to the security of the communication channel. Introduced by Scarani et al. in 2004 in the esteemed journal Physical Review Letters, SARG04 was proposed as a more efficient defense mechanism against PNS attacks, a prevalent form of quantum eavesdropping. Notably, when scrutinized from the perspective of quantum processing, SARG04 is essentially equivalent to BB84, underscoring its foundational principles in quantum cryptography. The operational framework of the SARG04 protocol is delineated as follows: Initially, Alice, the sender, endeavors to establish communication with Bob, the recipient. She initiates the process by generating two strings of bits denoted as x and y, each comprising n bits. Subsequently, a quantum state is created by encoding the two-bit string in a manner where the value of yi (the ith bit in the y string) dictates the basis in which xi is encoded. Specifically, if yi is set to 1, xi is encoded in the Hadamard basis, otherwise, xi is encoded in the computational basis. Upon encoding, Alice dispatches the quantum state through a quantum channel, with Bob serving as the intended receiver. Notably, an eavesdropper, Eve, could potentially gain access to the transmitted state. Nevertheless, the bit string y remains solely known to Alice, rendering both Bob and Eve incapable of discerning the qubit's state. Subsequently, Bob generates a random bit-string denoted as y0, also consisting of n bits. He proceeds to decode and measure the qubits, whereby y0i determines the basis for measuring xi. For each qubit, Alice selects one computational basis state and one Hadamard basis state. Bob's task entails distinguishing between the two potential states to extract the secret bits. By scrutinizing the measurement outcomes, Bob can ascertain whether they align with either of the possible states. In cases of consistency, the precise encoding basis remains ambiguous. Conversely, inconsistencies in the measurements empower Bob to deduce both the exact state and the secret bit. Following this deduction, Alice divulges a small fraction of the deduced bits to Bob publicly and evaluates their agreement against a predefined criterion. If the comparison proves successful, the communication proceeds; otherwise, the channel is discarded, and the process recommences.

In a practical implementation of the protocol, the operational dynamics are notably simplified. The experimental setup typically involves multiple iterations of the circuit (8192 runs), yielding a histogram of outcomes. Consequently, the intricate process of determining the exact state is obviated, as the state can be inferred from the probability distribution of—0i and—1i in the generated histogram. Noteworthy distinctions from the conventional BB84 protocol come to the fore, particularly in Alice's strategy. Unlike in BB84, where Alice discloses the chosen basis for each qubit, in SARG04, she refrains from announcing the basis of the bits. This strategic shift holds a distinct advantage in thwarting potential eavesdropping attempts, as Eve would be compelled to retain multiple copies of the qubit to ascertain the state accurately.

Our experimental validation of the scenario detailed in this section was conducted with the aid of the IBM QX quantum computing platform.

4.6 Advanced Encryption Standard (AES)

The Advanced Encryption Standard (AES) is a widely adopted encryption standard developed to replace the Data Encryption Standard (DES) [25]. It consists of three block ciphers: AES-128, AES-192, and AES-256, with key sizes of 128, 192, and 256 bits respectively, and has become a global standard for encryption, including for top-secret information. AES is utilized in various industry standards and commercial systems, such as IPsec, TLS, IEEE 802.11i, SSH, Skype, and more, highlighting its broad applicability and security features. The selection of AES followed a rigorous evaluation process, with Rijndael being chosen as the most suitable algorithm after competing with other candidates like MARS, RC6, Serpent, and Twofish. Its success stems from its efficiency, security, and ease of implementation, making it a cornerstone in modern encryption technologies [26] [(Fig. 6).

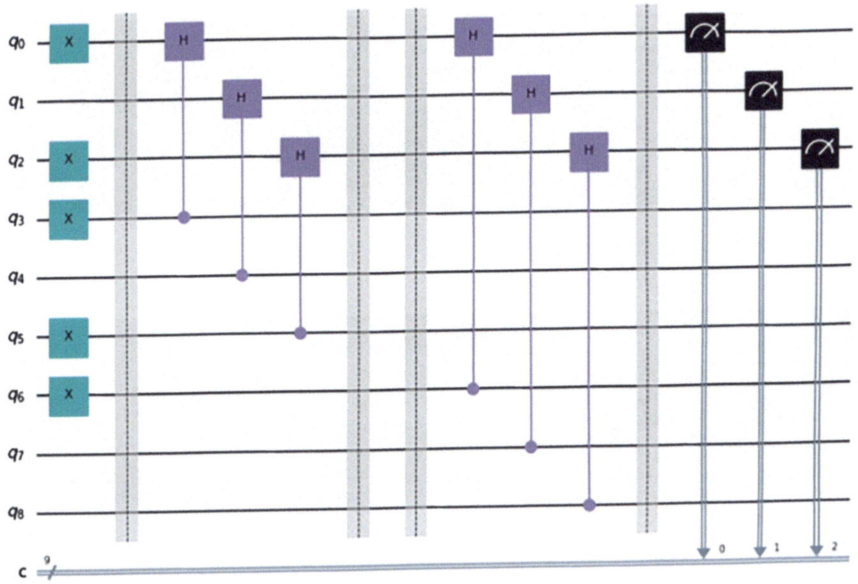

Fig. 6 SAG04 quantum circuit

4.7 Simulation Setup

All protocols employed Qiskit Aer [2] on an Intel i7 system. The simulations involved statistical key efficiency analysis, executing circuits for each protocol in 10,000 independent runs, with each run consisting of 2048 shots. The choice of 2048 shots ensured a comprehensive statistical sampling for robust analysis. Superposition states were transmitted without introducing noise or decoherence, maintaining a controlled environment for the comparison. The absence of noise and decoherence allowed for an isolated assessment of the inherent efficiency of each protocol. Post-processing involved determining the fraction of secured shared key bits over the total transmitted, yielding raw key efficiency metrics. The raw key efficiency metrics served as the basis for the quantitative comparison, offering insights into the performance of each protocol under the specified simulation conditions. The simulation design aimed at capturing both the theoretical strengths of the protocols and their practical efficiency in generating secure key bits. The resulting quantitative data allows for an informed evaluation and comparison of the BB84, E91, SAG04 and B92 protocols in the context of key distribution. We have successfully integrated all mentioned Qkd protocols with Advanced Encryption Standard (AES) (Figs. 7 and 8).

```
Enter in the 16 character passphrase to encrypt your text file
Inside your plaintext message is:
name,pokedex id,height,weight,type,secundary type,hp,attack,defense,sp atk,sp def,speed
bulbasaur,1,7,69,grass,poison,45,49,49,65,65,45
ivysaur,2,10,130,grass,poison,60,62,63,80,80,60
venusaur,3,20,1000,grass,poison,80,82,83,100,100,80
charmander,4,6,85,fire,None,39,52,43,60,50,65

The output hex value for the entire message is:
7f5271fcac809af71850b556b819231b6cef35e6ce3cc9abac4b8cd05c5c380509752945dfca9aa53d9281925b1ff5e26beb356e3703c8eb3
```

Fig. 7 Input data along with Ciphertext

```
Enter in the 16 character passphrase to decrypt your text file ciphertext.csv
Inside your ciphertext message is:
7f5271fcac809af71850b556b819231b6cef35e6ce3cc9abac4b8cd05c5c380509752945dfca9aa53d9281925b1ff5e26beb356e3703c8eb3

The decrypted message for the entire ciphertext is:
name,pokedex id,height,weight,type,secundary type,hp,attack,defense,sp atk,sp def,speed
bulbasaur,1,7,69,grass,poison,45,49,49,65,65,45
ivysaur,2,10,130,grass,poison,60,62,63,80,80,60
venusaur,3,20,1000,grass,poison,80,82,83,100,100,80
charmander,4,6,85,fire,None,39,52,43,60,50,65
```

Fig. 8 Output data after conversion from Ciphertext

5 Observations

The comparison between the BB84 and E91 quantum key distribution (QKD) protocols reveals a nuanced trade-off between simplicity and potential advantages in the presence of eavesdropping and for long-distance quantum communication. BB84, with its straightforward single quantum state preparation and measurement, offers ease of implementation and typically achieves higher efficiency in terms of key generation rate. This simplicity is advantageous in practical applications and may make BB84 more accessible for various quantum communication scenarios. On the other hand, E91, despite its more intricate setup, holds the promise of higher final key rates in the presence of potential eavesdropping activities. Its compatibility with quantum repeaters also positions E91 as a strong candidate for long-distance quantum communication, where signal loss traditionally poses a significant challenge.

However, it's crucial to note that realizing these potential benefits requires further research and successful reproduction of room-temperature superconductors like LK-99. Exploring the compatibility and integration of such materials with existing QKD setups would be a critical step towards unlocking the full advantages of room-temperature superconductors in quantum communication protocols. The SARG04 procedure's initial phase is the same as BB84's. Instead of disclosing their bases at the beginning of the second phase, in the SARG04 protocol, Client A and Client B identify which of their bases match. Client A reveals two non-orthogonal states, one of Which she used to encode her bit. By choosing the correct basis, Client B can accurately measure the state. Making an incorrect choice will result in the inability to measure any of Client A's states or determine the bit's location. In the absence of errors, the length of the key remaining after the sifting phase is a quarter of the original key. The level of security provided by the SARG04 protocol is similar to the security provided by the BB84 protocol in ideal single-photon setups (Figs. 9, 10 and 11).

The illustration provided offers a comparison of the Average key efficiency for BB84, B92, and E91 quantum key distribution protocols. It is observed that the B92 protocol presents a reduced number of states compared to BB84 and E91 protocols. The average raw key efficiency related to the BB84 protocol is not mentioned in the abstracts being reviewed. Conversely, the abstracts do note that the B92 protocol is characterized by an optimal maximum efficiency of 25 percent over the quantum channel, a trait it shares with the E91 protocol. The abstracts delve into an examination of the Quantum key distribution protocols BB84, B92, and E91. The focus is particularly on comparing the efficacy of BB84 and B92 protocols in free space quantum optical communication links between a ground station and a satellite orbiting in low earth orbit. Notably, the BB84 protocol has shown superior performance in enabling secure distribution of communication keys across long distances. The security foundations for both BB84 and B92 protocols have been established theoretically, although the ability to detect eavesdroppers depends on the amount of qubits being

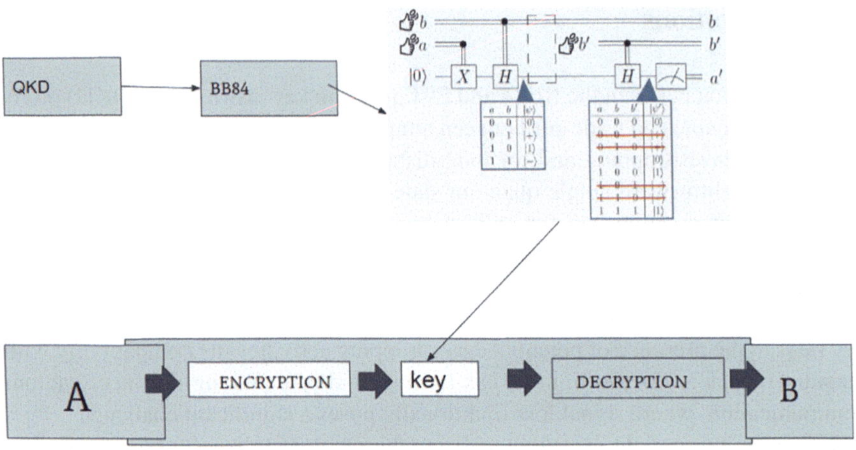

Fig. 9 Figure 4: architecture

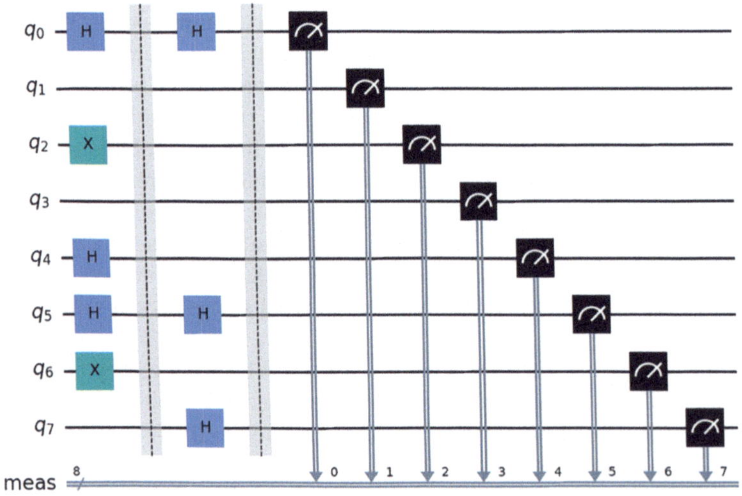

Fig. 10 BB84 quantum circuit

exchanged. Additionally, the abstracts emphasize the existence of commercial products utilizing quantum key distribution protocols, highlighting the practical applicability of these protocols in various fields. SARG04 provides the highest efficiency using 8 qubits because it is a robust protocol for weak coherent pulse implementation of Quantum Cryptography (QC) [1]. It is a variant of the BB84 protocol, which has been widely studied and implemented since its proposal in 1984 [2]. SARG04 tackles the practical challenges in implementing BB84 and presents enhancements

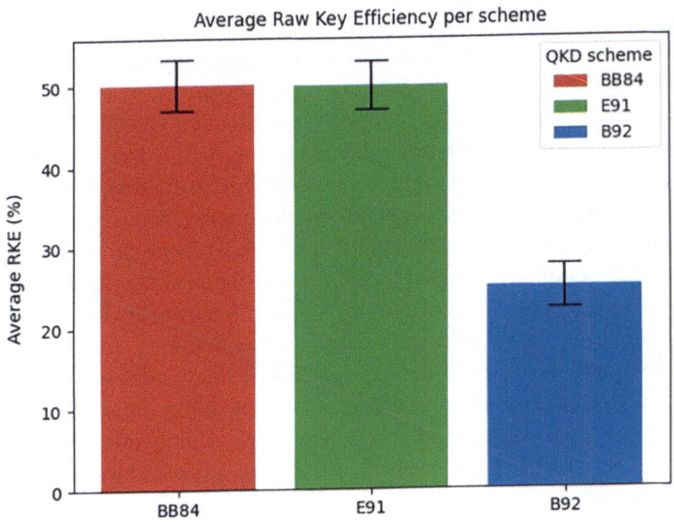

Fig. 11 Average raw case efficiency

in protocol efficiency, Quantum Bit Error Rate (QBER), and resistance to eavesdropping [3]. This protocol is tailored for integrating Quantum Cryptography (QC) into current wireless systems [4]. Through leveraging the advantages of SARG04, particularly its resilience and enhanced quantum communication range, significant efficiency is attained with a minimal requirement of 8 qubits. The QBER of the BB84 and BBM91 protocols plays a crucial role in assessing the security of QKD systems. In the BB84 protocol, the QBER is influenced by the strategies of the eavesdropper as well as the imperfections of the physical devices and the channel through which the quantum states propagate. However, in a polarization-encoded BB84 transmitter sourced by a SiGe light source, it has been shown that the QKD scheme can operate well below the QBER threshold at which secret keys can be established. This demonstrates the potential for a reliable and secure communication system using polarization encoding in QKD. BB84 achieve higher secret key rate and greater secure distance than SARG04.The execution of the BB84 protocol on the IBM Quantum Experience platform utilizing 8-qubits was conducted with meticulous attention to detail, employing a variety of three bases for quantum operations. Subsequently, a comprehensive analysis was performed to juxtapose the anticipated outcomes with the actual experimental results obtained from the implementation. The data sets were scrutinized meticulously, leading to the determination that the degree of overlap between the expected and acquired values fell within an acceptable range deemed suitable for further analysis and interpretation.

Following the successful completion of the BB84 protocol, the exploration and application of the B92 protocol were undertaken on the same quantum computing platform. This involved the intricate construction of non-orthogonal bases using a series of gate operations, enabling Alice and Bob to conduct measurements in strict adherence to the principles of the B92 protocol. The experimental outcomes were meticulously presented and subjected to in-depth scrutiny and analysis to extract meaningful insights.

Moreover, the implementation of the SAG04 protocol involved formulating bases through the utilization of quantum gates according to the predefined theoretical framework of this protocol. The average values of the expressions were meticulously measured and evaluated to draw insightful conclusions regarding the efficacy of these measurements in contributing to the establishment of a raw cryptographic key. These implementations served as a crucial tool in visualizing abstract theoretical concepts through the practical application of gate operations, thereby providing a deeper understanding of the foundational principles of quantum computing. Beyond the realm of QKD, the implications of room-temperature superconductors are expansive. The anticipated advancements in space travel, communication, navigation, propulsion systems, and energy storage could propel a new era of innovation and exploration. This marks a significant step forward in leveraging advanced technologies for the benefit of humanity.

However, it is imperative to emphasize the preliminary nature of these findings and projections. While grounded in plausible assumptions aligning with quantum mechanics and superconductivity principles, the actual feasibility and benefits of room-temperature superconductors in these applications necessitate rigorous testing and validation. This study serves as a foundational exploration, urging further research and empirical validation in the captivating frontier of room-temperature superconductors, exemplified by LK-99, and their transformative potential in quantum computing, communication, and broader technological landscapes (Figs. 12, 13, 14, 15, 16, 17, 18 and 19; Table 1).

6 Conclusion

This paper provides a thorough examination of various Quantum Key Distribution (QKD) protocols, specifically BB84, B92, E91, and SARG04, emphasizing their respective advantages and drawbacks in ensuring secure communication. The study highlights that BB84 is notably effective in secure key distribution over long distances due to its simplicity and efficiency in key generation. On the other hand, E91, despite its complexity, shows promise in higher final key rates and compatibility with quantum repeaters, making it suitable for long-distance communication where signal loss is a challenge. The paper also details the experimental implementation of these protocols on the IBM Quantum Experience platform, with a focus on practical application and theoretical validation. Moreover, the presence of commercial products utilizing QKD protocols is acknowledged, indicating the practical feasibility and growing recognition of these technologies in real world applications. The study concludes that while QKD protocols offer substantial cybersecurity advantages, their full potential will be realized through continued research and the successful incorporation of advanced materials and technologies. In summary, the exploration and

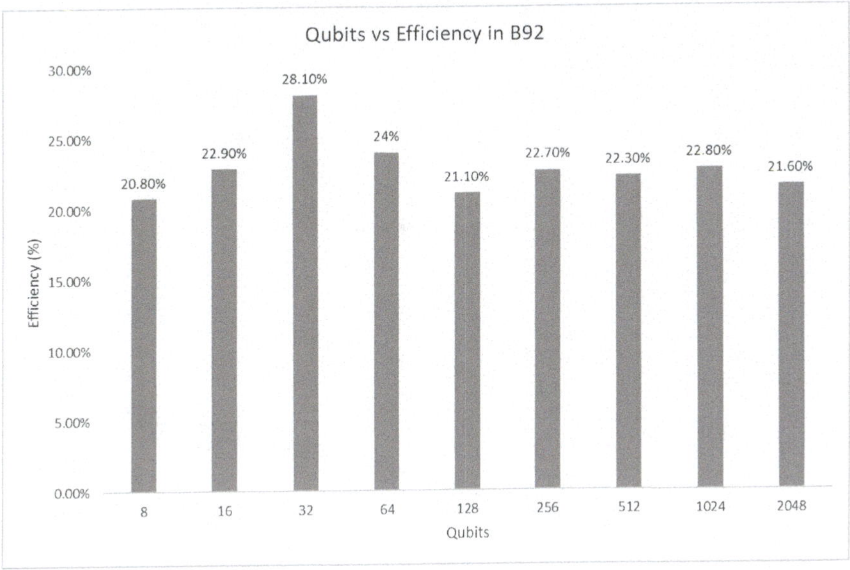

Fig. 12 Efficiency versus qubits of B92

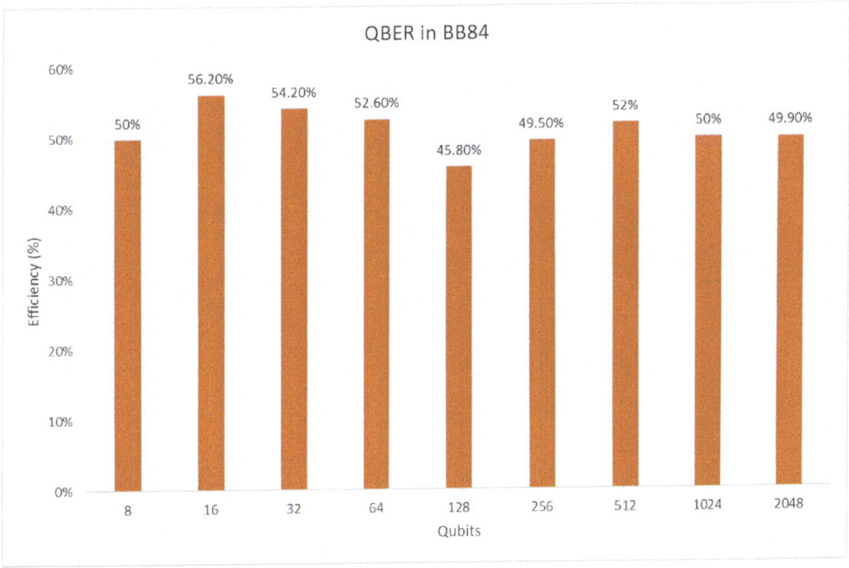

Fig. 13 Efficiency versus qubits of BB84

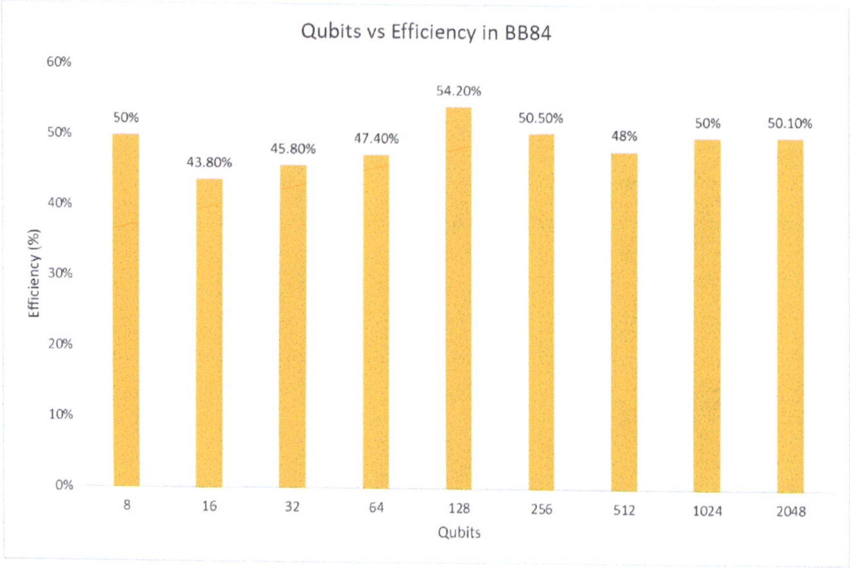

Fig. 14 Efficiency versus qubits of E91

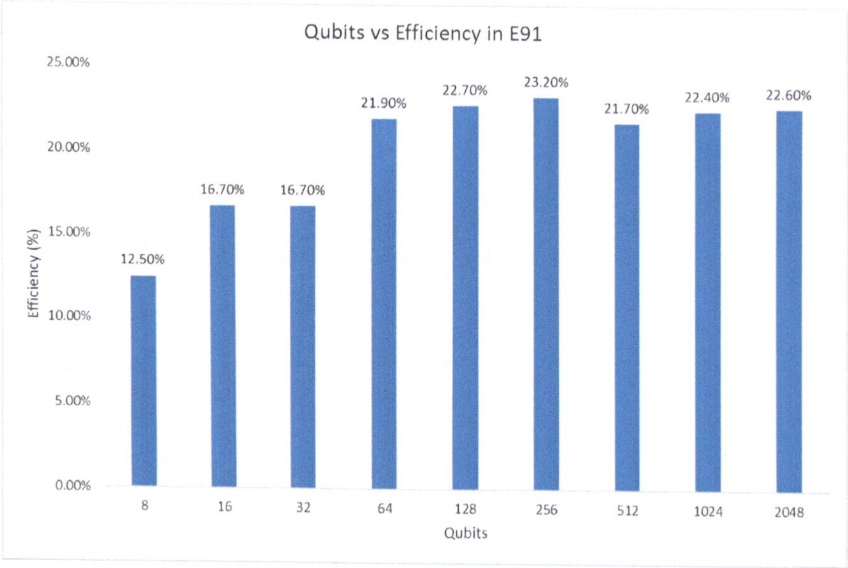

Fig. 15 Efficiency versus qubits of SAG04

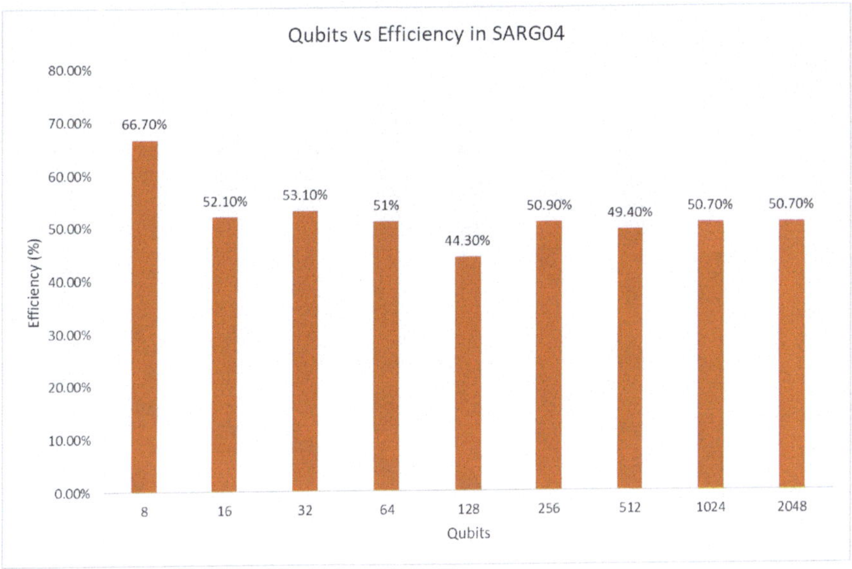

Fig. 16 QBER versus qubit of BB84

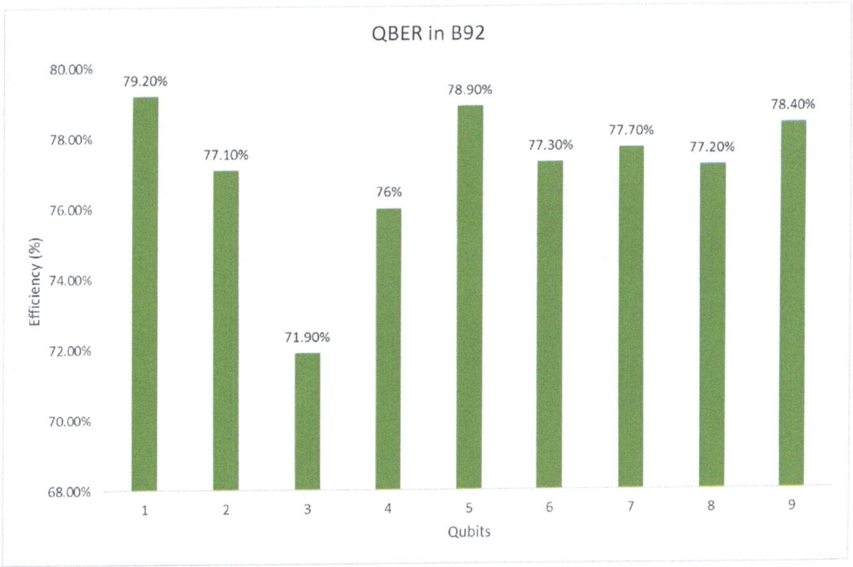

Fig. 17 QBER versus qubit of BBM92

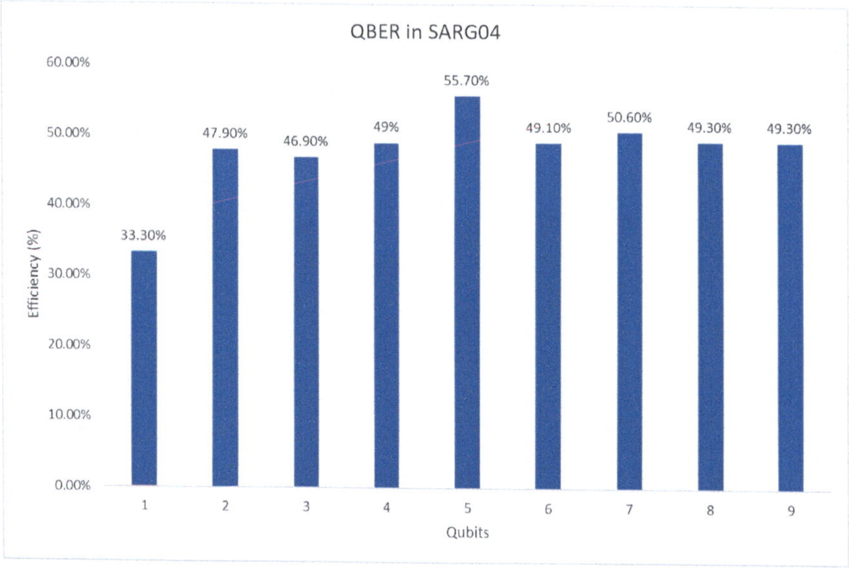

Fig. 18 QBER versus qubit of SAG04

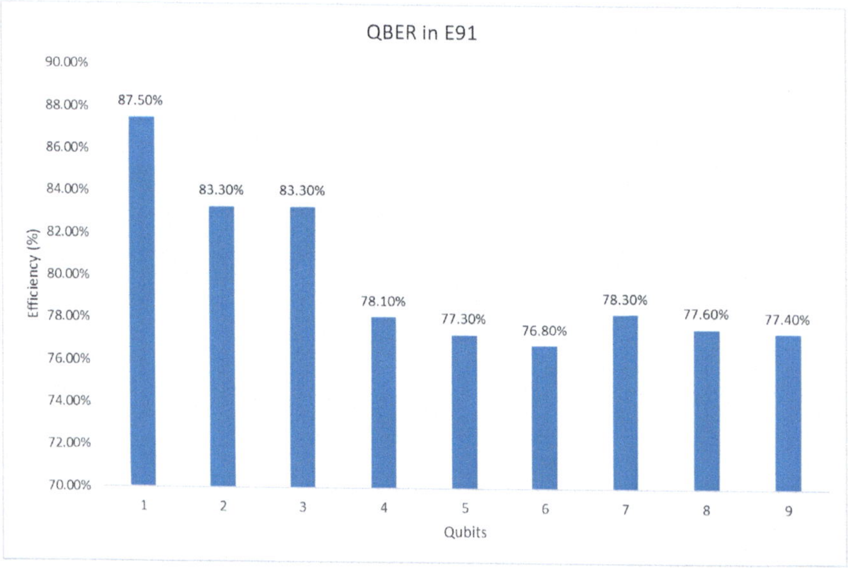

Fig. 19 QBER versus qubit of E91

Table 1 Comparison of QKD protocols

Protocol	Security	Efficiency	Implementation
BB84	High	Moderate	Widely implemented
B92	Moderate	High	Requires entanglement
E91	High	Low	Experimental
SARG04	High	Moderate	Asymmetric reconciliation

comparison of these QKD protocols, alongside the potential improvements from emerging technologies, underscore a promising future for quantum communication and its applications in enhancing global information security.

References

1. S. Lee, J.-H. Kim, Y.-W. Kwon. *The First Room-Temperature Ambient-Pressure Superconductor* (2023). arXiv preprint arXiv:2307.12008
2. J. Watrous, *Theory of Quantum Information*. Lecture Notes from fall 2008, Institute for Quantum Computing, University of Waterloo, Canada (2008)
3. Osada, R. Yamazaki, A. Noguchi, in *Quantum Technologies* ed. by T. Tanaka, Y. Yamamoto. Advances in Quantum Systems. Lecture Notes in Physics, vol. 1004 (Springer, 2022) pp. 235–243
4. C.H. Bennett, G. Brassard, Quantum cryptography: public key distribution and coin tossing, in *Proceedings of IEEE International Conference on Computers, Systems and Signal Processing* (1984), pp. 175–179.
5. A.K. Ekert, Quantum cryptography based on Bell's theorem. Phys. Rev. Lett. **67**(6), 661–663 (1991)
6. N. Gisin, G. Ribordy, W. Tittel, H. Zbinden, Quantum cryptography. Rev. Mod. Phys. **74**, 145 (2002)
7. S. Pirandola et al., Advances in quantum cryptography. Adv. Opt. Photonics **12**(4), 1012–1236 (2020)
8. K. Garapo, M. Mafu, F. Petruccione, Intercept-resend attack on six-state quantum key distribution over collective-rotation noise channels. Chin. Phys. B **25**(7), 070303 (2016)
9. W.K. Wootters, W.H. Zurek, A single quantum cannot be cloned. Nature **299**(5886), 802–803 (1982)
10. N. Gisin et al., Quantum cryptography. Rev. Mod. Phys. **74**(1), 145–195 (2002)
11. F. Xu et al., Secure quantum key distribution with realistic devices. Rev. Mod. Phys. **92**(2), 025002 (2020)
12. M. Mafu, M. Senekane, Security of quantum key distribution protocols, in *Advanced Technologies of Quantum Key Distribution* (IntechOpen, 2018)
13. C. Portmann, R. Renner, Security in quantum cryptography. Rev. Mod. Phys. **94**(2), 025008 (2022)
14. R. Liu et al., Towards the industrialisation of quantum key distribution in communication networks: a short survey. IET Quant. Commun. **3**(3), 151–163 (2022)
15. X.-B. Wang, Z.-W. Yu, X.-L. Hu, Twin-field quantum key distribution with large misalignment error. Phys. Rev. A **98**(6), 062323 (2018)
16. I.V. Puthoor, J. Dias, O. Dahlsten et al., Measurement-device-independent key distribution with three-party communications. Quant. Sci. Technol. **5**(2), 025001 (2020)
17. E.O. Kiktenko, A.S. Trushechkin, Y.V. Kurochkin, A.K. Fedorov, Post-processing procedure for quantum key distribution systems with imperfect quantum state discrimination. Phys. Rev. A **94**(6), 062305 (2016)

18. F. Xu, X. Ma, Q. Zhang, H.-K. Lo, J.-W. Pan, Secure quantum key distribution with realistic devices. Rev. Mod. Phys. **92**(2), 025002 (2020)
19. M. Lucamarini, Z.L. Yuan, J.F. Dynes, A.J. Shields, Overcoming the rate–distance limit of quantum key distribution without quantum repeaters. Nature **557**(7705), 400–403 (2018)
20. X.-S. Mao et al., Field test of twin-field quantum key distribution through sending-or-not-sending protocol. npj Quant. Inf. **5**, 80 (2019)
21. G.-J. Fan-Yuan et al., Optical-fiber-based quantum key distribution technology. Nat. Photon. **15**, 270–275 (2021)
22. Y. Pan et al., Towards practical continuous-variable quantum key distribution in real-world metropolitan optical fiber network. Nat. Commun. **13**, 1892 (2022)
23. Z.-Q. Yin, Y. Fu, Q. Liao, W. Chen, "Measurement-device- independent quantum key distribution with high key rate. Physical Re- view A **101**(6), 062327 (2020)
24. Q. Liao, Z.-Q. Yin, X. Wang, High-efficiency error correction for quantum key distribution. Opt. Express **29**(16), 25758–25767 (2021)
25. D. Selent, Advanced Encryption Standard. null (2010)
26. P. Frederic, M. Agnes, F. Vandome, J. McBrewster, Advanced Encryption Standard. null (2009)

Smart Factory Architecture, Infrastructure and Design of Manufacturing Systems Using Robotics

Kusum Yadav

Abstract Manufacturing has been transformed by technological advancements such as Cyber-Physical Systems (CPS), the Internet of Things (IoT), Big Data, and Cloud Computing, resulting in Industry 4.0 and smart factories. An examination of the architecture and infrastructure of smart factories is presented in this chapter, emphasizing their integration with these advanced technologies. Industry 4.0's smart factories increase productivity, flexibility, and quality through real-time data and autonomous systems. With short product life cycles, volatile demand, and customization requirements, modern manufacturing faces critical challenges. In addition to discussing the challenges of implementing smart factory systems, the chapter also discusses their modularity, adaptability, and real-time capabilities, along with their technological and economic challenges. In this chapter, enabling technologies and their benefits are discussed to provide a comprehensive overview of the transition to smart manufacturing.

Keywords Internet of Things · Smart factory · Industry 4.0 · Intelligent system · Manufacturing system

1 Introduction

In today's rapidly changing technology environment, manufacturers face a number of challenges, such as short product life cycles and volatile demand [1]. To compete in future markets, a company must be able to manufacture small batches or even a single item efficiently and cost-effectively. Furthermore, they ought to be scalable, functional, and connected to customers and suppliers [2]. The complexity and difficulty of monitoring and controlling these systems will also increase in response to such

K. Yadav (✉)
College of Computer Science and Engineering, University of Hail, Hail, Kingdom of Saudi Arabia
e-mail: y.kusum@uoh.edu.sa

R. Sharma and G. Jeon (eds.), *Data Analytics for Smart Robotics and Its Applications*, Intelligent Systems Reference Library 272,
https://doi.org/10.1007/978-3-031-87697-4_16

challenges. There will be a decline in the competitiveness of traditional manufacturing methods in the future (flexible manufacturing systems, dedicated production lines, etc.). CPS, IoT, and IoS, as well as Big Data, Cloud Computing, Semantic Web, and virtualization, are examples of recent technological advancements [3, 4]. The concept of Industry 4.0 (I4.0) has become widely recognized as a new industrialization model that leverages these new technologies to address the issues mentioned above. Smart factories play an important role in I4.0 [5]. Smart factories integrate these technologies to increase transparency, efficiency, and control over manufacturing processes. By using both physical and virtual information, a smart factory system makes it easier for people and machines to perform tasks. Several components in the system can negotiate with each other and with factory components to request or offer functions [6]. Figure 1 summarizes its layers.

Achieving high-quality, efficient, robust, and flexible products in the industry will require an increase in flexibility and automation [7]. As the market moves towards contemporary, customer-specific individual production, development and production cycles are being shortened, allowing different variations of products to be made quickly [8]. It is, however, time-consuming and expensive to adapt and convert future factories to the order situation with this flexible approach. It is essential to design parts of the plant in a modular and flexible manner in order to make such a complex project manageable. Innovative self-monitoring techniques will increase the efficiency of

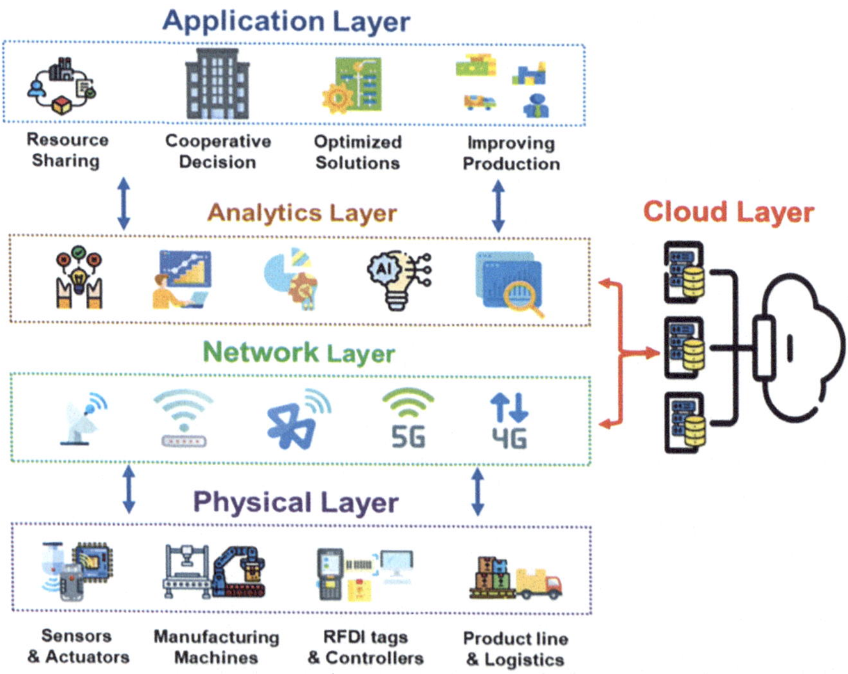

Fig. 1 An overview of a smart factory's architecture

manufacturing. Modern engineering is shifting away from mechanical processes and towards software-controlled processes that can be further optimized [9]. The Internet of Things (IoT) is a hybrid of the real world and the virtual world. CPS bridges the gap between mechanical, electrical, and computer sciences, thus contributing to this trend. According to this trend, Germany stands on the threshold of the fourth industrial revolution (Industry 4.0). Products manufactured in the future will have specific requirements. Increasing the efficiency of industrial production processes is expected to become an important competitive advantage due to new requirements for highly flexible production systems. In this approach, CPS is integrated with a real-world production environment, providing a solid foundation for Industry 4.0. The digital transformation and networking of production and products are at the core of Industry 4.0. Data is exchanged between machinery, equipment, and sensors as Industry 4.0 advances. A combination of the physical and virtual worlds is achieved in this way.

2 Related Work

Smart factories are at the centre of Industry 4.0. Using the Smart Factory, machines and systems can communicate intelligently with each other, allowing them to coordinate work steps automatically and with one another. This can only be achieved by using basic technologies. What is required technologically is outlined in this section [10]. A Smart Factory cannot be described without first describing the Cyber-Physical Systems (CPS) that underlie it [11]. A real system is also called a "physical system". In this sense, CPSs are autonomous so that they can be configured and extended independently [12].

Moreover, the CPS is able to connect to the Internet, as well as to other open and global information networks. Systems can be connected arbitrarily, changed their connections, terminated, and rebuilt while running. The CPS allows users to access data, information, and services anywhere [13–15]. An embedded system can be embedded in a product, a device, a building, a production facility, or a logistics component.

Manufacturing was implementing specific digital technologies at the beginning of the fourth industrial revolution. As a result of technology, a vast amount of information can be collected, processed, analyzed, and stored instantly. Information is collected, processed, analyzed, and stored in real time using technology [16]. Systems for Industry 4.0 are characterized by automated processes and real-time data exchange, allowing products to control systems in real time [17]. This capability disrupts the traditional method of controlling product flow by manufacturing systems [18]. Knowledge can be derived, analyzed, and stored from information retrieved in the field by using Industry 4.0 systemic skills. In smart factories, production, warehousing, and distribution facilities can be autonomously managed, thereby reducing human error and improving efficiency. When customer demands are uncertain or rapidly evolving, skilled, smart factories can adapt to these changes in a timely manner. As part of Industry 4.0, a variety of disruptive technologies and concepts are being utilized to bring intelligence into manufacturing processes. The transition to Industry 4.0 requires manufacturers to understand how to use technology to connect processes [19]. Machine learning, artificial intelligence, and big data analytics are key technologies for the internet of things [16, 20].

2.1 Smart Manufacturing: An Overview

Data collection can only be understood by understanding the field in which it takes place. Digitalization is primarily responsible for driving the fourth industrial revolution described in current literature. A manufacturing environment where equipment and devices are interconnected and internet-based scaling is enabled should provide maximum flexibility, optimum resource use, and optimum efficiency [21, 22]. In terms of smart manufacturing, there are a number of approaches that can be used. In 2011, the Smart Manufacturing Leadership Coalition recommended that coordinated manufacturing enterprises integrate "customer intelligence" throughout the entire manufacturing supply chain. The ability to achieve this definition requires new, flexible factories and supply chains driven by demand, which require data analytics, modelling, and simulation [21].

In addition to Germany, Industry 4.0 was also introduced by the country. As part of the manufacturing process, the product itself should have sensors and actors that actively participate in it. Industrial processes have been vertically integrated with economic software to maximize efficiencies and automate manufacturing processes. An interoperable production environment can be achieved through the exchange of information, triggers, and controls [23]. Recent, rapid technological advancements have enabled digital transformation by enabling enablers, which are technologies that have been evolving and improving. Smart manufacturing can be achieved through leveraging these enablers. In the next step, digitalization and smart manufacturing can be enhanced by identifying the so-called enhancers. To accomplish this, we must work together and have a common understanding [24] (Fig. 2).

A second key enabler of technology for smart manufacturing is Big Data, which emerges from the Internet of Things, artificial intelligence, and cloud computing. When data-driven smart manufacturing is combined with manufacturing, it has become a significant concept. Throughout the production process, AI technologies collect, store, process, and analyze data. Self-organizing systems that utilize historical data for learning and adoption enable self-organization, self-execution, self-regulation, and learning and adoption by themselves. Data-driven processes are controlled by the system, which enables customer-centric product development, self-organization, and transparency over manufacturing resources. As a result, manufacturing efficiency is increased, and product performance is improved [25]. There are many definitions of these terms in the literature, and a clear understanding of the next step requires distinguishing between them. In its broadest sense, the Internet of Things refers to the interconnection of smart consumer electronic devices. It is a

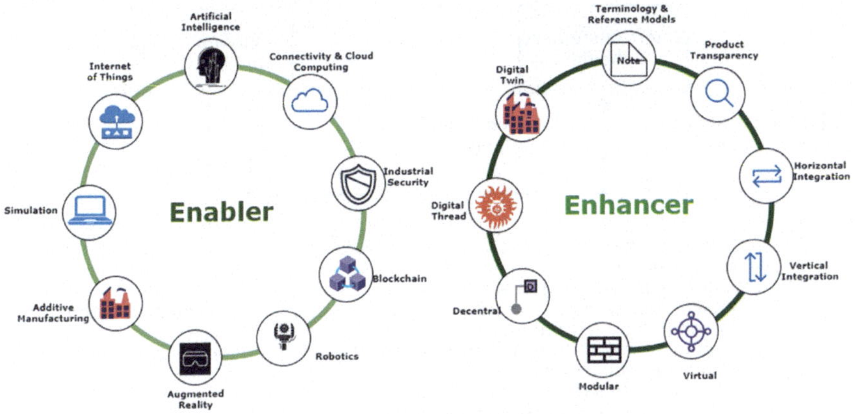

Fig. 2 Developing smart manufacturing technologies and enhancing them

device-to-device interaction and a human-to-human interaction as well. Interactions like this are similar to user-machine interfaces [26]. As opposed to the IoT, the IIoT focuses on the communication between machines in the manufacturing environment and industrial settings. IoT integrates operations technology with information technology. A number of key factors are incredibly important within the IIoT section, including resilience, precision, security, and operational efficiency [26].

3 Smart Factory

- **Smart Machines and Devices**: Factory devices, such as field devices, mobile devices, operating systems, and M2M communication, as well as communication between machines and humans, would be included in this definition.
- **Smart manufacturing processes**: A highly lively manufacturing environment fueled by IoT can be managed and controlled dynamically, efficiently, and automatically.
- **Smart Engineering**: Designing and developing products, engineering products, and manufacturing them. Consequently, manufacturing process data may be used for planning, and the machines (mechanical, electrical, etc.) may be optimized.
- **Big Data and the Cloud computing Data**: Algorithmic and analytical applications. With the increased popularity of big data analytics, factories will be able to improve their production processes and introduce new products to the market.

This system relies on iMCS for intelligent coordination of the smart elements shown in Fig. 3. Flexible [27], visible, and proactive characteristics should be present in an iMCS [28]. Compared to Industry 3.0, Industry 4.0 is characterized by flexibility [27], optimal decision-making, data-driven decisions, and remote monitoring with IoTs. A smart factory based on IoT is illustrated in Fig. 3:

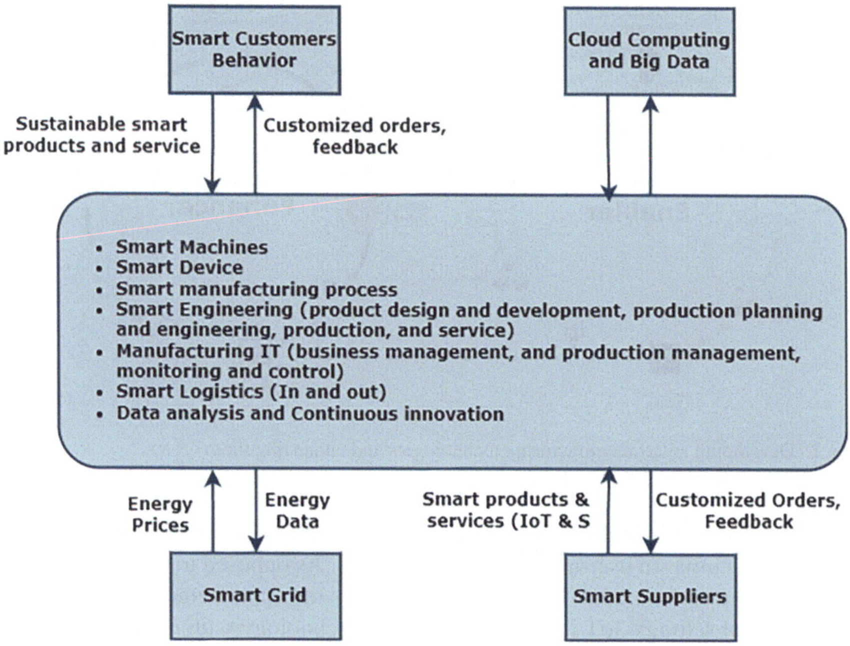

Fig. 3 Smart factories powered by IOT

For the factory of the future to be smart, we must equip manufacturing resources with IoT capabilities. Since iMCS uses a completely different technology than traditional SCADA systems, it is essential for the transition from Industry 3.0 to Industry 4.0.

3.1 Challenges for Implementing Smart Factory

An intelligent factory is based on standards, human-centred, and semantically uncoupled. A high degree of adaptability, reconfigurability, reliability, and precise traceability is required throughout and beyond the entire production chain [27]. Smart factories should be modular, interoperable, decentralized, virtualized, service-oriented, and real-time. Implementing smart factories should address people, cost, technology, structural, and security challenges. A common vision of smart factories and a common understanding of how to implement them often exist among factory staff [31]. It is unclear to them what smart factories can do for them. The second advantage of smart factories is that their production systems are highly automated and computer-driven [32]. Horizontal integration of intercompany value chains and information networks is needed, as well as vertical integration of production systems, customized production, and product life cycles [33]. The smart factory, therefore, requires several key elements, such as modular workstations or machine tools, a multisized workforce, reconfigurable fixtures, and standard communication and infrastructure. In addition, customization, real-time capabilities, and online monitoring and control are also required [11]. A huge investment is required for these elements. As a result, smart factories are expensive to implement. In addition, it will be necessary to solve many technical problems before the implementation of smart factories can be accelerated. The implementation of smart factories relies heavily on high-speed networks and communications [34]. Despite the rapid growth of intelligent equipment, industrial networks have a limited bandwidth, which poses a dilemma for smart factories [35, 36]. In addition, smart factories generate enormous volumes of unstructured sensor data, machine logs, and data related to manufacturing processes in real-time. In addition to their high dimensions, these data have variable metrics and a high level of noise, which makes them difficult to analyze directly [30]. The ability to store large amounts of data, acquire data efficiently, analyze it quickly, and connect to high-speed networks are all necessary components of a smart factory [34, 37, 38]. Integrated smart factories integrate supply chain, product, and logistics data. Industry big data analytics presents a significant challenge for improving the performance of smart factories [39].

Further, traditional manufacturers do not have the resources to develop software, especially small and medium-sized businesses. In the case of some manufacturers, growth involves acquisitions that result in a mixture of databases and software systems that are heterogeneous and intermingled. The fourth challenge is structural. Developing smart factories requires the systematic reuse of processes and improving understanding of complex structures [29]. Thus, smart factories require improved production, marketing, and controllability of production processes, as well as reduced manual intervention in workshops [40]. Digital transformation requires a change in traditional routines and processes [31]. Manufacturing characteristics must be taken into account in order to keep up with the rapidly changing demands of the market [41, 42].

Furthermore, we must take into account compound talents, multifield collaborations, and status quo, as well as multifield cooperation and manufacturing requirements [30]. A three-pronged approach is needed to integrate supply chain management with marketing, sales, and engineering. According to [43], contends that the temporal structure of machines, different processing speeds at parallel machines, and sudden arrivals of workers greatly complicate short-term supply chain planning in smart factories. The fifth reason is that smart factories are more vulnerable to security threats than traditional factories. The architecture of smart factories is limited and the underlying equipment is vulnerable, making a large amount of sensitive data vulnerable [44]. The security and privacy of smart factories must be improved urgently.

3.2 *Functional Architecture of Smart Factory*

Figure 4 illustrates how a manufacturing enterprise must fulfil several main functions during its processes. An organization's life cycle consists of several phases, including manufacturing, maintenance, education, and training. Planned investments are another phase. Data management becomes increasingly complex as architecture evolves. An example of a Smart Factory is displayed in Fig. 4, which shows how real-time information and control enable a manufacturing enterprise to execute manufacturing, maintain equipment, educate, and train employees.

Moreover, the Smart Factory can also be used for other functions within a manufacturing firm. In the Smart Factory, information can be useful for many functions throughout the factory's life cycles, such as process design and equipment planning. The Smart Factory gives context-aware information, unlike other existing applications. People can make informed decisions in production activities using a Smart Factory that canalizes information overflow. Application context awareness can optimize manufacturing processes and decrease search times for tools, for instance. I will briefly describe a maintenance scenario below as an example. In the scenario, a milling machine fails, demonstrating the benefits of using Smart Factory technologies.

When the machine breaks down, it sends an error message. An alert is sent to the shop floor manager and maintenance worker as a result of this information. This step analyzes the error information and assigns the appropriate workflow. As soon

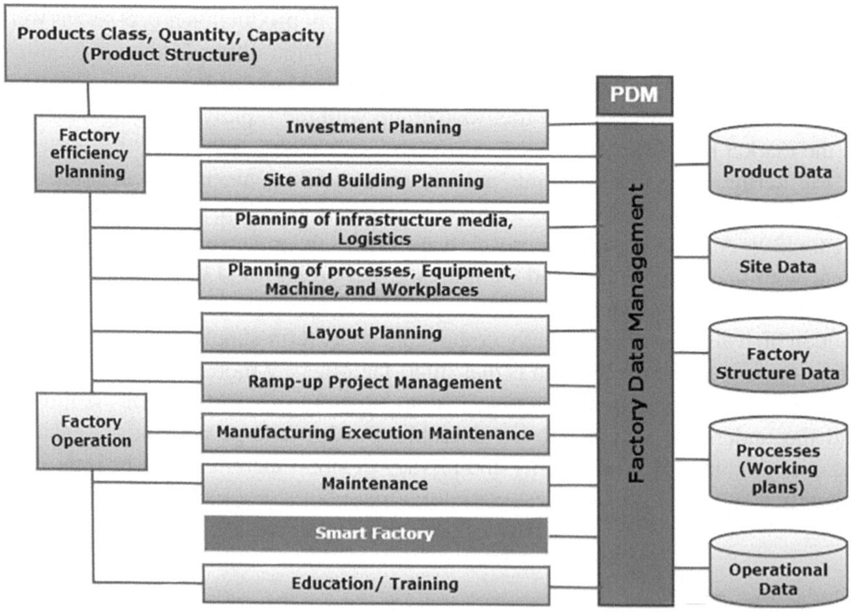

Fig. 4 An overview of a manufacturing system's functional architecture

as the error is documented, the worker receives the steps for repairing it as well as the necessary tools. In addition, the application displays the tool locations. When the workers get to the machine, they can order or pick up the tools themselves. As soon as he arrives at the machine, he can begin repairing it immediately. This reduces the time required to search for tooling. A workflow that incorporates the time, actions, and tools used to repair needs to be updated in order to store the repairing knowledge for future use. MESs can also adjust schedules in the event of breakdowns based on information about time. As an example of a smart factory, here is what it would look like. With the Nexus Platform, modules like MES or smart machines in the factory can exchange sensors, processes, jobs, geometric information, and more using the Augmented World Model (Fig. 5).

4 Methodology

Demonstrating manufacturing system architectures is dependent on enterprise requirements. Using conceptual data collection architecture, the researcher aims to make manufacturing more efficient and intelligent by managing it. Data mining processes are usually standardized across industries. This is an excellent starting point for data mining and can be used by entry-level manufacturing firms (Fig. 6).

To demonstrate the effectiveness of Data Enterprise Architecture, the research partner solved a real-life manufacturing problem, and a conceptual solution was then developed. For the purpose of demonstrating the effectiveness of Data Enterprise Architecture, a real-world manufacturing problem was solved, followed by the development of a conceptual solution. A cognitive solution based on functions and attributes was created through broader flowchart networks using database roles. An example is the division of functions according to the company's character in the process production configuration. The smart manufacturing users were provided with a comprehensive data collection architecture. This was done to ensure that the right tools were used in the correct environment and that the intended results were obtained. Figure 7 illustrates how the data architecture structure was conceptualized in this study to enable a smart system phase.

The importance of real-time information in smart manufacturing systems cannot be overstated. In this concept, data is analyzed and predicted using cell sensors, an external server, and an internal database. Using the structure. Enterprises can collect and analyze data and make decisions about manufacturing decisions, continuity, quality of products, and digital orders.

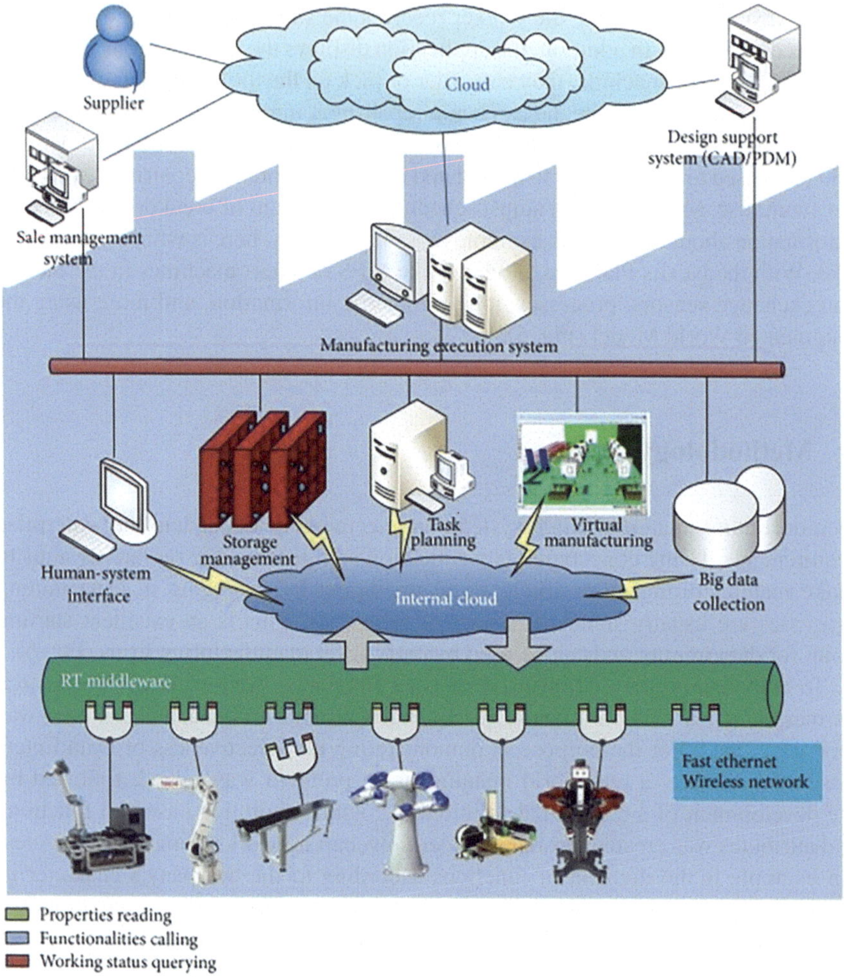

Fig. 5 Factory architecture based on smart technologies

5 Results and Discussions

Manufacturing has undergone a fourth industrial revolution with automation and data exchange, referred to as Industry 4.0. [45]. As a result of technological advancements in cyberphysical systems, the Internet of Things (IoT), cloud computing, and cognitive computing, this shift is taking place at an unprecedented pace. Smart factories can take advantage of these advancements in order to optimize their production processes and predict equipment failures, as well as adjust to changing conditions in the future. Industrial 4.0 was developed by the Germans in order to promote advanced manufacturing technology and methods [46]. Automated manufacturing, data exchange, and

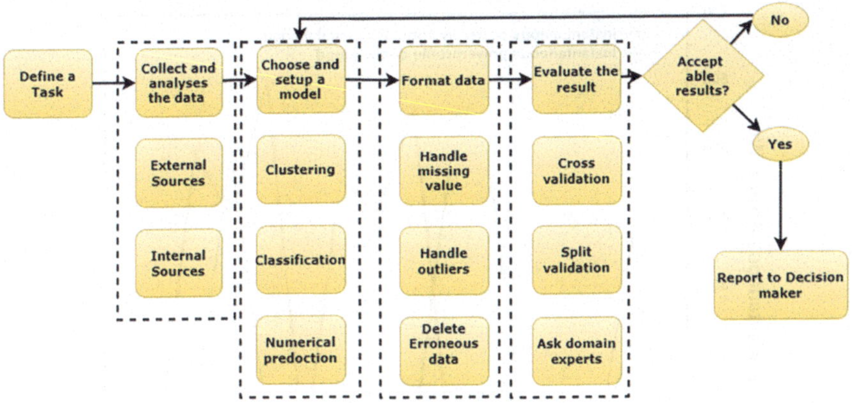

Fig. 6 Data mining process based on cross-industry standards

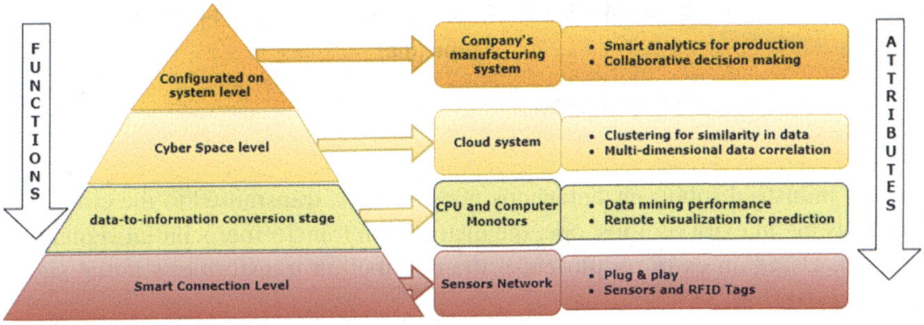

Fig. 7 Analytical data Architecture

data management are all part of this field. An organization's operational flexibility is enhanced as a result of Industry 4.0. Data is collected and processed in real-time throughout the organization, which facilitates decentralized decision-making. Industry 4.0 emphasizes smart manufacturing by integrating computational cores and networking into physical assets such as machinery and production lines. As a result of the use of these systems, it is possible to make localized adjustments automatically, which facilitates decentralized decision-making. Systems that are cyber-physical are capable of learning and adapting, using telemetry data to improve overall performance. The real-time monitoring of equipment and processes enabled by networked sensors reduces maintenance costs and detects early issues. Unplanned downtime and breakdowns can be reduced through this proactive approach. System disruptions can be identified swiftly and restored to normal operation using smart systems. As RFID tags, readers, and Internet of Things (IoT) protocols are widely integrated, factory assets, products, and machines can be tracked in real time. Production managers can monitor inventory and orders with precise visibility, and automated systems can reorder supplies based on usage patterns. Pick-and-place tasks can be performed

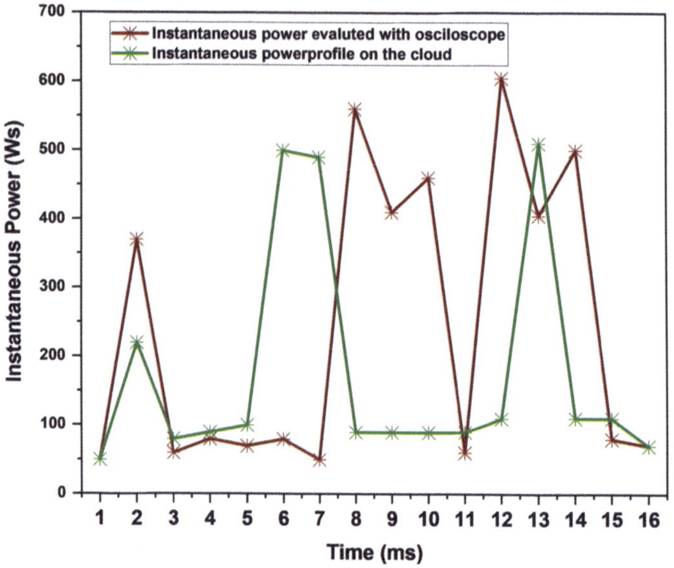

Fig. 8 Power versus time (Wm/ms) in Wm

using an industrial robot; instantaneous power data is transmitted to the cloud and stored. Using an extra oscilloscope, the distribution transformer's initial signal was observed. Using an extra oscilloscope, the distribution transformer's initial signal was observed (Fig. 8).

Collaborative robots, which integrate with humans and perform tasks including quality inspection and custom parts retrieval, are considered to be part of Industry 4.0 robotics. Robots replace workers rather than displace them, enhancing their cognitive capabilities. Sensors, equipment, and operations generate huge amounts of data that can be utilized by big data analytics. Using machine learning algorithms, we can gain insights into process optimization and find correlations that humans might miss. Inefficiencies, bottlenecks, and failures in the future are predicted by predictive analytics. Through the integration of cloud computing and industrial IoT, manufacturers can deploy decentralized and flexible production models that enable them to adjust production lines according to market conditions. A major benefit of cloud computing is that it provides remote access to systems, provides simulation resources, and facilitates big data analysis. In the future, Industry 4.0 will create smart factories that are connected, visible, decentralized, and capable of autonomous decision-making through the implementation of cutting-edge technologies [47]. As manufacturing capabilities advance, mass customization becomes possible at unprecedented scales. Industry 5.0 is the next evolution in advanced manufacturing after Industry 4.0, marked by greater flexibility, sustainability, and human-centred design [48]. As a result of this emerging concept, humans and machines are encouraged to collaborate and leverage technologies from previous industrial revolutions in order to advance

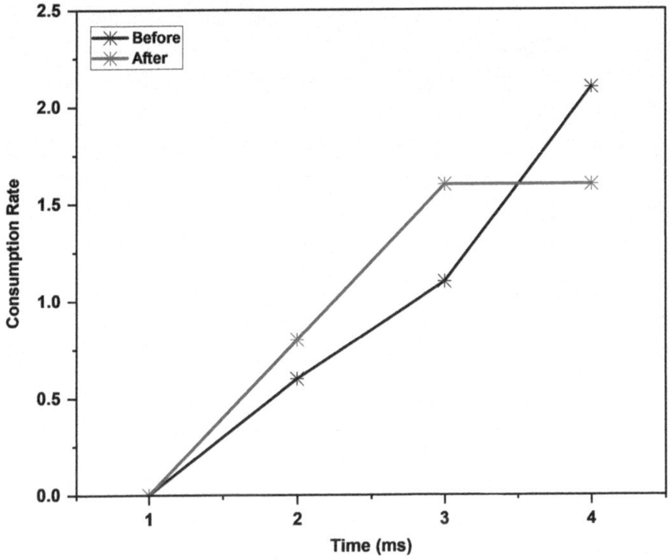

Fig. 9 Emissions from robots

society further. The purpose of Industry 5.0 is to enhance worker safety, productivity, and decision-making through human–machine collaboration. Both humans and machines benefit from wearable devices such as exoskeletons, which provide both cognitive and muscular enhancements. A comparison of energy consumption before and after the repair is shown in Fig. 9.

AI and advanced robotics are also key components of Industry 5.0, enabling flexible, reconfigurable production systems. Processes can be adapted seamlessly to accommodate product customizations or design iterations, allowing companies to manufacture small batches at mass production levels. 5.0 systems emphasize multidirectional communication, enabling robotic systems and organizational roles to make decisions in a timely manner. Predictive maintenance and self-adjusting operations can be performed using machine learning algorithms, providing actionable insights and making sense of the situation. A major focus of Industry 5.0 is sustainability and circular economies, including optimized manufacturing processes designed to conserve resources, reduce waste, and reduce environmental impact. Self-maintenance technologies such as 3D printing, predictive diagnostics, and automated inspections are employed in production systems. By utilizing additive manufacturing in local, flexible factories, decentralized production minimizes the gap between manufacturers and their end-users. It reduces transportation costs. Consumers and producers are becoming more connected with emerging technologies like drone delivery and augmented reality. Data transmission infrastructure must be expanded, computing power must be increased, and AI training systems must be improved in order for Industry 5.0 to be realized. Innovating will be driven by collaboration across multiple domains, such as biotech and robotics. By integrating an ethical framework into Industry 5.0, critical societal issues can be addressed. To advance equity, it places a high priority on maximizing human potential, regenerating the environment, and democratizing production. Human principles must be incorporated into technology, however [49]. Taking a holistic approach to manufacturing and embracing human–machine collaboration, Industry 5.0 will transform manufacturing as we know it. Though it has early effects, it signals a profound shift towards creating mutually beneficial environments and futures.

6 Conclusion

Manufacturing is transforming as Industry 4.0 technologies are incorporated into smart factories to become more efficient, adaptive, and data-driven. In addition to responding quickly to market changes, companies can improve the quality of their products and optimize their resources through this transformation. Nonetheless, implementing smart factories is challenging, requiring specialized workforce skills, high costs, and complex infrastructure requirements. Smart factories, despite these obstacles, are an essential component of the future manufacturing landscape thanks to benefits like enhanced production efficiency, predictive maintenance, and decentralized decision-making. A more collaborative relationship between humans and autonomous systems will increase manufacturing's sustainability and adaptability as Industry 4.0 becomes Industry 5.0.

References

1. H. Gaub, Customization of mass-produced parts by combining injection molding and additive manufacturing with industry 4.0 technologies. Reinf. Plast. **60**(6), 401–404 (2016). https://doi.org/10.1016/j.repl.2015.09.004
2. A. Schumacher, S. Erol, W. Sihn, A maturity model for assessing industry 4.0 readiness and maturity of manufacturing enterprises. Procedia CIRP **52**, 161–166 (2016). https://doi.org/10.1016/j.procir.2016.07.040
3. G. Adamson, L. Wang, P. Moore, Feature-based control and information framework for adaptive and distributed manufacturing in cyber physical systems. J. Manuf. Syst. **43**, 305–315 (2017). https://doi.org/10.1016/j.jmsy.2016.12.003
4. N. Hussain, P. Rani, Comparative studied based on attack resilient and efficient protocol with intrusion detection system based on deep neural network for vehicular system security, in *Distributed Artificial Intelligence* (CRC Press, 2020), pp. 217–236
5. J. Lee, Smart factory systems. Informatik Spektrum **38**(3), 230–235 (2015). https://doi.org/10.1007/s00287-015-0891-z
6. X.F. Liu, M.R. Shahriar, S.M.N. Al Sunny, M.C. Leu, L. Hu, Cyber-physical manufacturing cloud: architecture, virtualization, communication, and testbed. J. Manuf. Syst. **43**, 352–364 (2017). https://doi.org/10.1016/j.jmsy.2017.04.004
7. N. Kumar, P. Rani, V. Kumar, S.V. Athawale, D. Koundal, THWSN: enhanced energy-efficient clustering approach for three-tier heterogeneous wireless sensor networks. IEEE Sens. J. **22**(20), 20053–20062 (2022)

8. S. Bergweiler, Smart factory systems-fostering cloud-based manufacturing based on self-monitoring cyber-physical systems. Development **2**, 3 (2016)
9. M. Loskyll, I. Heck, J. Schlick, M. Schwarz, Context-based orchestration for control of resource-efficient manufacturing processes. Future Internet **4**(3), 737–761 (2012)
10. B. Holtkamp, A. Iyer, Industry 4.0: The Future of Indo-German Industrial Collaboration. Accessed 15 Oct 2024. [Online]. Available: http://aei.pitt.edu/102682/
11. M.M. Mabkhot, A.M. Al-Ahmari, B. Salah, H. Alkhalefah, Requirements of the smart factory system: a survey and perspective. Machines **6**(2), 23 (2018). https://doi.org/10.3390/machines6020023
12. T. Bauernhansl, M. Ten Hompel, B. Vogel-Heuser (eds.), *Industrie 4.0 in Produktion, Automatisierung und Logistik: Anwendung Technologien Migration* (Springer Fachmedien Wiesbaden, Wiesbaden, 2014). https://doi.org/10.1007/978-3-658-04682-8
13. J. Lee, E. Lapira, B. Bagheri, H. Kao, Recent advances and trends in predictive manufacturing systems in big data environment. Manuf. Lett. **1**(1), 38–41 (2013). https://doi.org/10.1016/j.mfglet.2013.09.005
14. N. Hussain, P. Rani, H. Chouhan, U.S. Gaur, Cyber security and privacy of connected and automated vehicles (CAVs)-based federated learning: challenges, opportunities, and open issues, in *Federated Learning for IoT Applications* (2022), pp. 169–183
15. J. Lee, B. Bagheri, H.-A. Kao, A cyber-physical systems architecture for industry 4.0-based manufacturing systems. Manuf. Lett. **3**, 18–23 (2015). https://doi.org/10.1016/j.mfglet.2014.12.001
16. C. Santos, A. Mehrsai, A.C. Barros, M. Araújo, E. Ares, Towards industry 4.0: an overview of European strategic roadmaps. Proc. Manuf. **13**, 972–979 (2017). https://doi.org/10.1016/j.promfg.2017.09.093
17. J. Butt, A strategic roadmap for the manufacturing industry to implement industry 4.0. Designs **4**(2), 11 (2020). https://doi.org/10.3390/designs4020011
18. J. Qin, Y. Liu, R. Grosvenor, A categorical framework of manufacturing for industry 4.0 and beyond. Proc. CIRP **52**, 173–178 (2016). https://doi.org/10.1016/j.procir.2016.08.005
19. Q. Li et al., Smart manufacturing standardization: architectures, reference models and standards framework. Comput. Ind. **101**, 91–106 (2018). https://doi.org/10.1016/j.compind.2018.06.005
20. A. Singh et al., Smart traffic monitoring through real-time moving vehicle detection using deep learning via aerial images for consumer application. IEEE Trans. Consumer Electron. 1–1 (2024). https://doi.org/10.1109/TCE.2024.3445728
21. S.M.L. Coalition, Implementing 21st century smart manufacturing, in *Workshop Summary Report*, pp. 1–36 (2011). Accessed 14 Oct 2024. [Online]. Available: https://www.academia.edu/download/28748242/smlcpfizer11911presv2.pdf
22. A. Schlemitz, V. Mezhuyev, Approaches for data collection and process standardization in smart manufacturing: systematic literature review. J. Ind. Inf. Integr. **38**, 100578 (2024)
23. H. Kagermann, W.-D. Lukas, W. Wahlster, Industrie 4.0: Mit dem Internet der Dinge auf dem Weg zur 4. industriellen Revolution. VDI nachrichten **13**(1), 2–3 (2011)
24. I. S. M. C. Committee, White Paper on Smart Manufacturing. Charlotta Johnsson (2021)
25. F. Tao, Q. Qi, A. Liu, A. Kusiak, Data-driven smart manufacturing. J. Manuf. Syst. **48**, 157–169 (2018). https://doi.org/10.1016/j.jmsy.2018.01.006
26. W.Z. Khan, M.H. Rehman, H.M. Zangoti, M.K. Afzal, N. Armi, K. Salah, Industrial internet of things: recent advances, enabling technologies and open challenges. Comput. Electr. Eng. **81**, 106522 (2020). https://doi.org/10.1016/j.compeleceng.2019.106522
27. D. Zuehlke, SmartFactory—towards a factory-of-things. Annu. Rev. Control. **34**(1), 129–138 (2010)
28. T. Iot, Building Smarter Manufacturing With the Internet of Things (IoT) (2014)
29. J. Cadavid, M. Alférez, S. Gérard, P. Tessier, Conceiving the model-driven smart factory, in *Proceedings of the 2015 International Conference on Software and System Process* (ACM, Tallinn Estonia, 2015), pp. 72–76. https://doi.org/10.1145/2785592.2785602
30. B. Chen, J. Wan, L. Shu, P. Li, M. Mukherjee, B. Yin, Smart factory of industry 4.0: key technologies, application case, and challenges. IEEE Access **6**, 6505–6519 (2018). https://doi.org/10.1109/ACCESS.2017.2783682

31. D. R. Sjödin, V. Parida, M. Leksell, A. Petrovic, Smart factory implementation and process innovation: a preliminary maturity model for leveraging digitalization in manufacturing, in *Moving to Smart Factories Presents Specific Challenges that can be Addressed Through a Structured Approach Focused on People, Processes, and Technologies. Research-Technology Management*, vol. 61, no. 5 (2018) pp. 22–31. https://doi.org/10.1080/08956308.2018.1471277
32. K. Jung, S. Choi, B. Kulvatunyou, H. Cho, K.C. Morris, A reference activity model for smart factory design and improvement. Prod. Plann. Control **28**(2), 108–122 (2017). https://doi.org/10.1080/09537287.2016.1237686
33. S. Park, Development of innovative strategies for the Korean manufacturing industry by use of the Connected Smart Factory (CSF). Proc. Comput. Sci. **91**, 744–750 (2016). https://doi.org/10.1016/j.procs.2016.07.067
34. S. Wang, J. Ouyang, D. Li, C. Liu, An integrated industrial ethernet solution for the implementation of smart factory. IEEE Access **5**, 25455–25462 (2017). https://doi.org/10.1109/ACCESS.2017.2770180
35. M. Chen, J. Wan, S. Gonzalez, X. Liao, V.C.M. Leung, A survey of recent developments in home M2M networks. IEEE Commun. Surv. Tutorials **16**(1), 98–114 (2014). https://doi.org/10.1109/SURV.2013.110113.00249
36. L. Kong, D. Zhang, Z. He, Q. Xiang, J. Wan, M. Tao, Embracing big data with compressive sensing: a green approach in industrial wireless networks. IEEE Commun. Mag. **54**(10), 53–59 (2016). https://doi.org/10.1109/MCOM.2016.7588229
37. Z. Song, Y. Sun, J. Wan, P. Liang, Data quality management for service-oriented manufacturing cyber-physical systems. Comput. Electr. Eng. **64**, 34–44 (2017). https://doi.org/10.1016/j.compeleceng.2016.08.010
38. J. Wan, D. Li, H. Yan, P. Zhang, Fuzzy feedback scheduling algorithm based on central processing unit utilization for a software-based computer numerical control system. Proc. Inst. Mech. Eng. Part B J. Eng. Manuf. **224**(7), 1133–1143 (2010). https://doi.org/10.1243/09544054JEM1690
39. X. Xu, Q. Hua, Industrial big data analysis in smart factory: current status and research strategies. IEEE Access **5**, 17543–17551 (2017). https://doi.org/10.1109/ACCESS.2017.2741105
40. Y. Lyu, J. Zhang, Big-data-based technical framework of smart factory. Comput. Integr. Manuf. Syst. **11**(22), 2691–2697 (2016)
41. S. Wang, J. Wan, M. Imran, D. Li, C. Zhang, Cloud-based smart manufacturing for personalized candy packing application. J. Supercomput. **74**(9), 4339–4357 (2018). https://doi.org/10.1007/s11227-016-1879-4
42. S. Wang, J. Wan, D. Zhang, D. Li, C. Zhang, Towards smart factory for industry 4.0: a self-organized multi-agent system with big data based feedback and coordination. Comput. Netw. **101**, 158–168 (2016). https://doi.org/10.1016/j.comnet.2015.12.017
43. D. Ivanov, A. Dolgui, B. Sokolov, F. Werner, M. Ivanova, A dynamic model and an algorithm for short-term supply chain scheduling in the smart factory industry 4.0. Int. J. Prod. Res. **54**(2), 386–402 (2016). https://doi.org/10.1080/00207543.2014.999958
44. J. Wan, J. Li, M. Imran, D. Li, Fazal-e-Amin, A blockchain-based solution for enhancing security and privacy in smart factory. IEEE Trans. Ind. Inf. **15**(6), 3652–3660 (2019). https://doi.org/10.1109/TII.2019.2894573
45. I. Podkolzina, A. Tenishchev, Z. Gornostaeva, H. Tekeeva, O. Tandelova, Ecological and food security in the conditions of the geopolitical situation in the worldglobal digital transformation trends in real sectors of the economy. SHS Web Conf. **172**, 02041 (2023). https://doi.org/10.1051/shsconf/202317202041
46. O. Dzhioeva, E. Magomadov, Green transformation in the context of sustainable development. BIO Web Conf. **63**, 04008 (2023). https://doi.org/10.1051/bioconf/20236304008
47. R. Gakaev, The world's forests are a stabilizing element of the climate system. BIO Web Conf. **76**, 06003 (2023). https://doi.org/10.1051/bioconf/20237606003
48. R. Gakaev, Functional classification of forests: study of carbon sequestration. BIO Web Conf. **76**, 06004 (2023). https://doi.org/10.1051/bioconf/20237606004

49. S.G. Shmatko, L.V. Agarkova, T.G. Gurnovich, I.M. Podkolzina, Problems of increasing the quality of raw material for wine in the stavropol region. Res. J. Pharm. Biol. Chem. Sci. **7**(2), 725–730 (2016)

The manufacturer's authorised representative in the EU is Springer Nature Customer Service Centre GmbH, Europaplatz 3, 69115 Heidelberg, Germany. If you have any concerns regarding our products, please contact ProductSafety@springernature.com

Printed and bound by CPI Group (UK) Ltd, Croydon, CR0 4YY
30/09/2026
02225562-0010